STANDARD RADIO COMMUNICATIONS MANUAL:

With Instrumentation and Testing Techniques

R. Harold Kinley, CET

Communications/Electronics Technician

South Carolina Forestry Commission

PRENTICE-HALL, INC.

Business and Professional Division

ENGLEWOOD CLIFFS, NEW JERSEY

Editora Prentice-Hall do Brasil Ltda., *Rio de Janeiro*
Prentice-Hall International, Inc., *London*
Prentice-Hall of Australia, Pty. Ltd., *Sydney*
Prentice-Hall of Canada, Inc., *Toronto*
Prentice-Hall of India Private Ltd., *New Delhi*
Prentice-Hall of Japan, Inc., *Tokyo*
Prentice-Hall of Southeast Asia Pte. Ltd., *Singapore*
Whitehall Books, Ltd., *Wellington, New Zealand*
Prentice-Hall Hispanoamericana, S.A., *Mexico*

©1985 *by*

PRENTICE-HALL, INC.
Englewood Cliffs, N.J.

Editor: George E. Parker

Library of Congress Cataloging in Publication Data

Kinley, R. Harold.
  Standard radio communications manual.

  Includes index.
  1. Radio—Apparatus and supplies—Testing—
Handbooks, manuals, etc. 2. Radio measurements—
Handbooks, manuals, etc. I. Title.
TK6553.K496 1985   621.3841′8   84-22893

ISBN 0-13-842394-6

Printed in the United States of America

DEDICATION

To my wife, Linda, who took over many of my routine chores and obligations so that I could devote maximum time to this project; to my children, Jennifer and Jack, who (for the most part) were very understanding, patient, and cooperative with me (and tolerant *of* me!) during this project; to my mother, Louise, who was always interested; to my brother, Jerry, whose courage serves as an inspiration to those who know him; to the memory of my father, Neal, who is missed; and to my uncle, Joel A. Waldrop, who encouraged my interest in the field of radio.

A BRIEF NOTE
FROM THE AUTHOR
ON THIS BOOK'S
PRACTICAL VALUE

A source of good practical, technical information is as much a valuable tool to the communications technician as is his or her voltmeter and soldering iron. The technician who is seriously interested in this work and has a desire to keep abreast is constantly looking for books that have a direct, practical application to the job and that will be of direct benefit in daily job performance. The technician doesn't want (and doesn't have time) to "wade and sift" through dozens of generalized books for specific information.

As a communications technician, much of your time is spent in test and measurement work either to evaluate the performance of equipment or to locate the area of trouble in the system. Over the years, a great number of test and measurement procedures have been developed by the industry. The Electronic Industries Association (EIA) has made great strides in *standardizing* many tests and measurement procedures in order to give real meaning to manufacturer's specifications, which often were based on improper and little-known test procedures.

Surprisingly, some technicians are still unaware of many test and measurement procedures that can help them to analyze the performance level of communications equipment more effectively. Others don't use the test and measurement procedures to their fullest advantage simply because the benefits are not completely recognized or understood. Perhaps part of the reason for this has been the lack of books that focus *complete* attention on this important, indeed vital, subject area.

In writing this book, it is my sole aim and purpose to provide you—the communications technician—with a complete source or directory of those tests and measurements relative to the servicing of AM, SSB, and FM radio communications gear, including antennas, transmission lines, and remote control lines. You will learn why these tests are necessary, step-by-step procedures for performing the tests and measurements, and proper interpretation and evaluation of the results. Where possible, more than one method of performing these tests and measurements (using different equipment) will be presented.

Chapter 1 tackles the *decibel* and how it is used in communications work. This chapter includes many graphs that you will find very useful in your day-to-day work. Chapter 2 describes, in great detail, the *spectrum analyzer*, a valuable piece of test equipment to the modern communications shop. Chapter 3 describes the modern, sophisticated instrumentation available to the communications shop. Chapters 4 through 9 cover tests and measurement procedures related to AM, SSB, and FM communications equipment. Chapter 10 provides good, practical coverage of those test and measurement procedures relative to antenna and transmission lines. The final chapter covers remote control line testing, various sweep techniques, and oscilloscope applications.

Finally, I wish to stress here that this book is intended as a *reference* manual. In this regard, extensive theoretical discussions are avoided in favor of practical, useful, *hard* information and data. The many graphs and illustrations serve to strengthen the book's practical value and to expand the book's usefulness. This book is not the type that lends itself to a *one-time* reading, then should be put away and forgotten. It will serve as a frequent reference in your work, giving you a real return on your *investment.* In summary, this book will help to make your job easier and more interesting and as a result will make you a better technician.

R. Harold Kinley

ACKNOWLEDGMENTS

Clemson University, Clemson, S.C., Research assistance.

Cushman Electronics, Inc., San Jose, CA. A special thanks to Michael J. Brehany (applications engineer), who offered assistance in the form of information, material, and checking a portion of the manuscript for accuracy.

Electronic Industries Association, 2001 Eye St. N.W., Washington, D.C. 20006, Phone (202) 457-4900. I am especially indebted to Stuart Meyer (Chairman of the Engineering Panel, Telecommunications Group) for answering many of the questions that arose during the preparation of this book. Thanks is also due the EIA for allowing me to adapt selected portions of their copyrighted publications for use in this book.

Helper Instruments Company, Indialantic, FL. The president of this company, William Detwiler, was called upon many times to answer questions of a highly technical nature and always responded positively with very good advice. Thanks, Bill.

Spartanburg Technical College, Spartanburg, SC. Thanks to George Bruce (Electronics Engineering Technology) for allowing me to use the laboratory to check out some of the test procedures presented herein, and to Becky Simpson (Librarian) for library assistance beyond the normal call of duty.

Many other companies offered assistance in the form of product information and photos of their equipment for use in the book. It would be impossible to list them all here, but their assistance is nonetheless appreciated.

CONTENTS

11 MISCELLANEOUS RELATED TESTS AND MEASUREMENTS   379

Remote Control Line Tests (379) Basic Sweep Techniques and
Tips (388) Miscellaneous Oscilloscope Applications (395)

1

THE DECIBEL IN COMMUNICATIONS WORK

The decibel is a widely used measuring unit, especially useful in communications work. Therefore, a good understanding of the decibel and the various forms in which it is used is essential to anyone working in the communications field. This chapter focuses complete attention on the decibel and the various forms in which you will find it used. The purpose of this chapter is not only to enlighten you on the use of the decibel but also to serve as a future reference for your on-the-job use from day to day. With this in mind, this chapter is loaded with useful charts, graphs, and formulas for your reference. Many practical examples in the use of the various formulas are presented here in the interest of demonstrating the practical value of the decibel and in the belief that an example of the application and execution of the formula is essential to a better understanding of it.

THE RESPONSE OF THE EAR TO SOUND

Scientific studies have concluded that the ear does not respond to *changes* in sound intensity in a *linear* manner. For example, if the power delivered to a loudspeaker is increased (or doubled) from 4 to 8 watts (W), the ear does not recognize this as a 2-to-1 increase in the sound intensity. Rather, the ear detects this doubling of power as only a *slight* increase in the sound intensity or volume. This is because the ear responds to changes in sound intensity in an approximately *logarithmic* manner.

1

THE BEL

A unit of measurement of the change in sound intensity was developed upon the principle of the human ear's logarithmic response to changes in sound intensity. This unit of measurement of change in power is called the *bel*, named in honor of Alexander Graham Bell—the inventor of the telephone. The relationship is shown in the formula:

$$bels = \log \frac{P1}{P2}$$

where P1 and P2 represent the two different power levels. In order to simplify the calculations, the larger of the two power levels usually is used as the numerator and the smaller as the denominator. Using the smaller value as the numerator results in a negative solution and requires a little more skill in the use and representation of logarithms (see Appendix A). P1 and P2 may be represented in watts, milliwatts, kilowatts, etc., as long as both are in the same unit of measure. That is, don't enter P1 in watts and P2 in milliwatts.

Let's now put the formula to work for us in calculating the gain (in bels) of an audio amplifier. Suppose the input to an audio amplifier is 1 W and the output is 10 W. We can calculate the gain of the amplifier in bels as follows:

$$bels = \log \frac{P1}{P2}$$

Allowing P1 to represent the larger of the two power levels and substituting, we have

$$bels = \log \frac{10}{1} = \log 10 = 1$$

Thus, we can see that a power ratio of 10 to 1 represents a power change of 1 bel. Thus, to say that a certain amplifier has a gain of 1 bel is the same as saying that it has a power gain of 10. In order to sound twice as loud to the ear, the gain of the amplifier would have to be increased to 2 bels. Calculate the necessary power gain required.

Solution: 2 (bels) = log *P*, where *P* is the power gain
P = antilog 2 = 100, the required power gain

Thus, it is shown that for each 1 bel increase, the power gain must be multiplied by 10. Figure 1-1 illustrates the relationship of power ratio to bels. Notice on the graph that a power ratio of 10 corresponds to 1 bel. A power ratio of 100 corresponds to 2 bels (as calculated in the example above). To double the volume of the sound to the ear (as going from 2 bels to 4 bels), the power ratio must be *squared.* This is shown on the graph. A

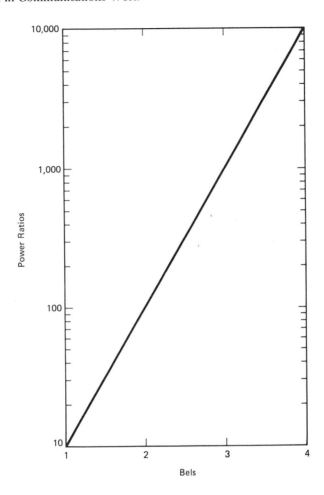

FIGURE 1-1. This graph illustrates the relationship of power ratios versus bels for ratios from 10 to 10,000.

power gain of 100 corresponds to 2 bels. To double the sound apparent to the ear (to 4 bels), the power ratio must be *squared.* As shown on the graph, this is the case—$100^2 = 10,000$, which corresponds to 4 bels. As you can see, the power ratio is quite misleading as to the actual change in sound to the ear. The bel serves to bring this in line so as to give a more *realistic* relationship of power ratios versus response of the ear. This enables us to analyze signal level relationships more effectively in terms of how they will be perceived by the ear.

It is important to keep in mind that the bel is based on power *ratios.* It has no *absolute* value unless it is *referenced* to some *specific* power level. For instance, the term 1 bel has no absolute value in terms of *actual* power level. However, 1 bel above 1 milliwatt (mW) does have an absolute value since a reference level is given. We know that 1 bel represents a power ratio of 10, so 1 bel above 1 mW is:

$$10 \times 1 \text{ mW} = 10 \text{ mW}$$

Since power (P) can be represented in terms of voltage (E), current (I), and resistance (R), so too can the bel be calculated from these terms. The formulas (and their derivation) for calculating bels from voltage and current ratios are presented here in complete detail to enable you to get a better insight into the various formulas.

The Bel and Voltage Ratios   Since the power in a load can be represented by the term E^2/R, we can calculate the bels from the voltage and resistance values by substituting the term E^2/R into the formula for P1 and P2:

1. $\text{bels} = \log \dfrac{\dfrac{(E1)^2}{R1}}{\dfrac{(E2)^2}{R2}}$ [Substitution]

2. $\text{bels} = \log \dfrac{(E1)^2 \, R2}{(E2)^2 \, R1}$ [Invert and multiply]

3. $\text{bels} = \log \left(\dfrac{E1}{E2}\right)^2 \left(\dfrac{R2}{R1}\right)$ $\left[\dfrac{E1^2}{E2^2} = \left(\dfrac{E1}{E2}\right)^2\right]$

4. $\text{bels} = \log \left(\dfrac{E1}{E2}\right)^2 + \log \left(\dfrac{R2}{R1}\right)$ $\left[\begin{array}{l}\text{The log of the product of} \\ \text{two values is equal to the} \\ \text{sum of the logs of the val-} \\ \text{ues.}\end{array}\right]$

5. $\text{bels} = 2 \log \dfrac{E1}{E2} + \log \dfrac{R2}{R1}$ $\left[\begin{array}{l}\text{The log of the square of a} \\ \text{value is equal to twice the} \\ \text{log of the value.}\end{array}\right]$

From step 2, we could also have done this:

2. $\text{bels} = \log \dfrac{(E1)^2 \, R2}{(E2)^2 \, R1}$

3. $\text{bels} = 2 \log \sqrt{\dfrac{(E1)^2 \, R2}{(E2)^2 \, R1}}$ $\left[\begin{array}{l}\text{The log of a value is equal} \\ \text{to twice the log of the} \\ \text{square root of the value.}\end{array}\right]$

4. $\text{bels} = 2 \log \dfrac{E1\sqrt{R2}}{E2\sqrt{R1}}$

You may use either formula—2 log $E1\sqrt{R2}/E2\sqrt{R1}$ or 2 log E1/E2 + log R2/R1—as you prefer. If the two load resistances, R1 and R2, are equal or if the voltages, E1 and E2, are taken across the same load resistor or impedance, the formula can be reduced to

$$\text{bels} = 2 \log \frac{E1}{E2}$$

The Bel and Current Ratios Since the power (P) in a load can be represented by the term I^2R, we can calculate bels from the current and resistance values by substituting the term I^2R for P1 and P2 into the formula as shown below:

1. $\text{bels} = \log \dfrac{(I1)^2 \, R1}{(I2)^2 \, R2}$ [By substitution]

2. $\text{bels} = \log \left(\dfrac{I1}{I2}\right)^2 \left(\dfrac{R1}{R2}\right)$

3. $\text{bels} = \log \left(\dfrac{I1}{I2}\right)^2 + \log \dfrac{R1}{R2}$ $\begin{bmatrix}\text{The log of the product of}\\\text{two values is equal to the}\\\text{sum of the logs of the val-}\\\text{ues.}\end{bmatrix}$

4. $\text{bels} = 2 \log \dfrac{I1}{I2} + \log \dfrac{R1}{R2}$ $\begin{bmatrix}\text{The log of the square of a}\\\text{value is equal to twice the}\\\text{log of the value.}\end{bmatrix}$

From step 1 above, we could have proceeded as follows:

1. $\text{bels} = \log \dfrac{(I1)^2 \, R1}{(I2)^2 \, R2}$

2. $\text{bels} = 2 \log \sqrt{\dfrac{(I1)^2 \, R1}{(I2)^2 \, R2}}$ $\begin{bmatrix}\text{The log of a value is equal}\\\text{to twice the log of the}\\\text{square root of the value.}\end{bmatrix}$

3. $\text{bels} = 2 \log \dfrac{I1\sqrt{R1}}{I2\sqrt{R2}}$

You may use either formula—2 log $I1\sqrt{R1}/I2\sqrt{R2}$ or 2 log I1/I2 + log R1/R2—as you prefer. As in the case of voltage ratios, if $R1 = R2$, or if the currents are taken through the same load resistor, the formula can be simplified to

$$2 \log \frac{I1}{I2}$$

Bel Formulas—Condensed Summary   All the formulas developed for use in calculating bels are listed below in their final condensed form.

1. bels $= \log \dfrac{P1}{P2}$

2. bels $= 2 \log \dfrac{E1}{E2}$, where the load resistances are equal

3. bels $= 2 \log \dfrac{E1\sqrt{R2}}{E2\sqrt{R1}}$, where the load resistances are not equal

4. bels $= 2 \log \dfrac{E1}{E2} + \log \dfrac{R2}{R1}$, where the load resistances are not equal

5. bels $= 2 \log \dfrac{I1}{I2}$, where the load resistances are equal

6. bels $= 2 \log \dfrac{I1\sqrt{R1}}{I2\sqrt{R2}}$, where the load resistances are not equal

7. bels $= 2 \log \dfrac{I1}{I2} + \log \dfrac{R1}{R2}$, where the load resistances are not equal

THE DECIBEL

In most practical applications, the bel proved to be too large a measuring unit for easy handling. As a result, the decibel (abbreviated dB) was introduced and has since gained wide acceptance in the field. The decibel is defined as one-tenth of a bel. Those formulas that were developed for use in calculating bels can easily be adapted to the decibel by simply multiplying the formulas by a factor of 10. These formulas for the decibel are listed as follows:

1. decibels $= 10 \log \dfrac{P1}{P2}$

2. decibels $= 20 \log \dfrac{E1}{E2}$, where the load resistances are equal

3. decibels $= 20 \log \dfrac{E1\sqrt{R2}}{E2\sqrt{R1}}$, where the load resistances are not equal

4. decibels $= 20 \log \dfrac{E1}{E2} + 10 \log \dfrac{R2}{R1}$, where the load resistances are not equal

5. decibels $= 20 \log \dfrac{I1}{I2}$, where the load resistances are equal

6. decibels $= 20 \log \dfrac{I1\sqrt{R1}}{I2\sqrt{R2}}$, where the load resistances are not equal

7. decibels $= 20 \log \dfrac{I1}{I2} + 10 \log \dfrac{R1}{R2}$, where the load resistances are not equal

As it works out, the smallest power change that can be detected by the ear is approximately 1 dB (for sine-wave signals). This enhances the practicality of the decibel. In dealing with complex signals such as human speech waveforms, where average power is low, the minimum power change detectable by the ear is approximately 3 dB.

Table 1-1 shows the relationship between power ratios, voltage ratios, and current ratios versus the decibel for a number of decibel values ranging from 0 to 100 dB. The chart can be used to find approximate dB values from voltage, current, and power ratios or to find the voltage, current, or power ratio corresponding to a given dB value.

dB	GAIN (+)		LOSS (−)	
	Power Ratio	Voltage Ratio	Power Ratio	Voltage Ratio
0	1.000	1.000	1.000	1.000
1	1.259	1.122	0.794	0.891
2	1.585	1.259	0.631	0.794
3	1.995	1.413	0.501	0.708
4	2.512	1.585	0.398	0.631
5	3.162	1.778	0.316	0.562
6	3.981	1.995	0.251	0.502
7	5.012	2.239	0.200	0.447
8	6.310	2.512	0.159	0.398
9	7.943	2.818	0.126	0.355
10	10.000	3.162	0.100	0.316
11	12.590	3.548	0.079	0.282
12	15.85	3.981	0.063	0.251
13	19.95	4.467	0.050	0.224
14	25.12	5.012	0.040	0.200
15	31.62	5.623	0.032	0.178
16	39.81	6.310	0.025	0.159
17	50.12	7.079	0.020	0.141
18	63.10	7.943	0.016	0.126
19	79.43	8.913	0.013	0.112
20	100.00	10.00	0.010	0.100
30	1000.00	31.62	0.001	0.032
40	10000.00	100.00	0.0001	0.01
50	100000.00	316.2	0.00001	3.16×10^{-3}
60	10^6	1000.0	10^{-6}	1.0×10^{-3}
70	10^7	3162.2	10^{-7}	3.16×10^{-4}
80	10^8	10000.0	10^{-8}	1.0×10^{-4}
90	10^9	31620	10^{-9}	3.16×10^{-5}
100	10^{10}	100000	10^{-10}	1.0×10^{-5}

TABLE 1-1. This chart shows the equivalent voltage and power ratios for decibel values from 0 to 100. The ratios are shown for both gains and losses.

The graph shown in Figure 1-2 is useful in determining *approximate* dB equivalents of power or voltage ratios. Conversely, the voltage or power ratio can be determined from a given dB value. The vertical scale covers power ratios ranging from 1 to 100. If the ratio is a *power* ratio, the equivalent dB value is read from the *top* horizontal scale labeled "power" (from 0 to 20). If the ratio is a *voltage* ratio, the equivalent dB value is read from the *bottom* horizontal scale labeled "voltage."

Although the ratio ranges from 1 to 100, we can use the same graph to convert higher ratios to dB values. We can do this by breaking down the higher ratio into factors that are within the range of the scale shown. For example, suppose we want to convert a power ratio of 150 into the equivalent dB value. Since 150 is beyond the range of the scale, we must break it down into smaller "bites." We could say: 75 × 2 or 50 × 3 or any other combination that, when multiplied, would yield 150. Let's use 75 × 2 here. First, find the dB value for a power ratio of 75. This corresponds to approximately 18.8 dB on the power scale. Now, find the dB value that corresponds to 2. This is 3 dB. The *sum* of the two dB values is the dB equivalent of a power ratio of 150. It is 18.8 + 3 = 21.8 dB.

In a similar manner, we could convert dB values outside the range of the values shown. For example, suppose we want to convert 55 dB to its equivalent voltage ratio. The top of the dB scale for voltage ratios is 40, which corresponds to a voltage ratio of 100. Now, find the ratio for 55 − 40, or 15 dB. (Remember to use the voltage scale.) It is 5.6. The voltage ratio that corresponds to 55 dB is equal to the *product* of these two ratios. It is 100 × 5.6 = 560.

Thus, we can convert between any value of voltage or power ratios to decibels by breaking down these values into smaller values that fall within the range of the graph.

Another useful graph is shown in Figure 1-3 on page 10. This is a plot of the percent of change in voltage or power versus change in decibels. To use this graph, you first must know the percent of change of power or voltage. For example, if P1 = 25 and P2 = 40, the percent of change in P1 is

$$\frac{40 - 25}{25} = 60\%$$

This correlates with approximately 2 dB on the graph.

Another graph is shown in Figure 1-4 on page 11. This graph is useful when the ratio of P1/P2 or E1/E2 is less than 1. For example, suppose that a voltage ratio E1/E2 is 45/60. This is equal to 0.75. From the graph, this correlates with approximately −1.2 dB.

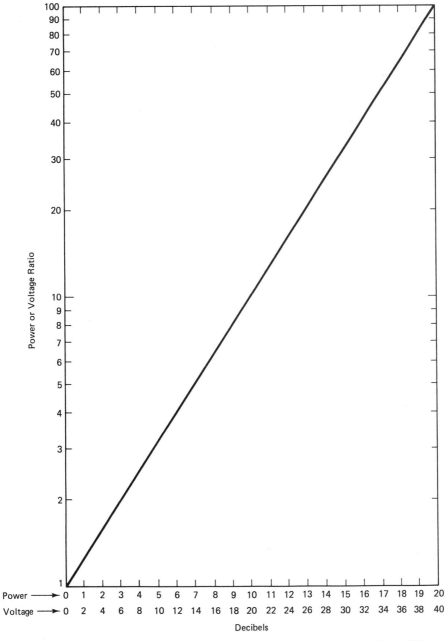

FIGURE 1-2. This graph converts voltage or power ratios from 1 to 100 to the equivalent decibel value. If the ratio is a voltage ratio, read the decibel value on the bottom horizontal scale. If the ratio is a power ratio, read the decibel value on the top horizontal scale.

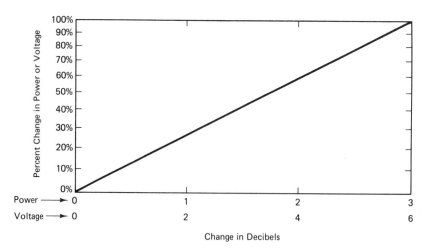

FIGURE 1-3. This graph converts the percentage of change of power or voltage to the equivalent decibel value. If the percentage of change refers to voltage, use the bottom scale. If the percentage of change refers to power, use the top scale.

All of these graphs are very useful and handy for calculating dB values quickly but will result in a small sacrifice in accuracy. You should use any tool available to you when working with dB to ratio conversions, including the charts, graphs, tables, etc., included here. However, there are times when you will need to be able to use the proper formula in solving these conversions. The following practical examples should help to clarify the use of the formulas in various practical situations.

In working out these formulas, there are several ways you can find the log of a number: (1) a slide rule, (2) an electronic calculator, (3) the table of logarithms in Appendix C, or (4) the graph in Appendix B. For most applications, accuracy to the nearest dB is sufficient.

Example 1: An audio amplifier is said to have a gain of 17 dB. The power output from the amplifier is measured and found to be 95 W. What is the power input to the amplifier?

Solution: We are given the dB gain and the power output, so the appropriate formula to use is dB = 10 log P1/P2. Using the output power for P1 and substituting the values into the formula, we have:

$$17 = 10 \log \frac{95}{P2},$$

$$1.7 = \log \frac{95}{P2},$$

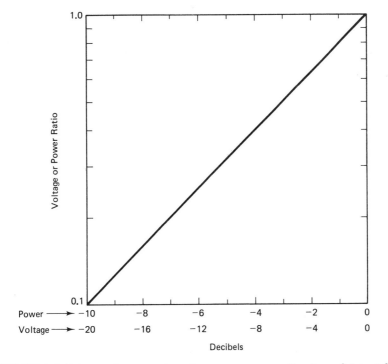

FIGURE 1-4. This graph converts voltage or power ratios from 0.1 to 1.0 to the equivalent decibel value. For voltage ratios, read the decibel value on the bottom horizontal scale. For power ratios, read the decibel value on the top horizontal scale. Notice that all the decibel values are negative. This is because the ratios are less than one, indicating a loss.

$$\text{antilog } 1.7 = \frac{95}{P2},$$

$$\text{antilog } 1.7 = 50 = \frac{95}{P2},$$

$$50 \text{ } P2 = 95,$$

$$P2 = \frac{95}{50} = 1.9 \text{ W}$$

Example 2: A transmitter is connected to a 50-Ω dummy load. With the transmitter keyed, the RF voltage across the dummy load is 15 volts (V). The loading of the transmitter is increased so that the RF voltmeter now reads 30 V. Calculate the change in transmitter power in decibels.

Solution: Since both the voltage readings are taken across the same resistance (50 Ω), we need not take the resistance into account. Thus, we can use the simpler formula for voltage ratios—dB = 20 log E1/E2. Allowing E1 to represent the larger of the two voltages and substituting, we have:

$$dB = 20 \log \frac{30}{15} = 20 \log 2 = 20(0.301) = 6.02 \text{ dB}$$

Example 3: A matching pad is used to match the 500-Ω output of an amplifier to a 16-Ω load. The voltage across the pad input (500 Ω) is 25 V and across the output (16 Ω) is 3.7 V. Calculate the insertion loss of the matching pad.

Solution: Since voltages across unequal resistances are involved here, we can do one of two things: (1) use the formula P = E²/R to convert the input and output of the pad to power units and then use the dB formula for power ratios, or (2) use one of the formulas developed for voltages in unequal resistances. The latter requires fewer computations because the first method would require converting the input and output to power and then using the dB formula for power ratios—resulting in three separate calculations. So, the logical choice here is decibel formula (3) or decibel formula (4) for voltages in unequal resistances. The solution for this example will be shown using both of these formulas for your benefit.

Let E1 represent the pad input and E2 the pad output. Using formula (3) first and substituting, we have:

$$dB = 20 \log \frac{25\sqrt{16}}{3.7\sqrt{500}} = 20 \log \frac{25(4)}{3.7(22.4)} = 20 \log \frac{100}{82.7} = 20 \log 1.2$$

$$= 20 (0.082) = 1.64 \text{ or } 1.6 \text{ dB}$$

Using formula (4) and substituting, we have:

$$dB = 20 \log \frac{25}{3.7} + 10 \log \frac{16}{500} = 20 \log 6.75 + 10 \log 0.032$$

$$= 20 (0.829) + 10 (0.505 - 2) = 16.58 + 5.05 - 20$$

$$= 21.63 - 20 = 1.63 \text{ or } 1.6 \text{ dB}$$

Notice the special manner in which the log of 0.032 is written—(0.505 − 2). This is because this log has a characteristic of −2. (See Appendix A.)

Example 4: The RF current at the end of a transmission line at the antenna is 4.7 amperes (A). The RF current at the input end of the transmission line (at the transmitter) is 5.6 A. (1) Calculate the dB loss of the transmission line. (2) Calculate the *percentage* of power loss in the transmission line.

Solution 1: Since the impedances in this case may be assumed to be equal, formula (5) is the appropriate one to use. Allowing *I*1 to represent the larger current value and substituting, we have:

$$dB = 20 \log \frac{5.6}{4.7} = 20 \log 1.19 = 20 \, (0.0755) = 1.5 \, dB$$

Solution 2: Since the impedance is not given, we can't calculate the power in terms of actual wattage. But by using the dB formula for power ratios, we can calculate the power ratio and then the power loss in percent. Substituting into the formula, we have:

$$1.5 \; (dB) = 10 \log \frac{P1}{P2}; \; \log \frac{P1}{P2} = \frac{1.5}{10}; \; \log \frac{P1}{P2} = 0.15, \frac{P1}{P2}$$

$$= \text{antilog } 0.15 = 1.41, \frac{P1}{P2} = \frac{1.41}{1}$$

Thus, the power at the end of the line is 1/1.41, or 0.708 times the power at the input side. This represents a power loss of approximately 30%, which sounds much worse than the 1.5 dB figure—but the 1.5 dB figure shows that the 30% loss really isn't very significant in practical terms.

Example 5: The RF current at the feedpoint of an antenna is 15 A. The antenna's feedpoint impedance is 72 Ω. A second antenna has a feedpoint impedance of 50 Ω and a feedpoint current of 22 A. Calculate the difference in power level in dB between the two antennas.

Solution 1: Since currents in unequal impedances are involved here, use formula (6)

$$dB = 20 \log \frac{I1\sqrt{R1}}{I2\sqrt{R2}}$$

or formula (7)

$$dB = 20 \log \frac{I1}{I2} + 10 \log \frac{R1}{R2}$$

The solution will be shown using both of these formulas. First, using formula (6) and substituting, we have

$$dB = 20 \log \frac{15\sqrt{72}}{22\sqrt{50}} = 20 \log \frac{15(8.48)}{22(7.07)} = 20 \log \left(\frac{127.2}{155.54} \right) = 20 \log 0.82$$

$$= 20 \, (0.9138 - 1) = 18.3 - 20 = -1.7 \, dB$$

The negative sign in front of the solution indicates that the *change* in power level is a *loss*. It worked out this way because the numerator was smaller than the denominator in our ratio. As mentioned before, it is common practice to write the ratio purposely so that the numerator is the larger of the two values. But in some cases (such as this one) it is difficult to determine just which of the two is larger until a certain point is reached in the calculations. It is at the point encircled in the calculations that it becomes clear that the numerator is smaller than the denominator. At this point, we could have turned the ratio upside down to get the larger quantity on top and the only effect it would have had on the result is that the solution would have turned out *positive* instead of negative. The point is to show that *only the sign of the answer is changed by inverting the ratio.* Still, it simplifies computations somewhat by having the larger quantity on top or as the numerator.

Solution 2: Using formula 7 and substituting, we have:

$$dB = 20 \log \frac{15}{22} + 10 \log \frac{72}{50}, \quad dB = 20 \log (0.68) + 10 \log (1.44),$$

$$dB = 20 (0.8325 - 1) + 10 (0.1584), \quad dB = 16.65 - 20 + 1.58,$$

$$dB = 18.23 - 20 = -1.77 \text{ or } -1.8 \text{ dB}$$

Converting Decibels to Power Ratios

Sometimes it is desirable to calculate the power ratio required for a specific dB figure. For instance, if we wish to achieve a dB increase of +7 dB in an amplifier, what must be the power gain of the amplifier? This is the reverse of working the dB formula, dB = 10 log P1/P2; so we must rearrange the formula to solve for P1/P2. The formula for P1/P2 is derived as follows:

$$dB = 10 \log \frac{P1}{P2}, \quad \frac{dB}{10} = \log \frac{P1}{P2}, \quad \frac{P1}{P2} = \text{antilog} \frac{dB}{10}$$

Now, let's put this formula to work in solving the example above. Substituting, we have:

$$\frac{P1}{P2} = \text{antilog} \frac{7}{10} = \text{antilog } 0.7 = 5.01 \text{ or } 5 \text{ to } 1 \text{ ratio}$$

If the dB figure had been negative, as in a loss, we could have made the calculation in the same manner and then simply inverted the ratio—that is, a dB figure of −7 dB translates to a power ratio of 1 to 5.

Converting Decibels to Voltage Ratios

This must be dealt with in two forms: (1) voltage in equal resistances or impedances, and (2) voltages in unequal resistances or impedances.

Voltages in Equal Impedances   This is the converse of solving for the dB value when the voltage ratio is given so that the formula used for calculating dB from voltages in equal impedances is rearranged to solve for the voltage ratio. The formula is derived as follows:

$$dB = 20 \log \frac{E1}{E2}, \quad \frac{dB}{20} = \log \frac{E1}{E2}, \quad \frac{E1}{E2} = \text{antilog} \frac{dB}{20}$$

Example: The output from a certain amplifier is increased by 12.5 dB. What is the voltage ratio required to produce this 12.5 dB increase?

Solution: Substituting into the formula, we have:

$$\frac{E1}{E2} = \text{antilog } \frac{12.5}{20} = \text{antilog } 0.625 = 4.2 \text{ or a 4.2-to-1 ratio}$$

Voltages in Unequal Impedances A modification of the formula used to calculate dB from voltages in unequal resistances will give us the formula for voltage ratios, given the dB value and the resistance values. The formula is derived as follows:

$$dB = 20 \log \frac{E1\sqrt{R2}}{E2\sqrt{R1}}, \quad \frac{dB}{20} = \log \frac{E1\sqrt{R2}}{E2\sqrt{R1}}, \quad \frac{dB}{20} = \log \left(\frac{E1}{E2}\right)\left(\frac{\sqrt{R2}}{\sqrt{R1}}\right),$$

$$\text{antilog } \frac{dB}{20} = \left(\frac{E1}{E2}\right)\left(\frac{\sqrt{R2}}{\sqrt{R1}}\right), \quad \frac{E1}{E2} = \text{antilog } \frac{dB}{20}\left(\frac{\sqrt{R1}}{\sqrt{R2}}\right),$$

$$\frac{E1}{E2} = \left(\text{antilog } \frac{dB}{20}\right)\sqrt{\frac{R1}{R2}}$$

Example: An amplifier has a power gain of 15 dB. The input imp edance of the amplifier is 500 Ω, the output impedance is 8 Ω. Calculate the voltage ratio.

Solution: Let E1 and R1 represent the output voltage and output impedance respectively and let E2 and R2 represent the input voltage and input imped-ance respectively. Substituting into the formula, we have:

$$\frac{E1}{E2} = \text{antilog } \frac{15}{20}\left(\sqrt{\frac{8}{500}}\right), \quad \frac{E1}{E2} = (\text{antilog } 0.750)\left(\frac{2.83}{22.36}\right),$$

$$\frac{E1}{E2} = (5.62)(0.1265) = 0.71 \text{ or a 0.71-to-1 ratio}$$

This indicates that the output voltage is less than the input voltage. On the surface, it may seem ridiculous that the voltage at the output of an amplifier is less than the voltage at the input, while at the same time a power gain of 15 dB occurs. The key word is *power* gain. Remember that the decibel is a measure of power ratio and power is a function of E^2/R. In our examp le, there was a tremendous change in the impedance from the input to the out-put. This accounts for the voltage loss/power gain relationship. Just to verify that the calculation is correct, let's use the voltage ratio to calculate the pow-er gain and then to calculate the dB from the power gain. If our calculations are correct, the solution will be 15 dB. Since E1/E2 is 0.71/1, let's assume the input voltage (E2) is 10 V. This makes the output voltage (E1) 7.1 V. The input power then is $(E2)^2/R2 = 10^2/500 = 100/500 = 0.2$ W input. The output power is $(E1)^2/R1 = 7.1^2/8 = 50.4/8 = 6.3$ W output. Now, to calculate the dB gain:

$$\text{dB} = 10 \log \frac{\text{P1}}{\text{P2}} = 10 \log \frac{6.3}{0.2} = 10 \log 31.5 = 10 \ (1.498) = 14.98 \ \text{dB}$$

This proves that our calculations were correct.

Using the Impedance Ratio Graph   When working with voltages in un-equal resistances, the graph in Figure 1-5 is very helpful. This graph corre-lates resistance or impedance ratios from 1 to 1,000 with the dB correction figure. The formula normally used for finding the equivalent dB value for voltage ratios in unequal resistances is dB = 20 log E1/E2 + 10 log R2/R1. Since the graph in Figure 1-5 solves the 10 log R2/R1 portion of the equa-tion, all you have to do is solve for 20 log E1/E2 and then *add or subtract* the correction factor taken from the graph. Let E1, R1 represent the output voltage and impedance and E2, R2 equal the input voltage and impedance. Then, if the output impedance (R1) is greater than the input impedance (R2), the correction factor will be negative. If the output impedance (R1) is smaller than the input impedance (R2), the correction factor will be positive. A couple of examples should clarify this for you.

Example: The voltage measured at the input to an amplifier is 3.5 V. The output voltage is 3.0 V. The input impedance is 500 Ω; the output imped-ance is 8 Ω. What is the dB gain of the amplifier?

Solution (Use graphs):   (1) E1/E2 = 3.0/3 = 0.857; from the graph in Fig-ure 1-4 for voltage ratios less than 1 find 0.857 on the vertical scale and read the dB value on the bottom horizontal scale. It is −1.4 dB. (2)R2/R1 = 500/8 = 62.5; from the graph in Figure 1-5, find 62.5 on the vertical scale and then read the dB correction factor on the horizontal scale. It is + 18 dB. We know the dB correction value is positive because the output im-pedance (R1) is lower than the input impedance (R2). (3) Now, simply add the results of steps 1 and 2 together to get dB = −1.4 + 18 = +16.6 dB.

Example: Suppose the output impedance (R1) of an amplifier is 600 Ω and the input impedance (R2) is 425 Ω. What dB correction value would you use if you are converting voltage ratios to decibels?

Solution: Since R2/R1 = 425/600 and this ratio is less than 1, we can *invert* the ratio to get a ratio greater than 1 *if* we remember to make the dB cor-rection factor *negative*. Thus, 600/425 = 1.41. This correlates to −1.5 dB on the graph.

Converting Decibels to Current Ratios

As for voltage ratios, the current ratios must be dealt with in two forms: (1) currents in equal resistances or impedances, and (2) currents in unequal resistances or impedances.

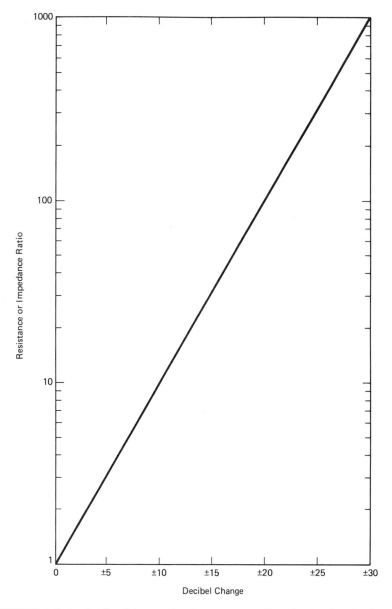

FIGURE 1-5. Basically, this graph gives a correction factor for dealing with voltages in unequal resistances or impedances. If the output impedance is lower than the input impedance the decibel value will be positive; if the output impedance is higher than the input impedance the decibel value will be negative.

Currents in Equal Impedances   The formula for calculating decibels from current ratios in equal resistances can be modified to solve for the current ratio when the decibel value is given. The formula is modified as follows:

$$\text{dB} = 20\log\frac{I1}{I2}\,;\,\log\frac{I1}{I2} = \frac{\text{dB}}{20}\,;\,\frac{I1}{I2} = \text{antilog}\,\frac{\text{dB}}{20}$$

Example: The gain of an amplifier is increased to give a 24-dB increase at the output. By what ratio did the current in the load change?

Solution: $I1/I2 = \text{antilog dB}/20 = \text{antilog } 24/20 = \text{antilog } 1.2 = 15.85$, or a 15.85 to 1 ratio.

Currents in Unequal Impedances   The formula for calculating decibels from currents in unequal resistances is modified as follows to yield a formula to solve for the current ratio $I1/I2$ when the dB and the resistance values are known. The formula is derived as follows:

$$\text{dB} = 20\log\frac{I1\sqrt{R1}}{I2\sqrt{R2}}\,;\,\frac{\text{dB}}{20} = \log\frac{I1\sqrt{R1}}{I2\sqrt{R2}}\,;\,\frac{\text{dB}}{20} = \log\left(\frac{I1}{I2}\right)\left(\frac{\sqrt{R1}}{\sqrt{R2}}\right)\,;$$

$$\text{antilog}\,\frac{\text{dB}}{20} = \left(\frac{I1}{I2}\right)\left(\frac{\sqrt{R1}}{\sqrt{R2}}\right)\,;\,\frac{I1}{I2} = \text{antilog}\,\frac{\text{dB}}{20}\left(\frac{\sqrt{R2}}{\sqrt{R1}}\right) = \text{antilog}\,\frac{\text{dB}}{20}\left(\sqrt{\frac{R2}{R1}}\right)$$

Example: An amplifier has a power gain of 15 dB. The input impedance is 500 Ω; the output impedance is 8 Ω. Calculate the current ratio required for this 15-dB power gain.

Solution: Let $I1$ and $R1$ represent the output current and impedance respectively and let $I2$ and $R2$ represent the input current and impedance respectively. Substituting into the formula, we have:

$$\frac{I1}{I2} = \text{antilog}\,\frac{15}{20}\left(\sqrt{\frac{500}{8}}\right)\,;\,\frac{I1}{I2} = \underset{\text{ratio}}{(\text{antilog } 0.75)(7.9)} = 44.4, \text{ or a 44.4 to 1}$$

This is quite a large current gain, which is caused by the large differential in the input and the output impedances.

Quick Estimations of Power and Voltage Ratios to Decibels and Vice Versa

Often, in field work, it is desirable to be able to convert power and voltage ratios to decibels or vice versa without the aid of log tables, charts, etc. Where accuracy is not too critical, this can be done fairly easily. The method presented here requires a minimum of memorizing. The fundamental thing to remember is that for each time the *power is doubled*, the dB scale increases by 3. Each time the power is multiplied by 10, the dB scale increases by 10. For each time the *voltage is doubled,* the dB scale increases by 6. Each time the voltage is multiplied by 10, the dB scale increases by 20. Thus, a voltage ratio produces twice as much change in the dB figure as the same power ratio. By simply remembering these basic rules of thumb, you can quickly calculate approximate dB values from power or voltage ratios. Just determine how many times the voltage or power was doubled for a given ratio. Then multiply the number of times it was doubled by 3 (for power ratios) or 6 (for voltage ratios). Stated another way: Determine the power (exponent) of 2 that yields a value closest to the given ratio and multiply that exponent by 3 (for power ratios) or 6 (for voltage ratios). For ratios of 10 or powers of 10, multiply the exponent of 10 by 10 (for power ratios) or 20 (for voltage ratios).

Example: What is the dB equivalent of a *power gain* of 20?

Solution: This is the same as 10×2, where the power is multiplied by 10 and then by 2. The multiplication of 10 gives us a dB value of 10 and the multiplication of 2 gives us a dB value of 3. The sum of the two (10 + 3) is the solution: 13 dB (approximately).

Example: In *equal* resistances, a *voltage gain* of 42 is equivalent to how many dB gain?

Solution: This ratio (42) is roughly equal to $10 \times 2 \times 2$ (or 40). For each time a voltage is multiplied by 10, we add 20 dB; for each time the voltage is multiplied by 2, we add 6 dB. In this example, this would be 20 + 6 + 6 = 32 dB (approximately). The exact solution is 32.5 dB, so we are not far off.

Example: What is the dB equivalent of a power gain of 34?

Solution: This is roughly the same as $2 \times 2 \times 2 \times 2 \times 2$, or 2^5, or 32. Thus, we multiply the exponent (5) by 3 to make 15 dB. By using the formula, the exact answer is 15.3 dB.

As you can see, this quick computing method is surprisingly accurate in most cases. If you desire even better accuracy in calculating dB values from power and voltage ratios and you don't have the aid of a chart, log table, etc., at hand, you can make up your own chart (from memory) as follows. Make a column for ratios, another column labeled power, and a third labeled voltage. Under the column labeled ratios, start at 1 and make a list of numbers, each entry being twice the previous one (1, 2, 4, 8, 16, . . .). Make this list out to as high a ratio as you need. Come back to the top under "power" column and write the dB value for the corresponding ratio. You will start at 0 and add 3 each time from top to bottom. Finally, under the "voltage" column, start at 0 at the top and add 6 each time in increments from top to bottom. The result will be similar to Table 1-2. As shown in Table 1-2, insert the 10, 100, 1,000, etc., at the appropriate places on the table along

Ratio	Decibels	
	Power	Voltage
1	0	0
2	3	6
4	6	12
8	9	18
(10)	10	20
16	12	24
32	15	30
64	18	36
(100)	20	40
128	21	42
256	24	48
512	27	54
(1000)	30	60

TABLE 1-2. This table shows the relationship between doubling the ratio and the effect on the decibel value for both power and voltage ratios. These values are approximate. The values for 10, 100, and 1,000 are exact.

with their equivalents under "power" and "voltage" columns. The ratios that are multiples of 10 are inserted because the dB values for these ratios are *exact* while the dB values for the other ratios are *approximate*. You should use this to advantage when estimating dB values for ratios of voltage or power.

For increased accuracy, Table 1-3 is helpful. It can be easily memorized by remembering that a 25% increase of power is approximately equal to a 1-dB increase and a 60% increase of power is approximately equal to a 2-dB increase; double this for voltage ratios to give 2 and 4, respectively. This will enable you to "fine-tune" your calculations to give an even better approximation.

Example: Convert a voltage gain of 180 to its equivalent value in decibels.

Solution: Within the 180 ratio, there is a multiplication factor of 100 that gives us exactly 40 dB. This leaves 80, which is an 80% increase over 100. From Table 1-3, under voltage ratios, we see that for a 60% increase the dB

PERCENT	ΔP	ΔE
25%	1 DB	2 DB
60%	2 DB	4 DB
100%	3 DB	6 DB

ΔP = Change in power
ΔE = Change in voltage

TABLE 1-3. This chart shows the relationship between the percentage of change and the decibel value. Memorizing this can help you when you need to mentally convert ratios to decibel values.

figure is 4 and for a 100% increase the dB figure is 6. Since 80% is between these two values we choose 5 dB. This gives us a total of 45 dB. The exact solution is 45.1 dB—close enough for any practical application.

Total System Gain or Loss in Decibels

Another advantage of working with decibels is that the individual gains and losses in a system (when represented by +dB or −dB values) are additive. As an example, refer to Figure 1-6. This represents a basic transmitter, a power amplifier, a transmission line, and an antenna with power gain. The dB values for the gains and losses incurred throughout the system are shown above each point in the system where the gain or loss occurs. The equivalent multiplication factor is also shown at the various points where a gain or loss occurs. A gain is indicated by a plus sign in front of the dB value, and a loss by a minus sign in front. In order to get the net dB gain of the system, simply add (algebraically) all the individual gains and losses throughout the entire system. In the example of Figure 1-6, the amplifier has a gain of +10.8 dB, the transmission line has a loss of −1.5 dB, and the antenna has an equivalent power gain of +8.8 dB. To get the net power gain of the system, add these values algebraically: +10.8 −1.5 +8.8 = 18.1 dB system net gain. If the algebraic sum of these values had been negative, this would represent a system net loss.

We could multiply together all the multiplication factors given to get the net system gain or loss in terms of a net multiplication factor. If the net multiplication factor amounts to less than 1, the system has a net loss. If the multiplication factor exceeds 1, the system has a net gain. However, this method isn't very indicative of the performance of the system in meaningful terms. The dB net result is more meaningful because it relates to how the gains and losses are perceived by the ear.

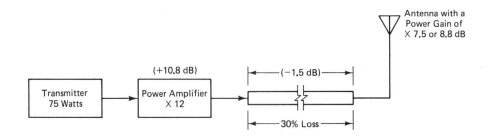

FIGURE 1-6. A typical transmitting system showing gains and losses in the system.

Let's calculate the net multiplication factor for Figure 1-6. The transmission line represents the only loss of the system with a loss of 30%. This translates to a multiplication factor of 0.7. Thus we have: 12 × 0.7 × 7.5 = 63 (net multiplication factor). This sounds like a great amount and it is a significant amount . . . but not that great! The dB figure of 18.1 dB gives a more *realistic* picture of the system gain.

THE dBm

As stated previously, the decibel is a *relative* unit of measure. For example, the value of +12 dB has no meaning in terms of *absolute* power level. If we wish to relate the dB to absolute value, we must establish a certain *reference* level. Many such reference levels are in common use. One of the most common reference levels is the *milliwatt* (mW). In the term dBm, the *m* signifies that the dB figure is based on a reference level of 1 mW. Thus, dBm values represent an absolute power level. Any dBm value can be converted to power in terms of watts by applying the formula:

$$P = \text{antilog}\left(\frac{\text{dBm}}{10}\right)$$

Example: What is the power of +12 dBm?

Solution: Substituting into the formula, we have:

$$P = \text{antilog}\frac{12}{10} = \text{antilog } 1.2 = 15.85 \text{ mW}$$

When the dBm value is negative, such as −12 dBm, there are two ways we could solve for *P*. One way is to keep the negative sign and simply substitute the negative value into the formula above. This will lead to a slight complication in the antilog but will give the solution if properly applied.

Another method is to drop the negative sign and use the formula:

$$P = \cfrac{1}{\text{antilog}\left(\cfrac{\text{dBm}}{10}\right)}$$

For your reference, an example of each method is given here. Use the method that suits you best.

Example: Find the power level of -12 dBm.

Solution 1: Using the first formula and substituting, we have:

$$P = \text{antilog}\left(\frac{-12}{10}\right) = \text{antilog} - 1.2$$

The figure -1.2 is a logarithm, but the negative sign in front indicates that the entire logarithm is negative. In logarithms, only the characteristic can be negative, so we must change this to get a positive fractional part (mantissa) and a negative whole number (characteristic). By the rules of algebra, $-1.2 = 2 - 1.2 - 2 = 0.8 - 2$. This form gives us the positive fractional part (mantissa) and the negative whole number (characteristic). Now, simply look up the antilog of 0.8 and write it down in standard notation as 6.31 and then move the decimal point two places to the left as indicated by the -2 characteristic. This makes the solution 0.0631 mW.

Solution 2: Using the second method and substituting, we have:

$$P = \cfrac{1}{\text{antilog}\left(\cfrac{12}{10}\right)} = \frac{1}{\text{antilog } 1.2} = \frac{1}{15.85} = 0.0631 \text{ mW}$$

With the proper application of these formulas, you can convert any dBm value, positive or negative, to the equivalent absolute power level.

Many voltmeters have a special dB scale on them. Usually the 0-dB reference point is 1 mW in a 600-Ω impedance. The impedance must be taken into account because the dB scale is based on the voltage reading and power $(P) = E^2/R$ or E^2/Z. The dB values are usually read directly from the scale when the meter is switched to the lowest ac range. When higher ranges are used, a correction figure must be added to the figure read from the scale. A correction chart is almost always printed somewhere on the meter face for easy reference.

A typical meter with a dB scale is shown in Figure 1-7. Notice that in the lower left-hand corner the 0-dB level is specified as "zero dB power level—0.001 watt 600 Ω," and in the lower right-hand corner a chart of correction factors is included for the three higher ac voltage ranges. The dB scale is read *directly* on the lower range.

When the meter needle is at the 0-dB mark on the scale, it also lines up over the 0.775-V mark on the lowest ac scale. This is the voltage required to give 1 mW in 600 Ω:

FIGURE 1-7. A typical multimeter with a decibel scale. (Courtesy of Simpson Electric Company, Elgin, IL 60120.)

$$P = \frac{E^2}{R}; \qquad 0.001 = \frac{E^2}{600}; \qquad 0.001 \times 600 = E^2; \qquad 0.6 = E^2;$$
$$E = \sqrt{0.6} = 0.775 \text{ V}$$

Voltmeters that have different voltage scales and ranges will have different correction charts, but the 0-dB mark will always be in line with the 0.775 volt mark as long as the zero reference is 1 mW in 600 Ω.

You can use the dB scale at impedances other than 600 Ω by adding a correction factor. The correction factor is determined by the formula:

$$dB \text{ correction} = 10 \log \left(\frac{600}{Z} \right)$$

As long as Z is less than 600 Ω, the dB correction factor will be positive. If Z is greater than 600 Ω, the dB correction factor will be negative. You can simplify your calculations by making the numerator larger than the denominator by inverting the term 600/Z to become Z/600 when Z is larger than 600 Ω. When you do this remember to *add a negative sign* to your correction factor. Of course, the formula can be used in the standard form (600/Z) if you desire. The dB correction factor will then *automatically* come out negative when Z is larger than 600 Ω. The solution by each method is shown for the following example.

Example: On the 10-V range of the meter in Figure 1-7, the meter is indicating -2 dB. The impedance of the circuit is 1,000 Ω. What is the actual dBm level?

Solution 1: According to the chart on the meter, we must add $+12$ dB to the meter reading for the 10-Volt range. This gives us: $+12$ dB $- 2$ dB $= +10$ dBm. To this, we must add another correction factor because the impedance is not 600 Ω, it is 1,000 Ω. Since the impedance is greater than 600 Ω, the correction factor will be negative. Keeping this in mind, we can simply invert the ratio in the equation to give:

$$\text{dB correction} = 10 \log \left(\frac{1,000}{600} \right) = 10 \log 1.667 = 10(0.222) = 2.22$$

Remember this is negative so the solution is $- 2.22$ dBm. We must now combine this with the figure from the previous step which is $+10$ dBm. This gives us:

$$+10 - 2.22 = +7.78 \text{ dBm}$$

Solution 2: This is the alternative to solution 1. Without inverting the ratio in the formula we have:

$$\text{dB correction} = 10 \log \left(\frac{600}{1,000} \right) = 10 \log 0.6 = 10 \ (0.778 - 1)$$
$$= 7.78 - 10 = -2.22 \text{ dBm correction}$$

Then just add this to $+10$ dBm to get 7.78 dBm.

A graph of 10 log (600/Z) is shown in Figure 1-8. This correlates the dBm correction factor for impedances from 10 Ω to 10,000 Ω.

The dBm versus the VU

A volume unit or VU is read from a *standard* volume indicator. A standard volume indicator is simply a specially designed meter used to monitor speech and/or music waveforms. These complex waveforms of high peak values and relatively low average power levels can't be adequately represented by ordinary types of voltmeters, and this requires a meter with specially designed ballistic characteristics. In our discussion, we are simply interested in the relationship of the dB and the VU. Often the two are confused and taken as meaning the same thing; however, they are not the same and cannot be interchanged except in special circumstances.

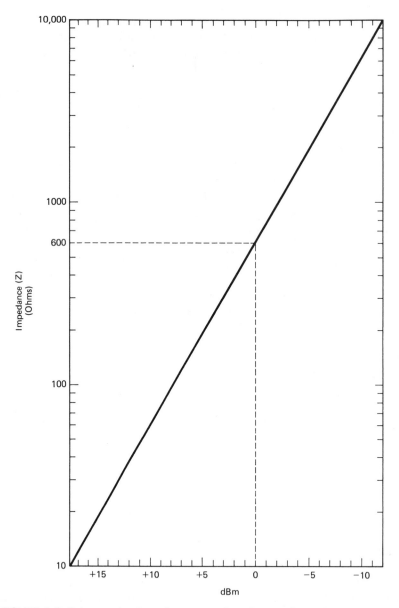

FIGURE 1-8. This graph gives the correction figures (in dB) for impedances other than 600 Ω. Add or subtract this value from the meter
reading on the multimeter or dB meter to get the true value.

A scale of a typical volume indicator or VU meter is shown in Figure
1-9. The 0-VU point is located at approximately two-thirds of full scale. The
lower two-thirds of the scale are calibration marks from 0 to 100, with 100

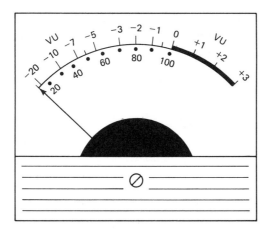

FIGURE 1-9. A typical VU meter showing the details of the scale.

coinciding with the 0-VU mark. The 0 to 100 calibrations represent modulation percentage.

With a steady sine-wave tone signal, the 0-VU point represents 0 dBm (1 mW in 600 Ω) also. The other calibration marks also represent dBm values as long as the signal is a sine-wave tone of steady duration. Thus, for sine-wave steady-state signals, the VU value is the same as the dBm value. However, for complex speech and/or music waveforms this relationship completely falls apart. For these complex waveforms, the VU-meter indication is constantly changing with the peaks of the signal and no definite mathematical relationship between the VU and the dBm can be established.

If, however, volume indicators are connected at several points within a system or chain, we can draw some conclusions as to dB difference from one point to another. For example, suppose that at the input to a line amplifier a volume indicator (with its range multiplier set to 0) is "peaking" often at approximately −1 VU on the scale. At the output of the line amplifier, another volume indicator (with its range multiplier set to +10 VU) is "peaking" often at approximately −2 VU. First, we add the multiplier setting of +10 VU to −2 VU to get +8 VU, which is the level at the output of the line amplifier. The level at the input of the line amplifier is −1 VU. The difference in the two signal levels is then: +8 VU − (−1 VU) = +9 VU. This difference also equals (or approximately so) the dB gain of the amplifier.

The VU indications are only valid when used across an impedance of 600 Ω. When the volume indicator is used across other impedances, a correction factor must be used to get the true volume level at that point. The formula for the correction factor is:

$$\text{VU correction factor} = 10 \log \frac{600}{Z}$$

Notice that this is the same as that used in the previous discussion (the dBm) for using the voltmeter across impedances other than 600 Ω and calculating the correction factor. As far as the correction factor is concerned, the examples discussed there also apply here. The graph in Figure 1-8 also applies here as well.

The dBm and Signal Generators

Professional types of signal generators have their attenuators and/or output level meters calibrated in dBm and/or microvolts (abbreviated μV). The 0-dBm level is the voltage level that will produce 1 mW of signal power. The output impedance of practically all signal generators used in communications work is 50 Ω. The voltage level which corresponds to 1 mW in 50 Ω can be calculated from the formula:

$$P = \frac{E^2}{R}$$

Substituting, we have:

$$0.001 = \frac{E^2}{50}; \qquad 0.05 = E^2; \qquad E = \sqrt{0.05} = 0.2236 \text{ V}, \qquad \text{or } 223{,}600 \text{ μV}$$

Thus, 0 dBm on a 50-Ω signal generator is 223,600 μV. Anything above this reference level would be specified as $+(x)$ dBm and anything below this level would be specified as $-(x)$ dBm. Most often, the levels encountered are below the reference level, or $-(x)$ dBm. The graph in Figure 1-10 correlates the dBm versus voltage levels (in μV) from 0.1 μV to 100 μV.

As you will see in the following chapters, it is a great advantage in having a dial calibrated in dBm as well as in microvolts. When working with signal generators that aren't calibrated in both dBm and microvolts use the graph in Figure 1-10 to convert between the two.

THE dBμV

The dBμV is sometimes used to indicate signal levels, usually RF signal levels. Occasionally, you will find this used in reference to signal generator levels. The dBμV "0" reference is 1 μV, while the dBm "0" reference is 223,600 μV. Be careful not to confuse the two.

The signal level in microvolts (μV) can be converted to dBμV by applying the following formula.

$$dBμV = 20 \log E$$

where E is the signal level in μV.

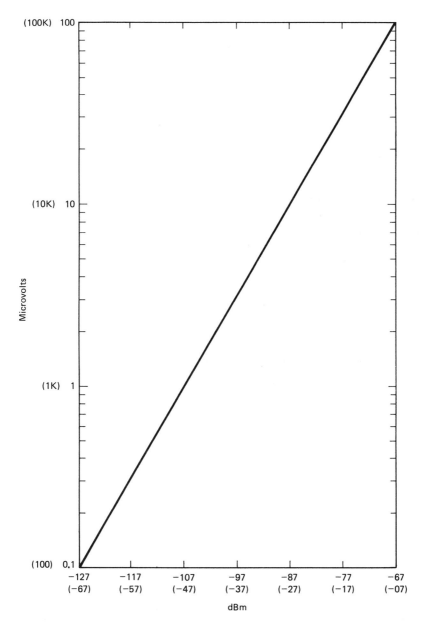

FIGURE 1-10. This graph converts microvolts in a 50-Ω load into the equivalent dBm value. Notice that two scales are used. The inside scales correlate microvolts from 0.1 to 100 to dBm values from −127 to −67 dBm. The outside scales correlate microvolts from 100 to 100K to dBm values from −67 to −07 dBm.

Example: Convert 0.3 μV to dBμV.

Solution: Substituting into the formula, we have:

$$dBμV = 20 \log 0.3$$
$$= 20 (0.477 - 1)$$
$$= 9.54 - 20$$
$$= -10.46 \text{ or } -10.5 \text{ dBμV}$$

THE dBW

The dBW is based on a 0-dB reference level of 1 W. Thus, the dBW is an *absolute* unit of measure. Just remember that the *W* signifies a reference of 1 W. Power levels greater than 1 W are expressed as + (x) dBW; power levels less than 1 W are expressed as − (x) dBW, 1 W being 0 dBW.

To convert watts to dBW, use the formula:

$$dBW = 10 \log P$$

where P is in watts.

Example: What is the dBW equivalent of 0.4 W?

Solution: Substituting into the formula, we have:

$$dBW = 10 \log 0.4$$
$$= 10 (0.6 - 1)$$
$$= 6 - 10$$
$$= -4 \text{ dBW}$$

To convert dBW to watts, use the following formula:

$$P = \text{antilog} \left(\frac{dBW}{10} \right)$$

where P is in watts.

Example: Convert 27 dBW into power in watts.

Solution: Substituting into the formula, we have:

$$P \text{ (watts)} = \text{antilog} \left(\frac{27}{10} \right)$$
$$= \text{antilog } 2.7$$
$$= 501 \text{ W}$$

THE dBk

The dBk is based on a reference level of 1 kilowatt (kW). It is an absolute unit of measure. The *k* signifies that the reference level, or 0 dBk, is 1 kW. Any level that is above 1 kW is expressed as $+(x)$ dBk; any level that is below 1 kW is expressed as $-(x)$ dBk.

To convert power in watts to dBk, use the following formula:

$$dBk = 10(\log P - 3)$$

where P is in watts.

Example: Convert 500 W to dBk.

Solution: Substituting into the formula, we have:

$$\begin{aligned}
dBk &= 10 \ (\log 500 - 3) \\
&= 10 \ (2.7 - 3) \\
&= 27 - 30 \\
&= -3 \ dBk
\end{aligned}$$

To convert from dBk to watts, use the formula:

$$P \ (watts) = antilog \left(\frac{dBK}{10} + 3 \right)$$

Example: Convert -2 dBk to watts.

Solution: Substituting into the formula, we have:

$$\begin{aligned}
P \ (watts) &= antilog \left(\frac{-2}{10} + 3 \right) \\
&= antilog \ (-0.2 + 3) \\
&= antilog \ +2.8 \\
&= 631 \ W
\end{aligned}$$

THE dBμV/m

The dBμV/m stands for dB (above or below) 1 μV per meter. The reference level (or 0 dBμV/m) is 1 μV per meter. Voltage levels greater than this are expressed as $+(x)$ dBμV/m and levels less than this are expressed as $-(x)$ dBμV/m.

This unit of measure is usually associated with field strength measurements (covered in a later chapter). If a piece of wire one meter long is oriented parallel to the lines of force and has a voltage of 1 μV induced into it, the strength of the field at that location is defined as 1 μV per meter or 1 μV/m. This is illustrated in Figure 1-11.

To convert from microvolts per meter (μV/m) to dBμV/m, use the formula:

$$dBμV/m = 20 \log μV/m$$

Example: Convert 17 μV/m to dBμV/m.

Solution: Substituting into the formula, we have:

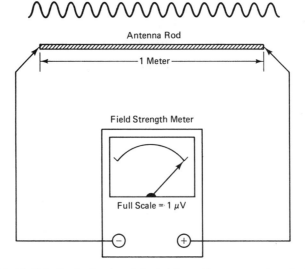

FIGURE 1-11. This illustrates in pictorial form the expression 1 μV/m (1 μV per meter).

$$\text{dB}\mu\text{V/m} = 20 \log 17$$
$$= 20 \ (1.23)$$
$$= 24.6 \ \text{dB}\mu\text{V/m}$$

To convert dBμV/m to μV/m, use the formula:

$$\mu\text{V/m} = \text{antilog} \ \frac{\text{dB}\mu\text{V/m}}{20}$$

Example: Convert 18 dBμV/m to μV/m.

Solution: Substituting into the formula above, we have:

$$\mu\text{V/m} = \text{antilog} \ \frac{18}{20}$$
$$= \text{antilog} \ 0.9$$
$$= 7.94 \ \mu\text{V/m}$$

BIBLIOGRAPHY

Andres, Paul G.; Miser, Hugh J.; and Ringold, Haim, *Basic Mathematics for Engineers.* New York: John Wiley and Sons, Inc., 1944.

Juszli, Frank L. and Rodgers, Charles A., *Elementary Technical Mathematics.* Englewood Cliffs, N.J.: Prentice-Hall, Inc., 1962.

Geiger, Darrell L., *Decibels.* Cleveland Institute of Electronics, 1967.

2
AN INTRODUCTION TO THE SPECTRUM ANALYZER

The spectrum analyzer is potentially the most valuable test and measurement aid available to the communications technician. In the past, spectrum analyzers were intended primarily for use in the laboratory. Now spectrum analyzers are being manufactured *especially* for use in the communications service shop. Their popularity has grown tremendously in just a few short years. A busy communications shop can hardly get along without the use of a spectrum analyzer nowadays, since this instrument not only saves valuable time but also enables the technician to do a better and more thorough job. Spectrum crowding, caused by the increased demand for two-way radio communications, has created some severe interference problems especially for users in highly populated areas. Such interference problems caused by harmonics, intermods, and other spurious signals caused by spectrum crowding can be best analyzed and solved through the use of a spectrum analyzer.

This chapter focuses full attention on the spectrum analyzer. Covered here are basics of the spectrum analyzer, operation of the instrument, various control functions, what to look for in spectrum analyzers, ways to maximize the use of the instrument, potential problems, and ways to avoid them. More specific uses of the spectrum analyzer are covered in many of the tests and measurements described in the following chapters.

BASICS OF THE SPECTRUM ANALYZER

The block diagram in Figure 2-1 shows a mixer and oscillator stage that can be found in any radio receiver. Suppose this is a standard AM broadcast receiver. If you were to connect the vertical input of a scope (through a demodulator probe) to the output of the mixer and manually tune the receiver across a portion of the band, every signal within that por-

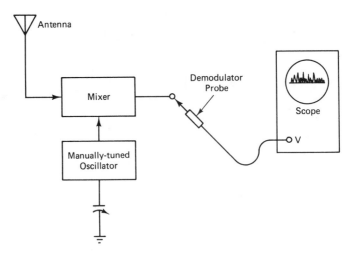

FIGURE 2-1. A typical front-end of a receiver being manually tuned to produce a display of signal on the scope.

tion of the band would register as a "pip" on the scope's screen. The problem is that there would be no order in how these "pips" appear on the scope. The traces of these various signals would be completely out of order, would fall on top of each other, and would be constantly shifting in position. To better understand this, look at Figure 2-2A. The sweep of the local oscil-

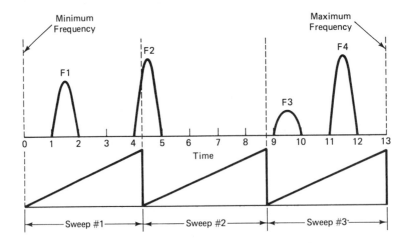

FIGURE 2-2A. The upper part of the figure shows that as the receiver is tuned from minimum to maximum frequency in the time period T-0 to T-13, four signals (F1-F4) will be detected. The lower part of the figure shows that the oscilloscope's horizontal sweep will make three complete sweeps during this time interval.

lator (effected by manually tuning it) will not be locked to the scope's horizontal sweep, which is swept by its own internal sawtooth generator. Thus, the scope will make many horizontal sweeps in the time required to tune the local oscillator manually from minimum to maximum frequency. Just for this example, let's simplify things a bit by assuming that in the time it takes to manually tune the local oscillator from minimum to maximum frequency (time 0 to time 13 in Figure 2-2A), the scope makes three horizontal sweeps across the screen. From Figure 2-2A, you can see that during the first sweep the scope will display F1 and a small part of F2. This is shown in Figure 2-2B. During the second sweep, the remaining portion of F2 will be displayed. During the third sweep, F3 and F4 are displayed. Since the three sweeps are superimposed, the resultant display will look like the one labeled "resultant" in Figure 2-2B. Notice how all the "pips" are displaced from

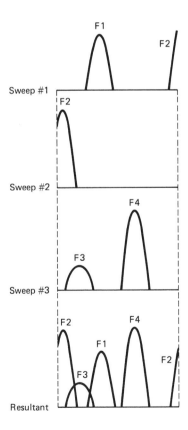

FIGURE 2-2B. The three separate sweep traces are shown here along with the resultant of all three traces as it actually appears on the scope.

their relative positions on the scale at A. In reality, the scope's horizontal sweep would probably have made many more than three passes during the time interval 0–13, which was required to tune the radio manually from the minimum frequency to the maximum frequency as shown at A. This faster sweep rate would cause the "pips" to look like "grass" on the screen. It would be of absolutely no value at all except to indicate that there is at least one signal in the band being scanned.

From this it is obvious that the sweep rate of the radio's oscillator must be exactly synchronized with the horizontal sweep rate of the oscilloscope so that during the time interval required to tune the radio across a certain portion of the band the scope's horizontal sweep will make one pass across the screen. A means of accomplishing this is shown in Figure 2-3. A sawtooth generator supplies a sawtooth waveform that is applied to a varactor diode, the varactor diode being a part of the frequency-determining circuit of the local oscillator. This same sawtooth waveform is applied to the horizontal input of the oscilloscope. The potentiometer, VR1, controls the tuning range

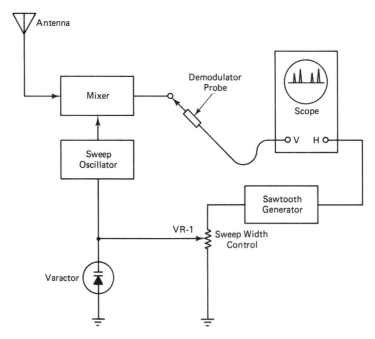

FIGURE 2-3. An arrangement in which a common sawtooth voltage is used to sweep the local oscillator and the scope's horizontal sweep in order to synchronize the two. This results in an orderly display on the scope in which the detected signals appear in proper sequence and spacing.

or sweep width of the oscillator by controlling the amplitude of the sawtooth signal applied to the varactor.

Figure 2-4 illustrates exactly what happens during the time that the radio oscillator is tuned or swept from the minimum to the maximum frequency. Between T-0 and T-13, the sawtooth voltage is rising, thus sweeping the oscillator frequency. This causes the receiver to "scan" the portion of the band that falls between the minimum frequency and the maximum frequency.

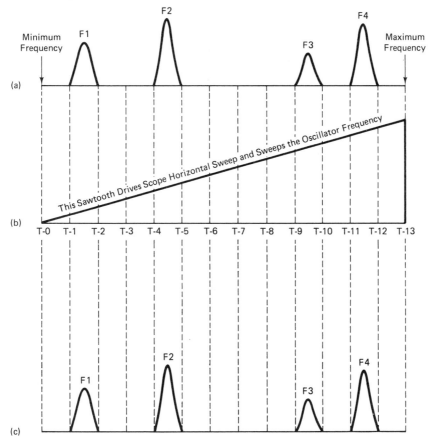

FIGURE 2-4. The top line shows the part of the spectrum scanned by the receiver in Figure 2-3. The drawing at (B) shows that the sawtooth voltage goes from 0 to maximum in the time interval T-0 to T-13, the same time interval in which the local oscillator is swept from the minimum to the maximum frequency. At (C) is the display as it will appear on the scope. Notice that it lines up properly with the display at (A) because the local oscillator is swept in "sync" with the scope's horizontal sweep.

Between T-1 and T-2, the sawtooth voltage level causes the receiver to be tuned to the frequency of F1. Thus, F1 is mixed with the local oscillator frequency (LO) to produce a signal at the mixer output. This signal is detected by the "demod" probe at the mixer output and then applied to the vertical input of the scope. This signal will cause a "pip" to occur on the scope. In the same manner, signals F2 through F4 are detected and seen as "pips" on the scope. Notice in Figure 2-4C that the "pips" on the scope (representing F1-F4) fall in proper sequence and spacing. Notice in Figure 2-4B that when the sawtooth voltage reaches its maximum point at T-13 it drops sharply back to zero. This causes two things to happen: (1) the scope's electron beam is returned swiftly to the other side of the screen, and (2) the local oscillator (LO) frequency of the radio is returned to the minimum or "resting" frequency. At this point, the sawtooth voltage starts rising again and the entire sequence is repeated.

The "pips" caused by F1-F4 will fall at exactly the same place on the screen during each subsequent sweep because the sweep of the oscillator is locked to the sweep of the scope by the common sawtooth voltage.

Figure 2-3 is a basic representation of a "crude" spectrum analyzer. Spectrum analyzers which are designed for communications work are a "bit" more complex although the basic working principle is the same.

A TYPICAL SPECTRUM ANALYZER

Figure 2-5 depicts a spectrum analyzer that is designed specifically for communications service work. This is a Cushman Electronics model CE-15 Spectrum Monitor™. The front panel operating controls have been minimized to simplify the use of the instrument. Figure 2-6 shows a simplified block diagram of the CE-15. Though the following discussion centers around the CE-15, the same principles apply to spectrum analyzers in general.

The CE-15 is a triple-conversion superheterodyne receiver. It is similar in operation to any superheterodyne type of receiver. Let's first discuss the individual stages of the CE-15 as shown in Figure 2-6.

The YIG Oscillator

The first local oscillator is called a "YIG" oscillator, so named because of the composition of the ferromagnetic coil cores used in the oscillator (yttrium, iron, and garnet). In a YIG oscillator, the frequency of oscillation is determined by the strength of the magnetic field. The YIG oscillator in the CE-15 operates over a frequency span of 2.1 to 3.1 GHz (1 GHz = 1,000 MHz).

FIGURE 2-5. The CE-15 Spectrum Monitor™. (Courtesy of Cushman Electronics, Inc.)

The YIG Driver

The YIG driver controls the frequency of the YIG oscillator by varying the current in the coils and hence the strength of the magnetic field. Notice on the block diagram that two outputs from the YIG driver feed the YIG oscillator. One of these outputs is a steady dc current used to set the center frequency of the YIG oscillator. This current level is set by the position of the "coarse" and "fine" frequency adjust resistors. The other YIG driver output to the YIG oscillator feeds a sawtooth or "ramp" signal to the oscillator. This is the signal that does the "sweeping" of the oscillator. The ramp signal is derived from the ramp generator and the amplitude of the ramp signal that is fed to the YIG oscillator determines the scan width or sweep range of the YIG oscillator. This ramp signal amplitude is set by the scan-width selector switch.

The steady dc voltage used to set the center frequency of the oscillator is used to drive a frequency-measuring circuit. This is fed to an *analog-to-digital* converter which changes the voltage level to a digital code which is then used to drive a display, thus giving a readout of the center frequency. The frequency readout is calibrated by the "Freq. Cal" resistor associated with the YIG driver.

There are four sweep ranges available as determined by the scan-width selector. These sweep ranges are: 100 kHz, 1 MHz, 10 MHz, and 100 MHz.

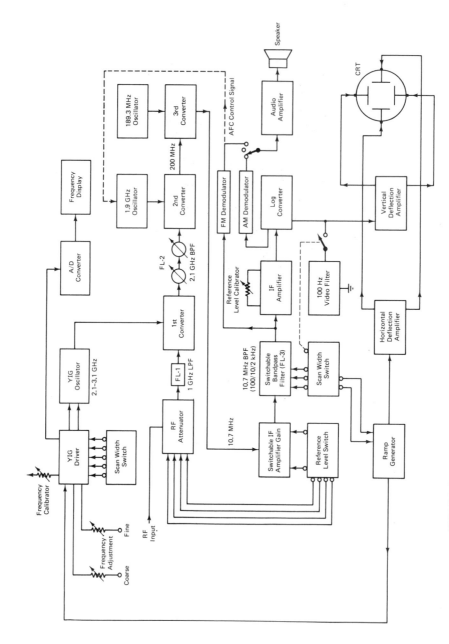

FIGURE 2-6. A simplified block diagram of the CE-15 Spectrum Monitor™.

Since the horizontal scale of the graticule is 10 centimeters (cm), we must divide the total sweep by 10 to get the hertz per cm or Hz/DIV. Thus, the four sweep widths are shown as 10 M, 1M, 100 K, and 10 K.

RF Attenuator

The RF attenuator reduces the level of the input signal to a level that can be handled properly by the first converter. The attenuator consists of a 20-dB attenuator and a 40-dB attenuator. The attenuators are switched in or out by biasing switching diodes. The reference-level switch determines whether one, both, or neither of the attenuators are in the circuit. The attenuation can thus be either 0 dB, 20 dB, or 40 dB.

1-GHz Low-Pass Filter (FL-1)

Since the frequency range of the CE-15 is listed as 1 to 1,000 MHz, it is desirable to block other frequencies from entering the analyzer. The 1-GHz low-pass filter passes frequencies below 1 GHz with little or no attenuation but greatly attenuates frequencies above 1 GHz.

The First Converter

Along with the input signal, the first local oscillator (LO) signal is fed to the first converter. The output from the first converter (the first IF frequency) is 2.1 GHz. Thus, the LO frequency must always operate 2.1 GHz above the RF input frequency. When the center frequency is set by the "coarse" and "fine" frequency adjust resistors on the front panel, the YIG driver causes the YIG oscillator to operate 2.1 GHz above this center frequency. For example, when the center frequency is set to 500 MHz, the LO frequency is set to 2.1 GHz + 0.5 GHz = 2.6 GHz. The scan-width setting controls the oscillator sweep above and below 2.6 GHz. If the scan-width setting is 10 MHz/DIV, the total sweep range is 100 MHz. Thus, the LO frequency swing is 2.6 GHz − 50 MHz (for the low) and 2.6 GHz + 50 MHz (for the high), or 2.55 GHz to 2.65 GHz.

Bandpass Filter (FL-2)

This is a cavity-type bandpass filter that has a center frequency of 2.1 GHz. As the name implies, the cavity filter passes a band of frequencies around 2.1 GHz.

The Second Converter

The second converter receives two input signals, the signal output of the bandpass filter (FL-2) and the 1.9-GHz oscillator signal from the second LO. The output of the second converter is the difference frequency, or 2.1 GHz − 1.9 GHz = 200 MHz. This is the second IF frequency.

The Third Converter

The third converter receives a 200-MHz signal from the second converter and a 189.3-MHz signal from the third LO. These two frequencies produce a difference frequency of 10.7 MHz at the output. This is the third IF frequency.

Switchable Gain IF Amplifier

The switchable gain IF amplifier contains two amplifiers that can be switched in or out depending upon the setting of the reference-level control. One of the amplifiers produces 20 dB of gain; the other produces 10 dB of gain. With only the 10 dB amplifier switched *out*, the attenuation is 10 dB. With only the 20-dB amplifier switched out, the attenuation is 20 dB. With both amplifiers switched out, the attenuation is 30 dB. Thus, the switchable gain IF amplifier can attenuate the signal in 10-dB steps from 0 to 30 dB, depending upon the position of the reference-level control. More about this later.

Switchable Bandpass Filter

The switchable bandpass filter has a bandwidth of 100 kHz, 10 kHz, or 2 kHz, depending upon the setting of the scan-width selector on the front panel. The center frequency of 10.7 MHz does not change, only the bandwidth of the filter changes. The reason for using this type of filter will be explained shortly.

IF Amplifier

This is an IF amplifier, the gain of which can be adjusted by the reference-level cal control to establish the proper reference level of the instrument.

The Log Converter

The purpose of the log converter circuit is twofold. First, it increases the *dynamic* range of the instrument; and second, it makes it possible to calibrate the vertical scale of the CRT graticule in dB units that are calibrated linearly up and down the scale. The display of an ordinary oscilloscope is directly proportional to the voltage of the input signal. If a 1-V signal were fed into the vertical input of the scope and the vertical gain adusted to give full-scale deflection on the screen and then the input voltage were halved (to 0.5 V) then the deflection on the CRT would also be half as much or half-scale. This is what is meant by *linear* display—the deflection changes in direct proportion to the input voltage change.

Figure 2-7A shows a linear display scale and Figure 2-7B shows a log display scale. Signals *A, B, C,* and *D* are displayed on the linear scale. The same signals are shown displayed on the *log* scale. The signal voltage required to produce 0 dBm in 50 Ω is approximately 223,000 μV.

Signal *A* represents a signal at the 0-dBm level. Notice in both Figures 2-7A and B that signal *A* reaches the full-scale deflection mark. The scale at Figure 2-7A is calibrated in μV while the scale at Figure 2-7B is calibrated in dBm, 0 dBm being the top calibration mark. The scale goes from 0 dBm at the top to -70 dBm at the bottom in -10 dBm increments. Signal *B* is one-half the voltage level of signal *A*. On the linear scale, the deflection with signal *B* is one-half that for reference signal *A*. Notice, however, that on the log scale the deflection drops only slightly, corresponding to -6 dBm on the log scale. Thus, signal *B* is 6 dB below signal *A*.

Signal *C* is one-tenth the voltage level of signal *A*. On the linear scale, the deflection for signal *C* is one-tenth the deflection for signal *A*, or one-tenth full-scale, while on the log scale the display for signal *C* is still high on the scale at the -20 dBm mark.

Signal *D* just barely appears above the baseline on the screen of the linear scale but still is high up on the log scale at the -26 dBm mark. Signal *D* is one-twentieth the voltage level of signal *A*. The gain of the instrument represented by the linear scale could be increased to allow signal *D* to give a

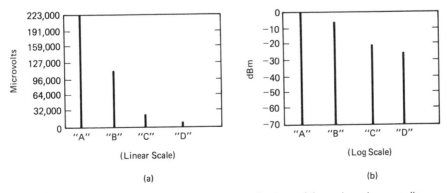

(Linear Scale)   (Log Scale)

(a)   (b)

FIGURE 2-7. The figure at (A) shows a display of four signals on a linear display. At (B), the same four signals are shown on a "log" display.

greater deflection, but then signal *A* would go out of sight. This is the disadvantage of the linear scale with such *limited dynamic range*. Since signal *D* is about as low as we can go and still get a useable deflection while limiting signal *A* to full-scale, the dynamic range of this instrument is about 20 to 1 in terms of voltage level; this is a 26-dB range (from full-scale to one-twentieth full-scale). By comparison, the dynamic range of the scale in Figure 2-7B is nearly 70 dB.

Vertical Deflection Amplifier

The vertical deflection amplifier amplifies the signal output of the log converter to a level sufficient to drive the vertical plates of the CRT.

Horizontal Deflection Amplifier

The horizontal deflection amplifier amplifies the ramp signal to a level sufficient to drive the horizontal plates of the CRT. This produces the horizontal sweep of the CRT.

100-Hertz Video Filter

As shown by the dashed line, the video filter switch is part of the scan-width selector. The 100-Hz video filter is switched in only when the scan-width selector is in the 10K (FLTR) position. This filter serves to increase the resolution by removing noise components. This is usually necessary only at very low signal levels.

The Ramp Generator

This circuit generates the basic sawtooth or ramp waveform that is used to supply the horizontal sweep of the CRT and to sweep the YIG oscillator. The time period of the waveform determines the sweep-rate of both the CRT horizontal sweep and the YIG oscillator.

AM and FM Demodulator

The AM and FM demodulators enable the operator to monitor AM or FM signals to which the CE-15 is tuned. In these two modes (MON-AM and MON-FM), the sweep of the local oscillator (LO) is stopped. The horizontal sweep of the scope is continued and when the instrument is manually tuned to a signal this will cause the baseline to shift, the signal being represented by a straight line across the CRT.

Scan-width Selector Switch

The horizontal scale of the CE-15 is divided into ten equal divisions of 1 cm each. Since the horizontal scale represents frequency, each division represents a certain frequency span. The particular frequency span depends upon the setting of the scan-width selector. The available scan widths are 10 MHz/DIV, 1 MHz/DIV, 100 kHz/DIV, and 10 kHz/DIV.

In addition to setting the scan width, the scan-width selector also selects the proper IF bandwith and the proper scan rate for each setting of the selector. This reduces the chance of operator error as might occur if the sweep rate, scan width, and IF bandwidth were selected independently. There is a proper combination of sweep rate and IF bandwidth which will give the best display. Generally, the greater the scan width, the higher the scan rate and the greater the IF bandwidth. Remember, scan width is the frequency range through which the first LO is swept and scan rate is how fast it is swept through this range. Table 2-1 correlates the IF bandwidth and sweep rate (scan rate) for each position of the scan-width selector.

Notice the extremely slow sweep rate (1 Hz) when the scan-width selector is in the 10-kHz (FLTR) position. Faster sweep rates would cause an ambiguous display with the 100-Hz filter in.

Reference-Level Control

The reference level on the CRT graticule is the top line as shown in Figure 2-7B. Although the reference line is marked 0 dBm on the scale, the actual reference level is determined by the setting of the reference-level control on the front panel.

Scan Width	I-F Bandwidth Filter	Sweep Rate
10 MHz/DIV	100 kHz Bandpass	20 Hz
1 MHz/DIV	100 kHz Bandpass	20 Hz
100 kHz/DIV	10 kHz Bandpass	20 Hz
10 kHz/DIV	2 kHz Bandpass	10 Hz
10 kHz/DIV (Filter)	2 kHz Bandpass 100 Hz video filter	1 Hz
MON FM	10 kHz Bandpass	No sweep
MON AM	10 kHz Bandpass	No sweep

TABLE 2-1

Making an Actual Measurement   When the reference-level control is set to one of the 0-dBm positions, the signal level can be read directly from the scale on the CRT. For other settings of the reference-level control, the actual signal level is equal to the algebraic sum of the reference-level setting and the scale calibration corresponding to the signal. For example, if the reference-level control is set to -30 dBm and a signal display reaches the -20 dBm calibration mark on the scale, the actual signal level would be $(-20) + (-30) = -50$ dBm.

RF and IF Attenuation   The reference-level control sets both the RF and IF attenuation, the total signal attenuation being the *sum* of the RF and IF attenuation. Table 2-2 shows the twelve positions of the reference-level control of the CE-15 and the RF and IF attenuation for each. The combination of RF and IF attenuation is such that at any reference-level control setting a signal input equal to the reference level will produce a specified signal level at the output of the switchable gain IF amplifier. To clarify this, suppose the reference-level control is set to -20 dBm and a signal level of -20 dBm is fed to the input of the instrument. The combination of RF/IF attenuation $(-30$ dBm) sets the output of the switchable gain IF amplifier to a specified level, which we will call X dBm. Now if the reference-level control is set to -30 dBm, this represents a reference level that is 10 dBm lower than for the -20 dBm setting. Thus, the RF/IF attenuation is reduced by 10 dBm to

Position	Reference Level	RF Attenuation	IF Attenuation
1	$+20$ dBm	-40 dBm	-30 dBm
2	$+10$	-40	-20
3	0	-40	-10
4	-10	-40	0
5	0	-20	-30
6	-10	-20	-20
7	-20	-20	-10
8	-30	-20	0
9	-20	0	-30
10	-30	0	-20
11	-40	0	-10
12	-50	0	0

TABLE 2-2

−20 dBm to compensate for this. This allows the −30-dBm signal level to produce the same IF level (X dBm) at the output of the switchable gain IF amplifier. It is the X-dBm level at the output of the switchable gain IF amplifier that causes full-scale deflection on the CRT. The RF/IF attenuator combinations simply determine the signal input level required to produce the X-dBm level at the output of the switchable gain IF amplifier.

Repeated dBm Positions

Notice in Table 2-2 that four of the reference-level positions are repeated. These are the 0 dBm, −10 dBm, −20 dBm, and −30 dBm positions. The total RF/IF attenuation is the same for these positions, but the combination has been changed. For example, in both 0-dBm positions the total RF/IF attenuation is −50 dBm, but in position 3 the RF attenuation is −40 dBm while in position 5 the RF attenuation is −20 dBm. The reason for this is that the input level to the first converter must not exceed −40 dBm for signals above 100 MHz or −50 dBm for signals below 100 MHz. This ensures that the distortion products will be at least 60 dB down. If signals exceeding these levels are allowed to reach the first converter, overloading will occur causing harmonic and intermodulation distortion to be generated. These distortion products can appear on the display and will look like "real" signals within the spectrum being analyzed. While the input to the first converter must not exceed the maximum specified levels to avoid overloading, it should be kept as high as possible for best signal-to-noise ratio. Table 2-3 shows the maximum signal level that should be applied to the CE-15 input for the various RF attenuator settings.

RF ATT	> 100 MHz MAX SIG LEVEL	< 100 MHz MAX SIG LEVEL
−40 dBm	0 dBm	−10 dBm
−20 dBm	−20 dBm	−30 dBm
0 dBm	−40 dBm	−50 dBm

TABLE 2-3

"Coarse" and "Fine" Tuning Controls

The "coarse" and "fine" tuning controls set the center frequency of the range being scanned. If the scanwidth selector is set to 1M and the center frequency to 100 MHz, the total range being scanned is 10 MHz, 5 MHz above and 5 MHz below the center frequency of 100 MHz. A 100-MHz signal would appear as a display at the center of the horizontal scale at "FO." A 105-MHz signal would appear on the last calibration mark at the extreme right, while a 95-MHz signal would appear at the last calibration mark to the extreme left.

Zeroing in for Monitoring   If it is desired to monitor a particular signal on a sweep display, simply adjust the tuning controls until the signal of interest is exactly over the center of the scale at "FO." Alternately, reduce the scan-width setting and retune the signal to the FO line. When you have reached the lowest scan-width setting and have retuned the signal to center scale, switch to MON-AM or MON-FM, depending upon the type of modulation. (The pattern associated with the signal display is a clue to the type of modulation.) Then carefully tune for the best sound reproduction.

OPERATIONAL HINTS AND "KINKS"

Identifying "False" Signals

As previously mentioned, overloading of the first converter can cause the generation of "false" signals due to intermodulation distortion products, harmonics, etc. False signals of this type can be identified by using the following procedure. Change the reference-level setting to a position that offers greater RF attenuation. Note the change in reference-level and the change in the level of the signals on the display. The amount of decrease in "real" signals will equal the "increase" in attenuation. The amount of decrease in "false" signals will be much greater than the increase in attenuation, thus identifying them as "false" signals. When changing the reference-level to identify false signals, pay close attention to the overall RF/IF attenuation change. It is the change in attenuation of the signal before it reaches the first converter that will have a dramatic effect on distortion-produced signals. If only the IF attenuation is changed, the RF attenuation remaining the same, the distortion-produced signal will change the same amount as the change in IF attenuation. The amount of RF attenuation used for any setting of the reference-level control is indicated by a red dot on the dial.

Refer again to Table 2-2. Assume you are checking a portion of the spectrum with the CE-15, and you suspect that some of the signals on your display are distortion-produced false signals caused by overloading in the front end. Furthermore, assume that the reference-level control is set initially to position 5, or 0 dBm. In this position, the RF attenuation is −20 dBm. Note the level of the signals on your display, paying particular attention to the signals that you suspect are false. Now switch to position 3, which is still 0 dBm, but now the RF attenuation has increased to −40 dBm. All real signals should remain at the same level on the display, but any distortion-produced false signals will change, probably much more than the change in RF attenuation or greater than -20 dBm.

The Zero-Reference Display

When the center frequency of a spectrum analyzer is tuned to one-half the total sweep width, the sweep will extend down to 0 frequency. For example, if the scan-width selector is set to 10M, the total sweep width is 100 MHz. If the center frequency is set to one-half this sweep width or 50 MHz, the low end of the sweep will reach the 0 frequency mark. At the 0 frequency mark, there will appear a display that serves to inform you that you are sweeping down to 0 frequency. This display also serves as a valuable reference at times. The 0 frequency display is actually the first LO signal. In the CE-15, the first local oscillator ranges from 2.1 GHz to 3.1 GHz. The first IF is 2.1 GHz, which is the difference between the LO frequency and the frequency being scanned:

$$f(\text{LO}) - f(\text{SIG}) = 2.1 \text{ GHz}$$

where $f(\text{LO})$ is the local oscillator frequency, and $f(\text{SIG})$ is the frequency being scanned. Thus, when the local oscillator frequency is 2.1 GHz, the signal frequency must be 0. The 2.1-GHz local oscillator signal feeds through the first converter and passes through the 2.1-GHz filter and all subsequent stages just as any other signal.

Precise Frequency Measurement Techniques

Frequency measurements on the spectrum analyzer's horizontal scale are limited in accuracy. Usually, precise measurement of the frequency of the signal display is not necessary. However, at times it may be desirable to measure the precise frequency of a signal display. Here are two methods used to measure the frequency of a signal on the display.

The Overlay Method If frequency measurement to within 1 kHz is adequate, the following method will work satisfactorily:

1. Tune the signal display to be measured to the center of the screen at FO.
2. Using an accurately calibrated frequency source, such as a synthesized signal generator, connect the signal generator output to the "Freq. Cal. In" connector on the CE-15. The display of the signal generator frequency should show up close to the signal to be measured.
3. Carefully fine-tune the signal generator frequency until the displays overlap exactly. The signal generator frequency is now equal to the previously unknown frequency, within 1 kHz or so. Figure 2-8 shows a

display of a marker frequency being used to determine the unknown frequency.

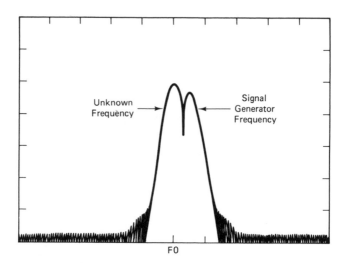

FIGURE 2-8. The "overlay" method of determining the frequency of a signal on the display. The signal generator is tuned until its signal is centered over the unknown signal, then the frequency of the signal generator is taken. (Courtesy of Cushman Electronics, Inc.)

The Zero-Beat Method   If even better accuracy is desired, use the following method:

1. Fine-tune the unknown signal to the FO line.
2. Switch to the MON position.
3. Adjust the signal generator frequency for "zero-beat." You can hear the "zero-beat" in the speaker, and you can view it on the display. There is no oscillator sweep in the MON modes, so the signal frequency will be a straight line across the CRT. However, when beating against another signal—in this case, the signal generator frequency— the resultant trace on the CRT will appear as a series of signals with peaks and nulls, as shown in Figure 2-9. Accuracy to within 100 Hz or so is possible using this method.

Increasing the Dynamic Range

The dynamic range of the spectrum analyzer is usually adequate for most purposes. However, there are times that it is desirable or necessary to

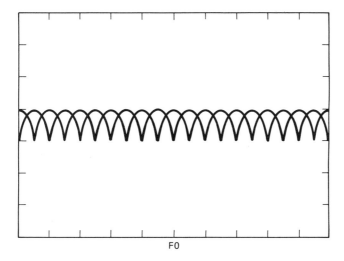

FO

FIGURE 2-9. The "zero-beat" method of determining the frequency of an unknown signal on the display. With the signal display tuned to FO and the instrument switched to MON, the signal generator is tuned for a zero beat. This will be audible from the speaker and will be visible on the screen. The display on the screen will look like this figure. (Courtesy of Cushman Electronics, Inc.)

expand the dynamic range of the instrument in order to measure a relatively weak signal in the presence of a strong signal. If you reduced the attenuation control of the instrument to give the weaker signal more "height" on the display, the strong signal would go "out-of-sight" and could also overload the front end. You could tune the strong signal out of the sweep range to get a better look at the weaker signal. But suppose you need to view the strong signal and the weak signal simultaneously, and the difference in the two signal levels is near or greater than the dynamic range of the instrument. This might be the case when you are trying to adjust a transmitter for minimum spurious signals and maximum carrier signal at the same time. You need to view both the strong signal and the weak signal at the same time on the display in order to know what affect your adjustment is having on both signals. You can extend the dynamic range of the analyzer by inserting a filter to attenuate the carrier frequency. The filter will also attenuate the spurious signal(s), but if the frequency separation between the carrier and the spurious is sufficient the carrier signal will be attenuated much more than the spurious signal. You will need to know how much attenuation the filter offers to both the carrier and the spurious signal. You can determine this by connecting the filter between a signal generator and the analyzer (see

Figure 2-10). First, tune the signal generator to a frequency outside the band-reject of the filter and establish a reference level, preferably at the 0-dBm line. Then tune the signal generator to the carrier frequency and measure the attenuation of the filter in dBm. Next, tune the signal generator to the spurious frequency in question and measure the attenuation. Note these two attenuation measurements for later use.

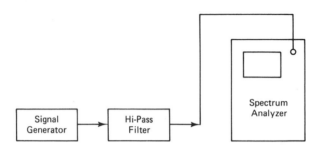

FIGURE 2-10. The setup for determining the bandpass or band-reject response of a filter.

Now hook the transmitter and spectrum analyzer up as shown in Figure 2-11. The carrier signal and spurious signal should now fall well within the dynamic range of the instrument. The display of the spurious signal might even be higher than the display of the carrier signal if the filter is sharp. It makes no difference, just as long as you don't get the two confused and tune up the transmitter for the greatest spurious output! After tuning for the best carrier/spurious relationship, carefully measure the levels of the two signals. Then apply the correction factors for the filter to get the actual level of the two signals.

Avoiding a Scan-Loss Condition

The CE-15 Spectrum Monitor™ is designed specifically for two-way radio communications work. Each setting of the scan-width selector sets the proper sweep width, sweep rate, and IF bandwidth combination for the best display. Many spectrum analyzers have separate controls for setting the sweep rate, sweep width, and IF bandwidth. This somewhat complicates the use of an instrument, but properly used it can give certain advantages. However, improper combinations of sweep rate, sweep width, and IF bandwidth can cause ambiguous displays. To see why, let's take a look at what happens when a spectrum analyzer is swept rapidly past an unmodulated carrier signal.

First, consider the case of using a sweep-frequency generator to check the filter bandpass of a receiver. The setup for checking the bandpass of a

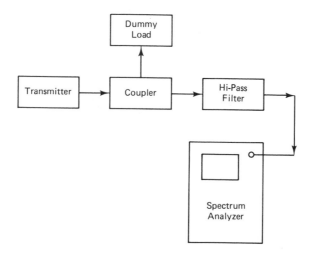

FIGURE 2-11. The setup for checking the transmitter output signal using a high-pass filter to expand the dynamic range of the instrument by reducing the level of the carrier. (Courtesy of Cushman Electronics, Inc.)

receiver's filter is shown in Figure 2-12A. (Note: The IF bandwidth is much more narrow than the preselector in the RF stage, so the width of the waveform will be determined primarily by the *width* of the *IF filter*, although technically the setup shown in Figure 2-12A gives the overall bandpass of the RF and IF stages.) The sweep-generator center frequency is set to the frequency of the receiver under test. An internal sawtooth generator "sweeps" the generator within a range of frequencies around the center frequency. This same sawtooth signal is fed to the scope's horizontal input. The output of the bandpass filter is detected and then fed to the vertical input of the scope. As the frequency of the sweep-generator is "swept" through the band, the output of the bandpass filter will vary in accordance with the response of the bandpass filter. This bandpass output then "draws" a graph on the scope that matches the shape of the bandpass filter's response curve.

The block diagram in Figure 2-12B shows a setup that is the converse of the situation in Figure 2-12A. In this setup, a fixed-frequency signal is fed to the input of the receiver while the frequency of the local oscillator is "swept" by a sawtooth signal from the sawtooth generator. The result is the same and the shape of the bandpass of the filter will appear on the scope just as it did in Figure 2-12A.

The setup in Figure 2-12B is the same in principle to a spectrum analyzer sweeping past a fixed carrier frequency. Thus, the bandpass characteristic gives a certain width to the signal on the display as the analyzer sweeps past it. It takes a certain or "finite" amount of time for the output of

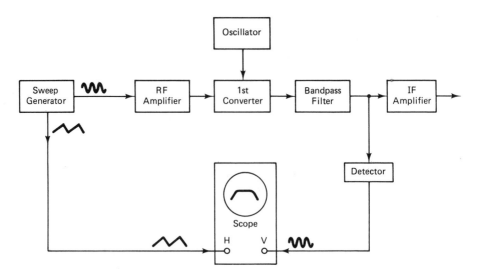

FIGURE 2-12A. Setup for using a sweep generator to check the bandwidth of the filter in the receiver.

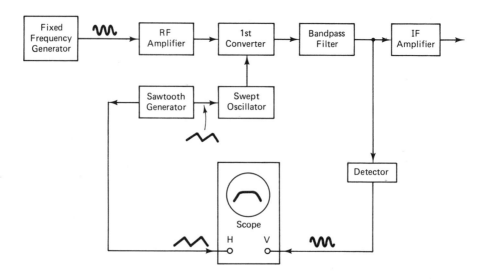

FIGURE 2-12B. Setup here is the converse of Figure 2-12A. Here the signal generator is fixed while the local oscillator of the receiver is swept. The result is the same and the bandwidth of the filter will show up as a display on the scope.

a filter to reach maximum amplitude after a signal "hits" the input. The more narrow the bandpass, the more time required to reach maximum amplitude after a signal "hits" the input. With high sweep rates on the spectrum analyzer, any signal picked up will be "hitting" the filter and quickly disappearing. If very narrow bandpass filters are used with these high sweep-rates, the output from the filter may not have had time to reach full amplitude before the signal is gone. This reduction in amplitude lowers the height of the display on the analyzer and causes the signal to appear wider than it should. The effect of too fast sweep-rates with narrow bandpass filters is shown in Figure 2-13. This type of distortion or loss is called *scan loss*. It can be avoided by using proper sweep rates with proper bandpass filters. Generally, for wide sweep ranges use fast sweep rates and wide passband filters. For narrow sweep ranges, use slow sweep rates and narrow passband filters. To check for scan loss, simply reduce the sweep rate or increase the bandpass. If the height of the displayed signal increases, this indicates that scan loss was occurring (and may still be). Reduce the sweep rate or increase the bandwidth until no further change in signal height is noted.

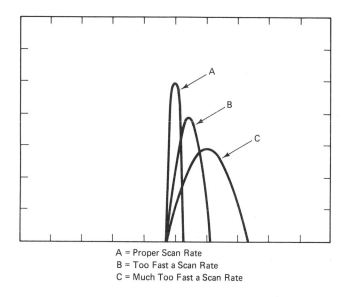

A = Proper Scan Rate
B = Too Fast a Scan Rate
C = Much Too Fast a Scan Rate

FIGURE 2-13. The result of different scan rates are shown here. (Courtesy of Cushman Electronics, Inc.)

Resolution

The resolution of a spectrum analyzer is defined as the ability of an analyzer to distinguish one signal from another on the display. When the frequency separation of two displays approaches the selected bandwidth of the spectrum analyzer, it becomes more difficult to resolve the displays. Two signals of equal amplitude must be separated by at least the selected bandwidth to be resolved effectively into two separate displays. For example, if the selected bandwidth is 10 kHz, two signals of equal amplitude must be separated by at least 10 kHz to be resolved effectively into two distinct displays (see Figure 2-14). Notice that signals A and B start to merge. Frequencies less than 10 kHz apart will merge as one display on the CRT. If one signal level is much less than the other, the required frequency separation must be much greater than the selected bandwidth in order for the spectrum analyzer to resolve the two signals.

Resolution can be improved by switching to a more narrow bandwidth and a slower sweep rate. Remember, fast sweep rates at narrow bandwidths can lead to scan loss, which will only cause further impairment of the resolution. In some cases, switching in a video filter can help to "clean up" the display, giving better resolution. Generally, when these video filters are switched in, the sweep rate should be lowered.

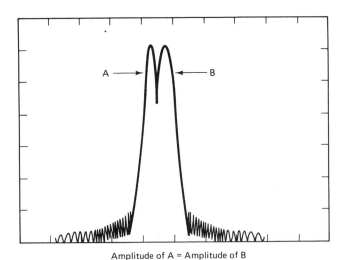

Amplitude of A = Amplitude of B

FIGURE 2-14. The separation of signals A and B is equal to the bandwidth of the instrument. This is the minimum separation for two equal-level signals that can still be resolved on the display. The greater the difference in signal levels, the greater the separation must be for the instrument to resolve the two signals into separate and distinct displays. Signals with close spacing and wide level differences tend to merge as one display. (Courtesy of Cushman Electronics, Inc.)

Bandwidth versus Stability

Obviously, from the previous discussion, the narrower the bandwidth, the greater the resolution possible. It would be possible to reduce the bandwidth easily enough. However, there are other factors that enter the picture when the bandwidth becomes very narrow.

First and foremost, the stability of the first local oscillator becomes a factor. There is always a small amount of residual FM on the local oscillator signal. Thus, the frequency of the local oscillator signal is constantly shifting. This shifting is only a very small amount and normally is not noticeable. But if the bandwidth were reduced below a certain minimum width, the residual FM on the local oscillator signal would cause the signal to move in and out of the bandpass of the filter, causing the display on the screen to become erratic. Obviously, such a narrow bandwidth would be worthless.

A second factor is the sweep rate. As the bandwidth is reduced, the sweep rate must be reduced to avoid scan loss. At very slow sweep rates, an objectional flicker is produced. Thus, the minimum sweep rate is limited except on expensive laboratory-type analyzers with special storage CRTs.

Baseline Shift

If a signal at one of the IF frequencies is present at the RF input position and of sufficient amplitude, it may cause an upward shift of the baseline from its normal location. This is caused by insufficient IF rejection. (The result is shown in Figure 2-15.) In well-designed instruments, this is rarely a problem. If a problem does arise where you must use an analyzer in the presence of an extremely strong signal on one of the IF frequencies, you could use a band-reject filter to attenuate the undesired signal. If the signal is so strong that it still causes a large shift, it is probably entering the IF section directly and there is probably little that can be done to alleviate the problem short of changing locations.

Images of the IFs Images of the second and third IF frequencies also can cause baseline shift, but such images would have to be extremely strong to cause much trouble. Still, you should be aware of this possibility. If the spectrum analyzer is in one of the scan modes, the "image" frequency of the first IF will vary with the sweep of the local oscillator. The image frequency of the CE-15's first IF would be 2.1 GHz above the first local oscillator frequency. Since the LO is swept from 2.1 GHz to 3.1 GHz, the image would range from 4.2 GHz to 5.2 GHz, which is far above the range of the instrument, since the input filter cuts off frequencies above 1 GHz. If a signal in this range were of sufficient amplitude to get through the bandpass filter, it could show up on the display as a regular in-band signal if the LO frequency happened to sweep through the frequency, which is exactly 2.1 GHz below the out-of-band signal. For example, if the center frequency of the CE-15

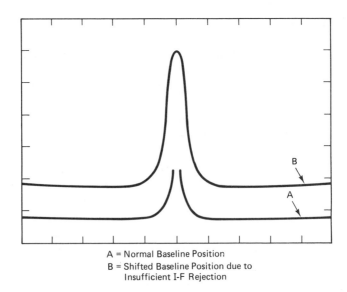

A = Normal Baseline Position
B = Shifted Baseline Position due to
    Insufficient I-F Rejection

FIGURE 2-15. The effect of "baseline" shift. (Courtesy of Cushman Electronics, Inc.)

were set to 100 MHz and the scan-width selector to 1M, the instrument would sweep from 95 MHz to 105 MHz. The LO frequency would be sweeping from 2.195 GHz to 2.205 GHz. Suppose a strong signal at 4.3 GHz were present. In order for this out-of-band signal to be converted to 2.1 GHz, the first LO frequency would have to be:

$$4.3 \text{ GHz} - 2.1 \text{ GHz} = 2.2 \text{ GHz}$$

With the CE-15 set up as described, the LO would sweep through 2.20 GHz; thus, the 4.3 GHz out-of-band signal would be converted to a 2.1-GHz signal and would pass through the 2.1-GHz filter and subsequent stages like an ordinary in-band signal. This type of situation would *not* cause baseline shift, but rather the out-of-band signal would appear on the display as a signal in the scan range. This is because the LO frequency is swept rather than fixed. Since the second and third local oscillators are fixed tuned, the second and third IF images would *not* change with the sweep of the first local oscillator and would therefore shift the entire baseline rather than just at one particular point. In the example above, the 4.3-GHz signal would show up at the same point on the horizontal scale as a 100-MHz signal (at center-scale). (see Figure 2-16.)

Problems such as this are usually not so severe that you can't work around them. They are purposely exaggerated here for the sake of illustration. Be aware, however, that these types of problems are possible.

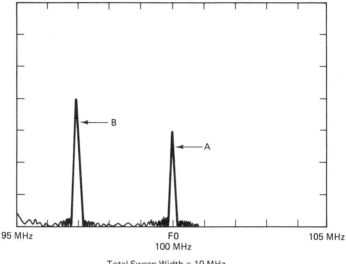

Total Sweep-Width = 10 MHz
Center-Frequency (F0) = 100 MHz
A = 4.3 GHz "Image" Signal
B = 97 MHz Real In-Band Signal

FIGURE 2-16. The signal A is an image signal that is located at 4.3 GHz but which will show up as a 100-MHz signal on the display when the analyzer is set up as described. The signal at B is a real in-band signal at 97 MHz.

TYPICAL DISPLAYS

If displays of unmodulated signals are "clean" signals—that is, not containing spurious signals of a high level—they will generally show up as a single vertical deflection or "pip" on the display. However, modulated signals appear differently, depending upon the type of modulation used. Each particular type of modulation produces its own special pattern or "signature" on the display. After a little experience, you will be able to identify the type of modulation on a signal by the pattern produced on the display.

AM Signals

Figure 2-17 shows a proper display pattern for a clean 27-MHz amplitude-modulated signal. The single-tone modulating frequency is 2 kHz. The sweep width is 1 kHz/DIV (total sweep width is 10 kHz). The higher the percentage of modulation, the higher the amplitude of the sidebands present, up to 100%.

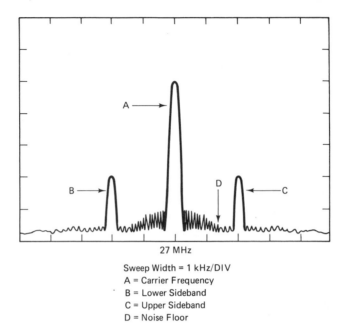

27 MHz

Sweep Width = 1 kHz/DIV
A = Carrier Frequency
B = Lower Sideband
C = Upper Sideband
D = Noise Floor

FIGURE 2-17. The display of a 27-MHz signal modulated by a 2-kHz tone.

SSB Signals

In the absence of modulation, there is supposedly no output signal from a single-sideband suppressed carrier-type transmitter. So if you were to use a spectrum analyzer to check the transmitter output, you wouldn't see any display until modulation was applied. In reality, you will probably see a small amount of carrier-leak, but it should be very small. If it isn't, the transmitter needs adjustment. (This is covered in Chapter 6.) If a 27-MHz SSBSC (single-sideband suppressed carrier) transmitter is modulated by a 1-kHz single tone, the proper output signal as displayed on a spectrum analyzer would look something like Figure 2-18. The signal in Figure 2-18A is the 27.1-MHz sideband. In Figure 2-18B is a small amount of carrier leakthrough.

Normally, SSBSC transmitters are tested with two-tone modulating signals. The two audio tones used should not be harmonically related. Let's assume that a 27-MHz SSBSC transmitter is modulated by two equal-level tones—a 500-Hz tone and a 2400-Hz tone. If the transmitter is properly adjusted, the pattern should closely resemble Figure 2-19. The displays at "A" and "B" of the figure represent the two modulating tones. The displays at "C" and "D" are third-order intermodulation products. The display at "E"

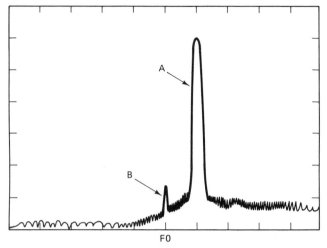

27 MHz Single Sideband Transmitter
Modulated by a 1 kHz Tone
Produces Signal at 27.1 MHz.
F0 = 27 MHz
Sweep Width; 1 kHz/DIV
A = 27.1 MHz Sideband
B = 27.0 MHz Carrier (Leak-Through)

FIGURE 2-18. The display of a 27-MHz single-sideband suppressed carrier signal modulated by a 1-kHz tone.

is a fifth-order intermodulation product. Another fifth-order intermodulation product ("F") is not shown because it lies off the screen beyond the sweep range of the analyzer for that particular sweep-width setting. It could be brought into view by increasing the sweep-width setting or by tuning the center frequency of the spectrum analyzer to a higher frequency. It would be located at 27.0062 MHz. The display at "G" is carrier leak-through.

FM Signals

When a carrier signal is frequency-modulated, something a little strange occurs. Carrier nulls will occur at certain modulation indexes. Modulation index is defined as *frequency deviation divided* by the *frequency of the modulating signal.* Thus, if the deviation is 5 kHz and the modulating frequency is 1 kHz, the modulation index is 5. It can be proven through a complex mathematical analysis using Bessel functions that at specified values of modulation index, the carrier is nulled out leaving only the sidebands. Figure 2-20 shows a typical spectrum analyzer display of a frequency-modulated signal. The center frequency is 100 MHz and the sweep width is 1 kHz/DIV. The

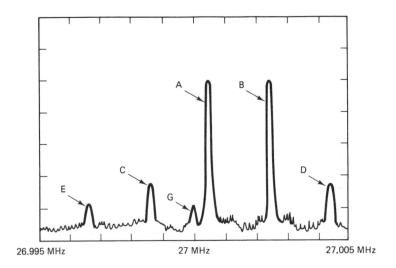

26.995 MHz   27 MHz   27.005 MHz

Sweep Width = 1 kHz/DIV
A = 27.0005 MHz (From 500 Hz Tone)
B = 27.0024 MHz (From 2400 Hz Tone)
C = 2A − B = 26.9986 MHz
D = 2B − A = 27.0043 MHz
E = 3A − 2B = 26.9967 MHz
F = 3B − 2A = 27.0062 MHz (Not Shown Because it Falls Off the Screen)
G = 27 MHz Carrier Leak-Through

FIGURE 2-19. The display of a 27-MHz single-sideband suppressed carrier using two-tone modulating signals.

modulating frequency is 1 kHz. The carrier frequency is at the center FO. Notice the large number of significant sidebands. The spacing of these sidebands is equal to the modulating frequency, or 1 kHz. If the deviation were increased (by increasing the level of the modulating signal), the carrier level would drop as the deviation increases. At a certain point, the carrier would disappear completely. With further increases in the deviation, the carrier would reappear, reach maximum amplitude, drop, and null out again. With continued increases in deviation, this would repeat itself at various values of modulation index. Figure 2-21 shows one such carrier null. The residual left at "C" in the figure when the carrier is nulled out is probably a third-order distortion product ($2A$ - B). Figure 2-22 shows a 150-MHz carrier frequency-modulated by a 1-kHz tone at 5-kHz deviation. This is very similar to the situation in Figure 2-19, except that the display is compressed horizontally due to the increased sweep width (10 kHz/DIV instead of 1 kHz/DIV). The second carrier null occurs at a modulation index of 5.5. The modulation for the display of Figure 2-22 is 5, so the carrier is approaching the second null. This explains the dip at the center of the display. The dip is actually much deeper than indicated by the display. A loss of resolution at this setting causes the dip to be somewhat masked.

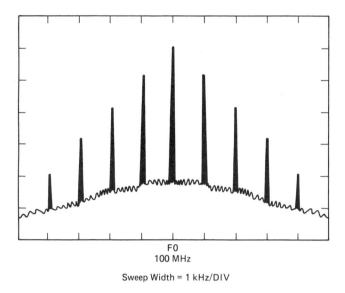

FO
100 MHz

Sweep Width = 1 kHz/DIV

FIGURE 2-20. A display of an FM signal modulated by a 1-kHz tone. Notice the number of significant sidebands produced.

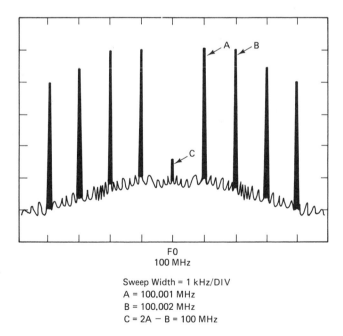

FO
100 MHz

Sweep Width = 1 kHz/DIV
A = 100.001 MHz
B = 100.002 MHz
C = 2A − B = 100 MHz

FIGURE 2-21. The display of an FM signal modulated by a 1-kHz tone. The modulation index is such that a carrier null (C) has occurred.

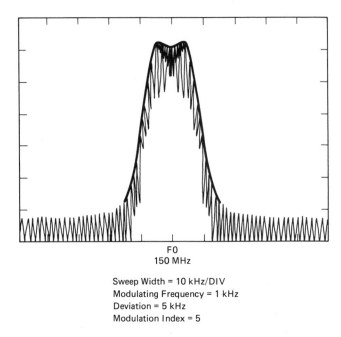

FO
150 MHz

Sweep Width = 10 kHz/DIV
Modulating Frequency = 1 kHz
Deviation = 5 kHz
Modulation Index = 5

FIGURE 2-22. This display is also that of an FM signal modulated by a 1-kHz tone except that the sweep-width has increased causing the display to be compressed horizontally. The dip in the center of the display is caused by a carrier null. (Courtesy of Cushman Electronics, Inc.)

SUMMING UP

As you can see, the spectrum analyzer can be an invaluable aid to the communications technician. It lets you see the individual "components" that make up a composite signal; that is, the carrier, its sidebands, intermod products, harmonics, etc. The technician who uses a spectrum analyzer to tune up a transmitter can do a much better job than the technician who simply watches the reading on a wattmeter and tunes the alignment coils, etc., for maximum meter reading. The wattmeter simply adds the sidebands, distortion products, etc., to produce the meter reading, so you really won't know what the make-up of the composite signal is really like. Remember: What you see is what you get.

Various uses of the spectrum analyzer are incorporated into the test and measurements that are covered in the following chapters. By no means is it possible to cover every application of such a diversified instrument. Its use is limited only by the imagination of the user. If you follow the guidelines given in this chapter, you should have little difficulty in understanding and using any spectrum analyzer.

BIBLIOGRAPHY

Titchmarsh, R.S.; C. Eng., M.I.E.R.E. MI Measuretest #13, "Modulation Measurements with Spectrum Analyser TF 2370." (An application note from Marconi Instruments.)

Cushman Electronics, Inc.; Staff. "CE-15 Spectrum Monitor™ Instruction Manual," San Jose, CA, 1977.

Cushman Electronics, Inc.; Staff. "Using the Spectrum Monitor™." San Jose, CA, 1978.

Texscan Corporation, "Spectrum Analysis, Parts I, II, & III." (An instrument application note.) Indianapolis, IN.

3
COMMUNICATIONS TEST AND MEASUREMENT INSTRUMENTS

This chapter presents the varied instrumentation used in communications tests and measurements procedures. Where appropriate, detailed analyses of the instruments are included. Most of the general-purpose types of instruments are not covered here because it is assumed that the reader is already familiar with these instruments. An understanding of the instrumentation used in test and measurement work is essential to getting the full use and benefit from the instruments and also in analyzing the results. The particular brands of test instruments presented here in no way represent an endorsement of these brands to the exclusion of others.

RF WATTMETERS

In transmitter testing, usually the first test or measurement that comes to mind is the measurement of the RF output power. Instruments that are used to measure the output power of a transmitter are usually referred to as *wattmeters*. The most commonly used types of wattmeters are covered here.

Plug-in-Element Wattmeters

Professional-quality directional wattmeters are designed around the "traveling-wave" concept. The wave on a transmission line can move in both directions; that is, from the source (transmitter) through the transmission line toward the load (antenna) and from the load through the transmission line to the source. Traveling waves that move from the transmitter toward the antenna or load are called *forward waves*. Traveling waves that move from the antenna or load toward the transmitter are called *reflected waves*. In an ideal situation, in which the load or antenna impedance perfectly matches the source or transmitter impedance, there will only be a forward wave on the transmission line. Since practical situations are almost never ideal, there will always be a forward wave and a reflected wave present on a transmission line (see Figure 3-1). The strength of the forward wave will depend upon the transmitter power, while the strength of the reflected wave will depend upon the *severity* of the impedance mismatch between the transmitter and the antenna.

The power actually *accepted* or *dissipated* by the load or antenna is equal to the *difference* between the forward power and the reflected power. A directional RF wattmeter that works on the traveling wave concept is shown in Figure 3-2. To measure RF power, this instrument is inserted in the trans-

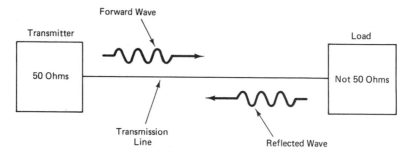

FIGURE 3-1. Forward and reflected power on a transmission line.

FIGURE 3-2. The Bird model 43 directional wattmeter. (Courtesy of Bird Electronic Corporation.)

mission line. The instrument itself contains a built-in section of 50-Ω air-line (tranmission line), which is connected into the transmission line by connectors on either end of the instrument. Plug-in coupling elements are used with this type of wattmeter. These plug-in elements can be rotated 180° to measure either the forward power or the reflected power. To measure forward power, the arrow on the element should be pointing *toward the load*. To measure reflected power, the arrow should point *toward the source*.

Notice that there are three scales on the wattmeter shown in Figure 3-2. These are: 0 to 25 watts, 0 to 50 watts, and 0 to 100 watts. The plug-in elements used with this wattmeter specify the full-scale power for that particular element. The plug-in element also determines the frequency range over which the measurements will be accurate. A variety of different elements are available for different frequency and power ranges. Figure 3-3 shows a diagram of a Bird Thruline™ directional wattmeter. The basic sensing circuit is shown by the simplified drawing in Figure 3-4. The capacitance C is shown as a lumped constant but is actually the capacitance between the bottom of the resistor and a portion of the loop parallel to the axis. Also, M is shown as a lumped constant but is actually the mutual inductance between the center conductor of the line section and the loop. In Figure 3-4, two traveling waves are present on the line. The forward wave is represented by the solid-line arrow, while the reflected wave is represented by the dashed-line arrow. The current through resistor R and inductance M will consist of

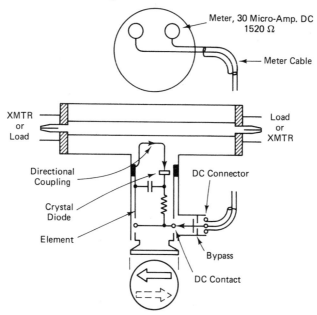

FIGURE 3-3. A schematic diagram of the Bird model 43 wattmeter. (Courtesy of Bird Electronic Corporation.)

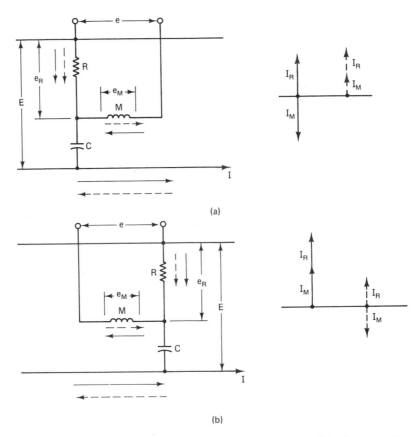

FIGURE 3-4. A simplified diagram of the sensing circuit in the coupling element of the Bird model 43 wattmeter. (Courtesy of Bird Electronic Corporation.)

two components. One current component is caused by the forward wave, while the other current component is caused by the reflected wave. The reflected wave current components are represented by the dashed-line arrows beside R and M, while the forward wave current components are represented by the solid-line arrows.

The current components that flow through R are caused by the capacitance between the loop and the center conductor of the line section. These current components will not be affected by the direction of the traveling wave. Notice that the arrows of both components point in the same direction. The current components that flow through M will be affected by the direction of the traveling wave. Notice that the two current components through M are pointed in opposite directions. The element is designed so that the current component through R (caused by C) is approximately the same amplitude as the current component through M (caused by C). This is true of both the forward wave components and the reflected wave components. This is illustrated by the fact that the dashed-line arrows beside R and M are equal in length, with the length representing the amplitude of the current component. The solid-line arrows are also equal in length. In this case, the current components of the forward wave (solid lines) will completely cancel each other since they are of opposite direction (180° out of phase). The current components of the reflected wave will *add* since they are in phase. Thus, the resulting voltage e will be due to the reflected wave components only, since the forward wave components completely cancel each other. The vectors to the right of Figure 3-4A serve to illustrate this point.

In Figure 3-4B, the coupling element is turned in the opposite direction. In this case, the reflected wave current components *cancel* while the forward wave current components add. Thus, the voltage e represents only the forward wave. This is illustrated by the vectors to the right of Figure 3-4B. The difference in the wattmeter readings in the forward and reflected positions is the actual power dissipated by the load.

Built-in-Element Wattmeters

Another approach using the traveling wave principle is to use *fixed* coupling loops that are permanently placed inside the line section. There are two coupling loops: one for the forward wave and the other for the reflected wave. One such instrument is shown in Figure 3-5. This is a Telewave mod-

FIGURE 3-5. The Telewave Model 44A broadband RF wattmeter. (Courtesy of Telewave, Inc.)

el 44A Broadband RF Wattmeter. The diagram of this wattmeter is shown in Figure 3-6. The wattmeter actually consists of two identical sections: one for the forward wave and the other for the reflected wave. The meter is switched between the two sections to make forward and reflected power measurements. An "OFF" position grounds the meter, thus damping it.

A disadvantage of this type of design is that a frequency-correction graph or chart must be used in certain frequency ranges. The advantage is that a wide range of power levels can be measured by simply flipping a switch rather than changing out elements. The frequency range is 20 MHz to 1,000 MHz.

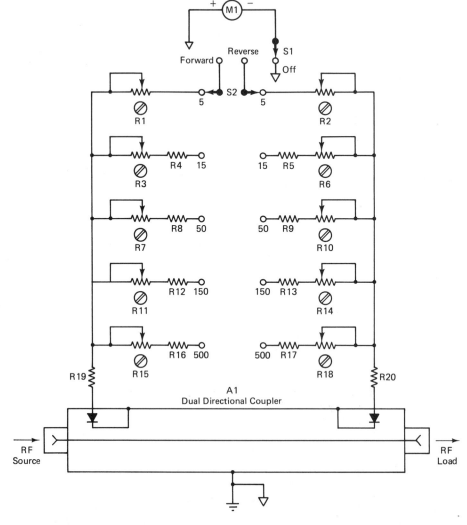

FIGURE 3-6. A diagram of the Telewave Model 44A wattmeter. (Courtesy of Telewave, Inc.)

The Toroid Coil Wattmeter

A typical wattmeter of this type is shown in Figure 3-7. Notice that the instrument consists of two separate parts. The smaller box with the two RF connectors on top is called the *remote sensor.* This part of the kit comes prealigned from the factory as a sealed unit. A simplified partial schematic diagram of the remote sensor is shown in Figure 3-8. Although it isn't obvious from the schematic, the line section passes through the center of L101; L101 is a *toroid coil.* The operation of this sensing circuit is really quite similar to the wattmeter previously discussed. This is a symmetrical circuit; the left side responds only to the reflected wave while the right side responds only to the forward wave. In Figure 3-8, the dashed-line arrows represent the reflected wave; the solid-line arrows represent the forward wave. Notice that the reflected-wave current components are series-aiding through R103 and R101, while the forward wave current components are series-opposing through R103 and R101. Therefore, the forward-wave components will cancel out and the reflected-wave components will add. The reflected wave components are detected by D101.

In the same manner, the forward wave current components through R103 and R102 add, while the reflected wave components through R103

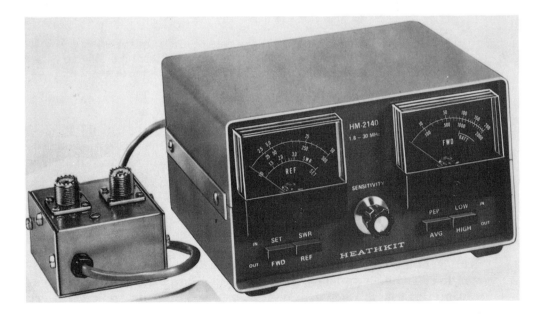

FIGURE 3-7. The Heathkit® Model HM-2140 wattmeter, which uses a toroid coil-type sensing circuit. The sensing circuit is located in the small box at the left in the photo. (Courtesy of Heath Company.)

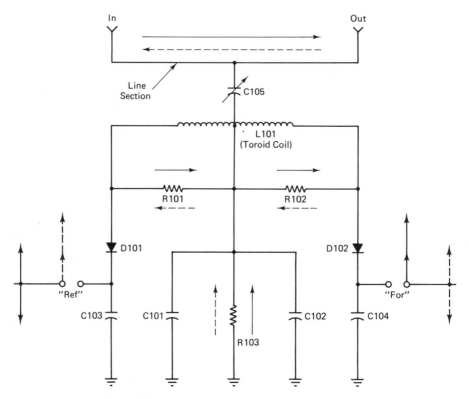

FIGURE 3-8. Schematic diagram of the sensing circuit of the Heathkit®
Model HM-2140 wattmeter, showing the relationships of the forward
and reflected wave current components in the sensing circuit.

and R102 cancel. Thus, the signal detected by D102 represents the forward
wave. The vectors on each side of the diagram show the phase relationships
of the forward and reflected components. This type of sensing circuit has
become very popular, especially in the lower-priced instruments.

Termination-Type Wattmeter

The *termination* type of wattmeter gets its name from the fact that the
instrument itself provides a termination or load impedance for the transmission line. The built-in dummy load is designed to provide an excellent impedance match to the transmission line so that any reflected wave
component will be negligible. This being the case, a directional type of coupling device is not needed.

Figure 3-9 shows a typical circuit of a termination wattmeter. The 50-Ω
load resistor dissipates the power. Since power is equal to E^2/R, the voltage

across the 50-Ω load resistor can be converted to watts. The capacitive voltage divider across the load resistor samples the RF voltage across the load resistor. A small portion of the voltage appears across the lower capacitor and is rectified and filtered before being applied to the meter. The meter is calibrated in watts.

The termination wattmeter in Figure 3-10 has full-scale power ranges of 3, 15, 50, and 150 W and a dummy load power rating of 150 W. The frequency range is from 20 MHz to 520 MHz. This type of wattmeter makes an excellent "workbench" wattmeter.

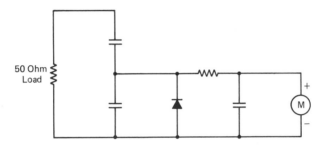

FIGURE 3-9. Typical sampling circuit used in the termination-type wattmeter.

FIGURE 3-10. A typical termination-type wattmeter is the Coaxial Dynamics, Inc., Model 85. (Courtesy of Coaxial Dynamics, Inc.)

WATTMETER INDICATIONS

Wattmeters can generally be classified as "peak reading" or "average reading." Some wattmeters are capable of making peak or average measurements by simply flipping a switch; others are manufactured specifically for average- or peak-reading applications.

Average-Reading Wattmeters

Average-reading wattmeters are used for measuring unmodulated carrier power. To understand how the average reading wattmeter responds, study Figure 3-11. At part A of the figure is an RF detector circuit. Suppose the signal at part B of the figure, representing an unmodulated carrier, is applied to the detector circuit. Diode D1 rectifies the RF signal, causing only the positive half-cycles to appear across R1 in the form of positive pulses. Capacitor C1 filters these pulses to produce a *steady* dc level as shown in Figure 3-11C. This dc level is applied to the meter, which is calibrated in watts. Suppose now that the modulated signal in Figure 3-11D is applied to the detector circuit. The waveform in Figure 3-11E will appear across the meter. Since C1 is too small to filter out the audio modulating component, this audio component will appear as a signal riding on the average dc level. Since the positive peaks are equal to the negative peaks, the wattmeter will respond only to the average level as shown by the dashed line in Figure 3-11E.

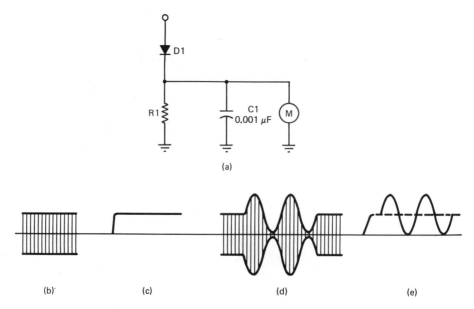

FIGURE 3-11. An average-reading wattmeter circuit.

Peak-Reading Wattmeters

Peak-reading wattmeters are designed to respond to modulation peaks so that the peak power in a modulation envelope can be determined. Peak-reading wattmeters are especially useful in such applications as single sideband. To understand how the peak-reading wattmeter works, refer to Figure 3-12. The RF detector circuit shown in Figure 3-12A is identical to that in Figure 3-11, except that capacitor C1 is much larger (1 μF). If the modulated signal in Figure 3-12B is applied to the detector circuit in part A, the waveform in part C will appear across the meter. Notice that the dc level of the waveform in part C corresponds to the peak of the modulation envelope. This is because the larger capacitor charges to the peak value of the modulation envelope and discharges very little between peaks. Thus, the meter indicates peak power in the modulation envelope.

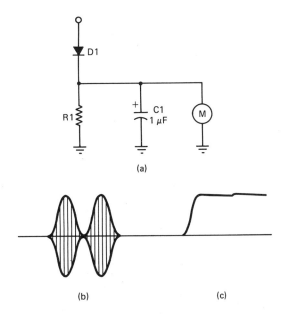

(a)

(b)   (c)

FIGURE 3-12. A peak-reading wattmeter circuit.

DUMMY LOADS

In order to obtain accurate power measurements and to minimize interference to others, the transmitter should be terminated into a *noninductive nonradiating*, purely resistive load for testing. This is usually referred to as a dummy load or dummy antenna. It consists of a special resistor or combinations of resistors arranged to present a resistance that is equal to the output impedance of the transmitter. In communications work this impedance is almost always 50 Ω.

Dry Dummy Loads

Dry dummy loads do not contain a liquid to dissipate the heat that is generated during transmitter testing. Instead, a large heat sink with radiating fins is used to keep the resistor(s) from overheating. The obvious advantage of the dry load is that it is lighter, more portable, and can be used in any position. The disadvantage is that dry loads cannot handle as much power. A typical dry load is shown in Figure 3-13. This particular dry load is rated at 200 W in any position.

A simple, low-power dry load can be constructed by connecting carbon resistors of appropriate resistance and wattage in series/parallel combinations to yield a 50-Ω resistance at the desired power rating. Suppose you need a 20-W, 50-Ω dummy load. Ten 2-W resistors connected in any combination of series/parallel circuits will yield a total power capacity of 20 W. The trick lies in determining what resistance value to use, assuming they are all to be equal. Using ten 2-W resistors of X resistance value must yield 50 Ω. Ten 510-Ω resistors connected in parallel would yield 51 Ω at 20 W. This would make an excellent low-power dummy load.

The resistors should be housed in a small aluminum box with holes drilled in it to allow sufficient ventilation to keep the resistors from overheating. An SO-239 coaxial connector should be used to facilitate connection of the load box to commonly used PL-259 coaxial plugs.

FIGURE 3-13. A typical "dry"-type dummy load. (Courtesy of Coaxial Dynamics, Inc.)

Oil-Filled Dummy Loads

When a higher power-handling capacity is needed, an *oil-filled* dummy load is generally used. An economical oil-filled dummy load is shown in Figure 3-14. This dummy load has a 50-Ω noninductive resistor element submerged in transformer oil. The transformer oil serves to keep the resistor element cool. The oil level should never be allowed to drop below the top of the resistor element. If it does, the resistor element may be damaged from overheating. The value on top of the can is a relief valve to allow venting of the vapors caused by heating.

Some transformer oils contain the chemical PCB (polyclorinated biphenyl), which improves the heat-resistance properties of the oil. Extreme care should be taken when handling transformer oil with PCB. Also, the disposal of transformer oil with PCB must meet proper guidelines for environmental protection. Mineral oil or PCB-less transformer oil is preferred for safety although a slight derating of the power-handling capacity of the load will be necessary. Motor oil should *never* be substituted.

FIGURE 3-14. An economical oil-filled dummy load. (Courtesy of MFJ Enterprises, Inc.)

(a)

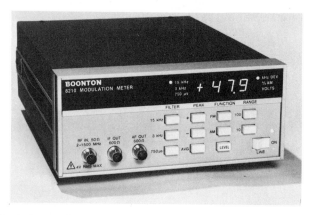

(b)

FIGURE 3-15. (A) A professional modulation meter of the analog type. (Courtesy of Marconi Instruments.) (B) A professional modulation meter of the digital type. (Courtesy of Boonton Electronics Corporation.)

MODULATION MEASURING INSTRUMENTS

Two high-quality professional instruments used to measure both amplitude and frequency modulation are shown in Figure 3-15. The instrument in Figure 3-15A is the *analog* meter type, while the one in 3-15B features *digital* readout. Both instruments feature automatic tuning and automatic leveling. Both IF and AF outputs are available from both instruments. The IF output is a nominal 400 kHz. A general-purpose scope can be connected to the IF output terminal to study AM modulation envelopes. The audio output terminal can be used to measure distortion of the modulated signal. The instrument in Figure 3-15A can be operated by a *nickel-cadmium* battery that recharges as the instrument is operated from an ac power line.

A Simple AM Modulation Meter

Figure 3-16 shows a portable multipurpose CB-type tester that features a simple but surprisingly accurate circuit for measuring amplitude modulation. The specific circuitry used for measuring amplitude modulation, redrawn and simplified from the full schematic, is shown in Figure 3-17. The pushbutton switch, SW1, and the MOD CAL control, R3, are the only *operator* controls necessary for the use of the modulation meter. The carrier set control, R5, is an internal adjustment that is used to calibrate the instrument. The MOD CAL adjustment simply allows the instrument to be used over a wide range of RF power levels.

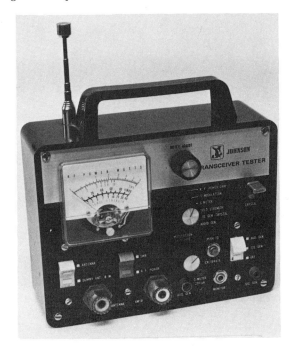

FIGURE 3-16. A multipurpose transceiver tester featuring a very accurate AM modulation meter. (Courtesy of E. F. Johnson Company.)

The RF output of the transmitter is sampled by R1, rectified (detected) by D1, and filtered by C1. This results in a negative dc voltage at "A," the junction of D1 and C1. This dc component at "A" is relative to the carrier level. This dc component is blocked from the circuitry at the right by C2. With the transmitter keyed and no modulation applied, the pushbutton, SW1, is pressed while R3 (MOD CAL) is adjusted for full-scale deflection on the meter, M1. With the pushbutton released, the instrument is now set to read modulation percentage.

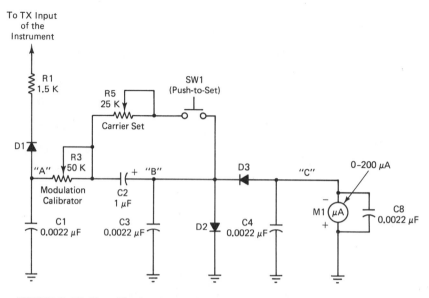

FIGURE 3-17. Simplified schematic diagram of the modulation meter circuit used in the E. F. Johnson Transceiver Tester pictured in Figure 3-16. Courtesy of E. F. Johnson Company.

Figure 3-18 serves to illustrate just how the signals are processed. Figure 3-18A shows how the unmodulated carrier signal is processed. The signal appearing at point "A" has been rectified by D1 and filtered by C1, resulting in a negative dc voltage with some RF ripple. This does not appear at point "B" unless the pushbutton, SW1, is pressed. Thus, the signal at point "B" is zero, as indicated in Figure 3-18. When the pushbutton, SW1, is pressed, the signal at point "A" appears at point "B" and is used to calibrate the meter to full scale by adjusting the MOD CAL control, R3.

Figure 3-18B shows what happens when modulation is applied. Capacitor C1 filters out most of the RF but offers considerable impedance to the audio component of the modulation envelope. The audio component then passes through C2 and appears at point "B," as shown in Figure 3-18. The positive half-cycles are clipped by D2, which shunts the positive half-cycles to ground. The negative half-cycles that remain are passed by D3 to the meter, M1. Some filtering is provided by C4 and C8 to smooth out these negative pulses so that the pulsations don't cause the meter to pulsate (see point "C" in the figure). This voltage component that appears across the meter is directly related to the percentage of modulation.

To summarize, the audio modulation component is separated from the envelope and compared with the unmodulated carrier component, which was used as a reference to set the meter to full scale.

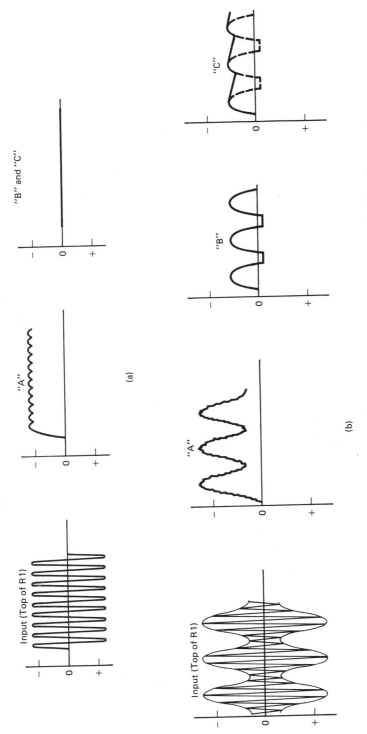

FIGURE 3-18. Waveforms as they appear at various points in the circuitry of the modulation meter of the E. F. Johnson Transceiver Tester.

Frequency Modulation Meter

Figure 3-19 shows a Heathkit® model IM-4180 FM modulation or deviation meter. Tuning and level setting are done manually but with a little
practice tuning presents little difficulty. The frequency range of this particular instrument is 25 MHz to 1,000 MHz. Deviation ranges are: 0 to 2 kHz, 0
to 7.5 kHz, 0 to 20 kHz, and 0 to 75 kHz. The deviation reading is the *peak*
deviation. An audio output jack is provided for monitoring the signal. A
scope output terminal is also provided for checking the deviation on the
scope. Switchable de-emphasis is also provided for proper response to
broadcast stations or to two-way radio services.

FIGURE 3-19. An FM deviation meter, the Heathkit® Model IM-4180 FM
Deviation Meter. Courtesy of Heath Company.

Figure 3-20 is a simplified block diagram of the Heathkit® IM-4180.
The FM signal is converted to a 200-kHz IF signal. The oscillator is tuned
200 kHz above or below the carrier to measure negative or positive deviation. The IF signal is amplified, limited, and then fed to a waveshaper. The
output of the waveshaper is a square-wave signal the frequency of which
varies with the deviation (modulation) of the FM signal at the input. This
square-wave signal feeds a trigger circuit which in turn triggers a one-shot
multivibrator. The output of the one-shot multivibrator is normally low, but
when a trigger pulse is applied to the input a positive pulse of approximately
1.6 μs duration is produced on the output. As the frequency of the IF signal
increases due to deviation the spacing of the pulses from the one-shot
multivibrator decreases (meaning more pulses per second). A decrease in

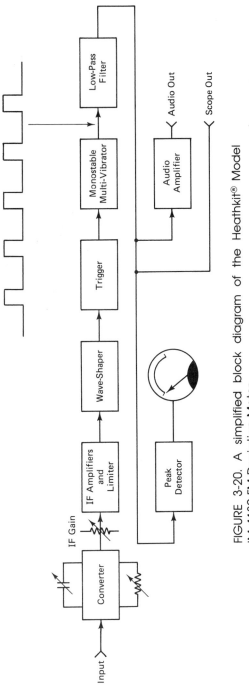

FIGURE 3-20. A simplified block diagram of the Heathkit® Model IM-4180 FM Deviation Meter.

the IF signal frequency due to deviation will increase the spacing of the pulses (fewer pulses per second).

Figure 3-21 illustrates the relationship of positive and negative deviation to the output of the one-shot multivibrator and the output of the low-pass filter. Figure 3-21A represents a carrier (Fc) that is modulated $+/-5$ kHz by a sine-wave audio tone. Figure 3-21B represents the instantaneous pulse rate at the output of the one-shot multivibrator. Figure 3-21C represents the output of the low-pass filter circuit.

In this analysis, the deviation meter is tuned to $Fc - 200$ kHz. When tuned to this point, the deviation meter will measure the peak deviation of the positive (high side) deviation. At time 0 (center), the audio-modulating tone is at the 0 level. The frequency of the IF at this instant will be

$$Fc - (Fc - 200 \text{ kHz}) = 200 \text{ kHz}$$

The output of the one-shot multivibrator at time 0 is shown in Figure 3-21B. It is this 200-kHz signal that produces the average level at the output of the low-pass filters as shown in Figure 3-21C. At time 1, the audio-modulating tone is at its maximum positive level, which is sufficient to cause the carrier to deviate $+5$ kHz. The instantaneous IF frequency at time 1 is:

$$Fc + 5 \text{ kHz} - (Fc - 200 \text{ kHz}) = 205 \text{ kHz}$$

This causes the pulse rate of the one-shot multivibrator to increase, as shown at time 1 in Figure 3-21B. The output of the low-pass filters rises to the maximum as shown in Figure 3-21C. At time 2, the audio-modulating tone is at 0, the IF frequency is 200 kHz, the pulse rate of the one-shot multivibrator is back to the steady-state rate, and the output of the low-pass filters is at the average level. At time 3, the audio-modulating tone is at its maximum negative level, sufficient to cause the carrier to deviate -5 kHz. The instantaneous IF frequency at time 3 is:

$$Fc - 5 \text{ kHz} - (Fc - 200 \text{ kHz}) = 195 \text{ kHz}$$

This causes the pulse rate of the one-shot multivibrator to decrease, as shown in Figure 3-21B. The output of the low-pass filters falls to the minimum level shown in Figure 3-21C. At time 4, the audio-modulating tone is back at the 0 level, the IF frequency is at 200 kHz, the pulse rate of the one-shot multivibrator is back to the steady-state rate, and the output of the low-pass filters is back to the average level.

The peak detector responds to the peak of the waveform at part C. Since this peak is caused by the peak positive deviation ($Fc + 5$ kHz), the deviation meter will read 5 kHz. The peak negative deviation can be measured by retuning the IM-4180 to $Fc + 200$ kHz. The frequency relationships will then be reversed and the peak of the waveform at part C will represent the peak negative deviation.

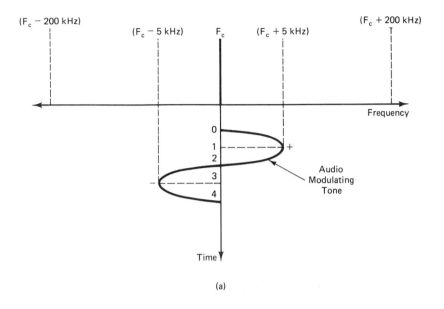

(a)

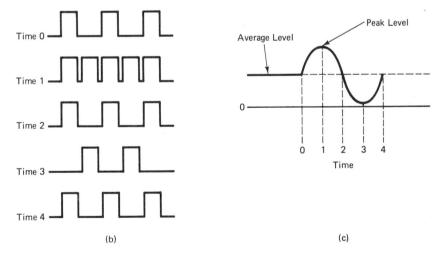

(b)   (c)

FIGURE 3-21. (A) This shows the frequency relationships corresponding to the audio modulating tone. (B) Pulses as they appear at the output of the one-shot multivibrator. (C) The output of the low-pass filters.

FREQUENCY MEASURING INSTRUMENTS

Advances in technology have yielded test equipment that is capable of measuring the frequency of radio and audio signals to a degree of accuracy unheard of several years ago. This technology has led us from the old tube-type micrometer zero-beat frequency meter to the digital IC frequency counter that reads out the frequency directly. A typical frequency counter is shown in Figure 3-22. Another popular type of frequency meter utilizes a digital frequency synthesizer, which serves as a local oscillator in an arrangement that closely resembles a typical FM receiver. This type of instrument and the frequency counter are more fully discussed in the following sections.

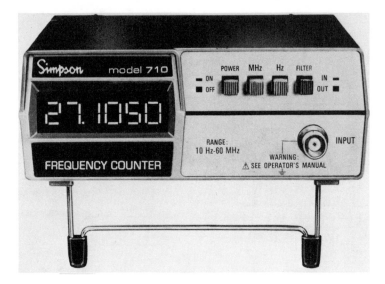

FIGURE 3-22. A typical frequency counter which may be used in communications work is the Simpson Model 710. Courtesy of Simpson Electric Company, Elgin, IL 60120.

The Frequency Counter

The frequency counter has become very popular with communications technicians. This is probably due to several factors: relatively low-cost, simplicity of operation, high degree of accuracy, portability, etc. A block diagram of a simple frequency counter is shown in Figure 3-23. This is a five-digit counter. Display #5 represents the most-significant-digit. The frequency to be counted is first amplified and then shaped by the Schmitt trigger. The

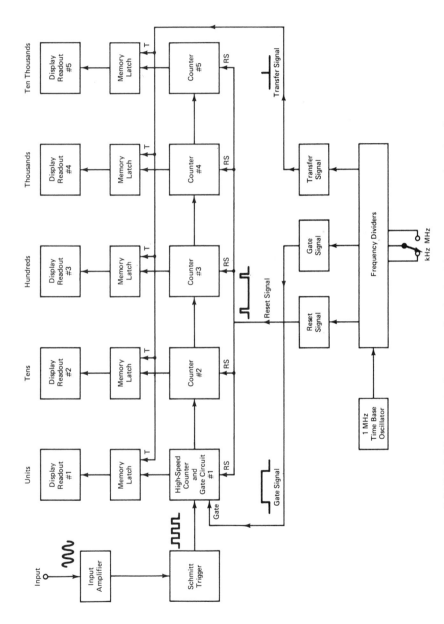

FIGURE 3-23. A block diagram of a typical frequency counter circuit.

Schmitt trigger produces a square-wave pulse for each cycle of the input signal. This pulse has the characteristics necessary for proper triggering of the high-speed counter. The high-speed counter only counts the pulses during a small time interval when the gate is open or *enabled*. During the time interval that the gate is open, every pulse that reaches the counting input is counted. If the time duration of the gating pulse is one millisecond (1 ms), this means that the gate of the high-speed counter will be open for 1 ms and all pulses entering the counter during this time interval will be counted.

Notice in Figure 3-24 that a reset pulse is generated just prior to the gating signal. This resets each of the counters to zero before each count is started. Also notice that a transfer pulse is generated at the trailing edge of the gating pulse. This is applied to each of the memory latches. At the end of the counting pulse, this transfer pulse signals the memory latches to accept and store the data from the associated counter, the previously stored data being "thrown out" at this time. When a 1-ms gating pulse is used, a transfer pulse is not sent after each gating pulse because such fast changes could not be followed by the human eye due to the persistence of vision.

Let's analyze what happens during a counting cycle. Assume the input frequency to be 27,125,417 Hz. For this count, assume the gating pulse is 1 ms in duration. First comes the reset signal that resets all the counters to zero. The 1-ms gating pulse is then applied to the first counter gate. During this 1-ms time interval, all the pulses at the counter input will be counted. Since it would take a full second for all 27,125,417 pulses to get through the gate, in 1 ms only 1/1,000 of these pulses (27,125) will get through. Each of the counters is a *decade*-type counter. This means it can only count to 10 (0 to 9). On the tenth count, the counter returns to 0 and the next counter to the right advances by 1. At the end of the 1-ms gating interval, 27,125

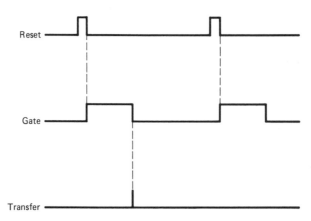

FIGURE 3-24. Various pulses that are used to control the frequency counter.

pulses will have been counted. Counter #5 will register 2; #4,7; #3,1; #
2,2; and #1,5. At this instant, a transfer pulse is sent to the memory latches.
Upon receiving the transfer signal, the memory latch accepts the input from
its associated counter. The memory latch holds this data (number) and produces an output that causes the proper segments of the display to light up,
thus displaying the number in the memory latch. The memory latch holds
this count until another transfer signal is received, which will be after 100
counting intervals have passed.

Notice that three of the digits are lost: 27,125,417 is the input, but only
27,125 is displayed. If the frequency is in MHz, the decimal point will be
placed three places to the left (between the 7 and 1). Since the three digits
to the right are lost, we can only measure the frequency to to the nearest
1,000 Hz. There is a way we can recover the three lost digits, however. If
the gating signal is increased to a one-second duration, every one of the
27,125,417 pulses will make it through the gate during the counting interval.
Again, three digits will be lost, but this time it will be the three higher digits
(2, 7, and 1) and we already know that they are from the 1-ms count. This
time, the display will be 25,417 (2 on Counter #5, 5 on #4, 4 on #3, 1 on
#2, and 7 on #1). If eight counters were used (with their associated memo-
ry latches and displays), all the digits could be displayed simultaneously with
a 1-sec gating interval. However, the three higher digits have been lost because there were not three more counters to the right to "catch" them. With
this one-sec time interval, the readout will be in kHz, with the decimal point
three places to the left (same as before). By making a measurement with the
1-ms gating pulse and then the 1-sec gating pulse, all eight digits can be determined (see Figure 3-25). A switch is provided on the front panel of such
counters to switch from a 1-ms gate pulse to a 1-sec gate pulse. The switch
position relating to the 1-ms gate pulse is labeled MHz, while the position
relating to the 1-sec gate pulse is labeled kHz.

The basic reference frequency of this counter is 1 MHz. The 1-MHz
signal is divided by 1,000 to get a 1-kHz pulse. This 1-kHz pulse is used to
start and stop the gating pulse with the switch in the MHz position. Thus,
the gating pulse duration is 1 ms.

In the kHz position, the 1-MHz signal is divided by 1,000,000. This
produces a 1-Hz pulse. This 1-Hz pulse is used to start and stop the gating
pulse, resulting in a 1-sec pulse duration.

The 1-MHz reference oscillator can be checked against another frequency standard such as WWV to calibrate the instrument. Another excellent calibration method is to couple the counter to the "color" oscillator of a
color TV receiver. The TV should be "color-locked" to a TV transmitter.
Then the 1-MHz oscillator in the frequency counter is adjusted until the
counter display reads 3.579545 MHz. It will be necessary to switch to the
kHz range for the last three digits.

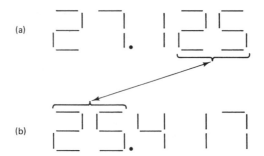

FIGURE 3-25. (A) The readout as it appears when the switch is in the MHz position. (B) The readout as it appears when the switch is in the kHz position.

Prescalers   In order to extend the upper frequency limit of a frequency counter, a prescaler can be connected to the input of the counter. The signal, the frequency of which is to be measured, is fed to the input of the prescaler. The input frequency is divided by ten in the prescaler and then applied to the input of the frequency counter. The actual frequency is ten times the frequency displayed on the counter.

Discriminator-Type Frequency Meters

The block diagram in Figure 3-26 shows a simplified version of a typical discrimination-type frequency meter. As you can see, this is very similar

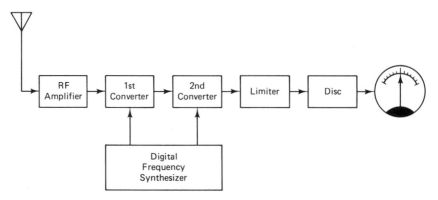

FIGURE 3-26. A simplified block diagram of a typical discriminator type of frequency meter.

to an FM receiver. In fact, it is an FM receiver. The frequency of the digital synthesizer is set by thumbwheel switches to the frequency to be measured. The output of the limiter is an amplitude-limited signal at a low IF frequency (75 kHz or so). This low IF is applied to the discriminator. If the signal at the input to the RF amplifier is equal in frequency to that set up by the thumbwheel switches, the discriminator meter will remain at zero (center scale). If the input frequency is higher or lower than the programmed frequency, the discriminator meter will move to the right or left of zero. The meter is calibrated in kHz. The exact frequency can be determined by adding or subtracting the meter reading from the frequency programmed by the thumbwheel switches. This type of frequency meter is commonly used in multifunction transceiver test instruments called *service monitors*. The digitally synthesized oscillator is also part of the signal generator.

RF SIGNAL GENERATORS

RF signal generators that are used in communications work must meet rigid specifications. Frequency drift must be minimal in order to minimize resetting during testing procedures. An accurate method of setting the frequency of the generator is essential. Dial-calibrated generators are generally very difficult to set to an exact frequency without the use of some external means of frequency measurement. Some signal generators feature a digital readout of the generator frequency to allow precise frequency setting. Other generators use a digital frequency synthesizer. These generators usually provide thumbwheel switches for precise frequency setting. Where the signal generator is required to operate over a relatively narrow range of frequencies, a channel switch is sometimes used; i.e., generators designed for CB work use a switch to cover the forty channels. This simplifies frequency setting and the phase-locked-loop circuitry of these signal generators provide extremely accurate frequency control.

In addition to tight frequency control, generators used in measuring the sensitivity of very sensitive receivers must have the RF output level calibrated down to 0.1 μV or so. In order for this calibration to be valid at such low levels the RF "leakage" from the signal generator must be kept to a minimum. The output level is usually controlled by a step attenuator in conjunction with a variable control and level meter. The output level of the generator is usually calibrated in both microvolts (μV) and dBm where the 0-dB reference level is 1 mW in 50 Ω.

It is essential to be able to modulate the signal generator; the type of modulation used depends upon the type of receiver being tested. AM receivers would require an AM signal while FM receivers would require an FM signal.

Figure 3-27 shows a professional-quality signal generator manufactured by LogiMetrics. This generator features digital frequency readout for precise

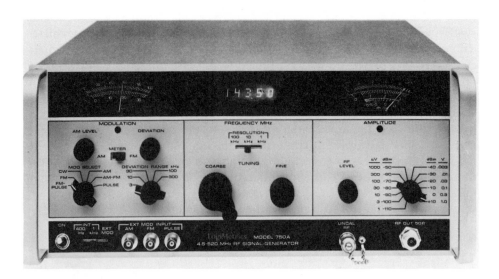

FIGURE 3-27. The LogiMetrics Model 750A professional quality signal generator, capable of AM and FM modulation. Courtesy of LogiMetrics, Inc.

frequency setting and monitoring. It can be AM or FM; the modulation level is indicated by the modulation meter. The output level is adjusted by a step attenuator and RF level control. The precise level is determined by the position of the step attenuator and the meter reading. Notice on the meter that two voltage scales (0 to 1 and 0 to 3) and a dBm scale (−15 to +3) are provided. When measurement of the level is desired in microvolts, use the 0 to 1 scale for attenuator settings of 1, 10, 100, etc. For attenuator settings of 3, 30, 300, etc., use the 0 to 3 scale. For example: If the attenuator is set to 1,000 μV/−50 dBm and the meter reading on the 0 to 1 scale (0.89) is multiplied by the attenuator setting (1,000 μV) to get the level in microvolts, this would be: 0.89 × 1,000 = 890 μV. The level in dBm is equal to the algebraic sum of the meter reading on the dBm scale and the attenuator setting. In this case:

$$-50 \text{ dBm} + 2 \text{ dBm} = -48 \text{ dBm}$$

RF SWEEP SIGNAL GENERATORS

In a sweep generator, the frequency of the signal generator is not constant but rather it is varied within a certain frequency range above and below its center frequency. The range of frequencies covered is called the *sweep width*. It is important that the output level remain constant over the sweep range.

The heart of the RF sweep generator is the *swept-tuned oscillator*. A popular method of sweeping the oscillator is to apply a sawtooth voltage across a varactor diode, the varactor diode being part of the oscillator's resonant circuit. The sawtooth voltage serves to vary the reverse bias of the varactor that in turn varies the capacitance of the varactor. The sweep width depends upon the amplitude of the sawtooth voltage while the sweep rate depends upon the repetition rate of the sawtooth voltage. Generally, slower sweep rates are preferred when sweeping sharp filter circuits. But if the sweep rate is too slow, the flicker on the scope will be objectional. This is because the horizontal sweep of the scope is synchronized with the sweep of the sweep generator.

In effect, an FM signal generator is also a sweep generator since the frequency is varied above and below the carrier frequency. The sweep width depends upon the amplitude of the audio-modulating signal while the sweep rate depends upon the frequency of the audio-modulating signal.

The basic purpose of the sweep generator is in determining the frequency response of radio frequency circuits of transmitters, receivers, antennas, etc. Typical uses and methods of interconnecting sweep generators with scopes, etc., are covered in the following chapters.

AUDIO GENERATORS

For maximum versatility in communications work, audio generators should be tuneable over a wide frequency range with a wide range of output levels. Low distortion in the output signal is imperative if the generator is to be used in making distortion tests. The instrument shown in Figure 3-28 is very suitable for communications work. This is a ViZ model WA-504B/44D audio generator. It covers the frequency range of 20 Hz to 200 kHz. Output levels up to 10 V peak-to-peak are available from the 600-Ω unbalanced output. The output signal is available as a sine wave or a square wave.

AUDIO DISTORTION ANALYZERS

Distortion analyzers are used to measure the distortion content of audio signals directly in percent or as so many dB down. Some distortion analyzers are designed to measure distortion at one fixed audio frequency or a narrow range around a fixed frequency while others can be tuned to measure distortion within a wide range of frequencies.

A simplified block diagram of a fixed-frequency type of distortion analyzer is shown in Figure 3-29. In this arrangement, the reject filter is designed to reject a 1-kHz signal while passing other frequencies with little attenuation. Initially, the function switch is placed in the set-level position as indicated by the solid-line arrow. The level–adjust control is then used to set the meter reading to the reference mark. The function switch is then set to the measure-distion position, as indicated by the dashed-line arrow. In this

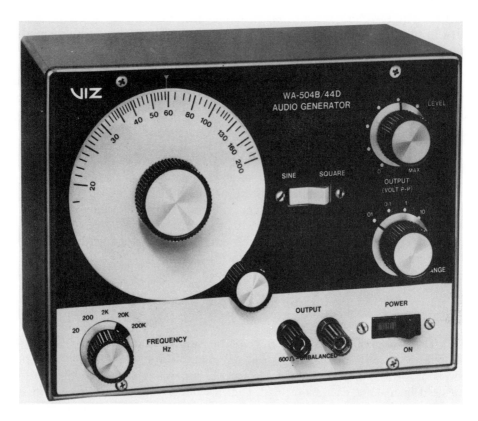

FIGURE 3-28. A typical variable-frequency audio generator. Courtesy of Viz Test Equipment.

position, the null-adjust control is adjusted for minimum meter indication. The remaining meter reading is caused by the distortion components that pass through the filter.

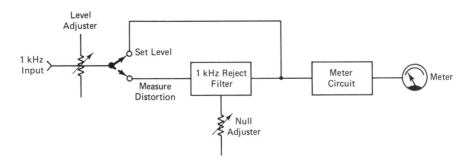

FIGURE 3-29. Simplified block diagram of a distortion analyzer.

Figure 3-30 shows a distortion analyzer that is tuneable over a frequency range of 5 Hz to 100 kHz. A special feature of this distortion analyzer is that an automatic null circuit can be used to locate the exact null once the null is tuned below the 10% distortion level. This is very helpful because fine-tuning of the null can sometimes be very difficult, especially at very low distortion levels. Accurate distortion measurement depends upon precise null tuning.

FIGURE 3-30. The Heathkit® IM-5258 Harmonic Distortion Analyzer, which is tuneable over a wide frequency range. Courtesy of Heath Company.

AUDIO LOADS

Audio loads are useful in providing proper load impedance to an audio output of a receiver. An audio load provides a uniform impedance over a wide range of frequencies and also reduces ear strain, especially when measuring power output levels. Commercial audio loads are available or a simple audio load may be constructed in a small box in the same manner described under "RF Dummy Loads." Noninductive resistors of appropriate wattage and resistance should be used. The most commonly encountered impedances in communications work are 3.2 Ω and 8 Ω.

The *rms* voltage across the load can be used to compute the audio power in the load. A graph in Appendix D correlates *rms* voltage with audio power. Curves are shown for both 3.2- and 8-Ω loads.

DIP METERS

The modern dip meter is the solid-state version of the old "grid dip" meter, so named because the meter was in the grid circuit of a vacuum-tube oscillator. Figure 3-31 shows a popular dip meter. This instrument basically consists of an oscillator and a detector with a meter in the detector circuit. The set-meter control is used to set the meter to a convenient level. The det-osc control is used to control the level of oscillation and to control the regeneration of the oscillator circuit when it is used as an absorption-type wavemeter. A headphone jack is provided to listen for beat notes. Headphones used with this instrument should be of the dynamic type in order to provide dc continuity; otherwise, the metering circuit will be interrupted. This dip meter covers the frequency range of 1.7 MHz to over 300 MHz using 7 plug-in coils.

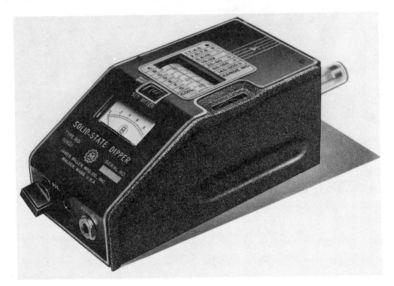

FIGURE 3-31. A popular solid-state dip meter, the Millen Solid State Dipper. Courtesy of Caywood Electronics, Inc.

Four modes of operation are possible with this dip meter. They are described below.

1. *Dip Meter*: This is the basic operation for which the instrument was designed. In this mode, the "DET-OSC" control is set fully clockwise to "OSC." The oscillator is now set for strong oscillation at a frequency that is determined by the plug-in coil and the setting of the tuning control. The "Set Meter" control is then adjusted to produce a reference reading at midscale or above. To determine the resonant frequency of a de-energized circuit, the plug-in coil is held near the circuit under test while the dip meter is tuned through its range. When the dip meter is tuned across the frequency of the resonant circuit under test, there will be a noticeable dip in the meter reading. The dip meter should be tuned for minimum meter reading. The resonant frequency

of the circuit under test is now determined by reading the dial of the dip meter. When the dip meter is tuned to the resonant frequency of the circuit under test, the circuit under test extracts energy from the dip meter's resonant circuit. This accounts for the dip.

2. *Signal Generator*: To set the dip meter up as a signal generator, it is set up as a dip meter as in (1). The plug-in coil is used to couple the signal to various circuits.

3. *Detector-Oscillator*: The same setup as for the dip meter (#1) is used here except that a pair of headphones is plugged into the phone jack. The purpose is to determine the presence of fundamental, harmonic, or parasitic frequencies of energized RF circuits. When the dip meter is tuned to the frequency of the signal in the circuit (fundamental, harmonic, or parasitic), a beat note will be heard in the phones.

4. *Absorption-Type Wavemeter*: In this mode, the "DET-OSC" control is first set to "OSC" and the "Set Meter" control adjusted for nearly full scale. The DET-OSC control is then slowly turned counter-clockwise until the oscillation stops. This occurs at the point where the meter reading stops falling as the DET-OSC control is turned counter-clockwise toward DET. The DET-OSC control should be set just below the point at which oscillation stops. The set-meter control is then adjusted for approximately one-fifth scale. The oscillator is not oscillating, but the circuit is regenerative. A signal at the resonant frequency of the dip meter coupled to the pickup coil will cause a sharp increase in the meter reading. The dial of the dip meter then indicates the frequency of the signal causing the meter reading.

Figure 3-32 shows a setup that can be used to simulate the action of a dip meter. Although not as versatile as the dip meter, it is nevertheless very useful. The signal generator substitutes for the oscillator. The meter should be a sensitive microammeter. As the signal generator is tuned through the frequency at which the circuit is resonant, the meter will show a dip. The frequency is then read from the dial of the signal generator. For higher accuracy, a frequency meter or counter may be used to measure the frequency of the generator.

NOISE BRIDGE

The noise bridge is useful in determining unknown impedances at RF frequencies. Basically, the noise bridge is a broadband noise generator. The noise output is applied to a bridge circuit that is adjusted to achieve a null or balanced condition. A low-cost noise bridge is shown in Figure 3-33. A zener diode is used to generate the noise that is then amplified by a three-stage broadband amplifier. This amplified noise is then fed to the input of a

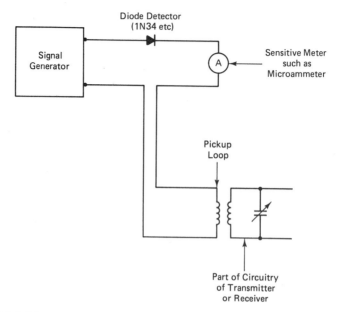

FIGURE 3-32. A method of connecting a signal generator and detector circuit for use as a dip meter.

FIGURE 3-33. A typical low-cost RF noise bridge. Courtesy of MFJ Enterprises, Inc.

bridge circuit through a transformer. The schematic of the bridge as it appears in the instruction manual is shown in Figure 3-34A. The bridge is redrawn at Figure 3-34B in order to show the bridge arrangement more clearly.

In order to use the bridge, a receiver is connected across the bridge as shown and tuned to the frequency at which the measurement is to be made. The unknown quantity to be measured is connected as shown to become part of one leg of the bridge. If R1 and C1 are properly adjusted to balance out the resistance and/or reactance of the unknown quantity, the bridge will be balanced and a null will be heard in the receiver. The dial readings are then used in formulas given in the instruction book for various calculations.

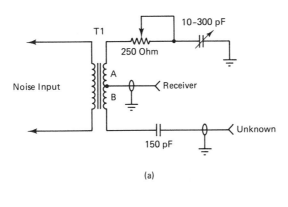

(a)

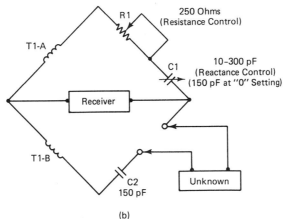

(b)

FIGURE 3-34. (A) The bridge circuit of the MJF Model 202 noise bridge as it appears in the instruction manual. Courtesy of MFJ Enterprises Inc. (B) The same noise bridge redrawn to show the basic makeup of the bridge components.

RF VOLTMETERS

RF voltmeters are valuable aids in the testing of various stages of receivers and transmitters. RF voltage measurements can be made with an electronic voltmeter or *VOM* using a special RF probe. Many voltmeter manufacturers make a special RF probe for their meters. They are very similar to the demodulator probes used with oscilloscopes. A simple RF detector probe that can be used with many electronic voltmeters is shown in Figure 3-35. This arrangement can cause a severe loading effect on the circuit under test and can cause the signal to disappear completely when trying to measure it. It can also cause an oscillator to stop oscillating. Generally, the RF probe/voltmeter combination is not very sensitive, but where adequate RF levels are present and loading isn't a problem this method can be quite satisfactory.

Sensitive and highly accurate RF voltmeters are available that can measure RF voltages in the millivolt and even in the microvolt range. One such instrument, manufactured by Helper Instruments Company, is the RF Millivolter shown in Figure 3-36. Full-scale voltage ranges are from 1 mV to 100 V. Lowest usable reading is said to be 300 μV. The instrument is usable over a wide frequency range. A dB scale is provided for dBm measurements or stage gain measurements.

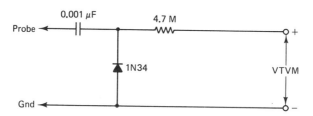

FIGURE 3-35. A schematic of a detector probe that can be used with an electronic voltmeter to measure RF voltages.

REGULATED POWER SUPPLIES

In order to bench-test mobile radio transceivers, power amplifiers, etc., it is necessary to have a dc power supply that can meet the voltage and current requirements of the equipment under test. Modern high-power transmitters and/or power amplifiers draw high currents while the current drain of modern receivers is quite low. As an example, when a modern transceiver is operating in the standby mode, current drain may be as low as 200 mA or less (including the indicator lamps). Conversely, when the transmitter is

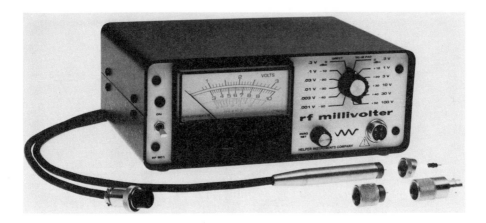

FIGURE 3-36. A very sensitive RF voltmeter, Helper Instruments Company's *RF Millivolter*. Courtesy of Helper Instruments Company.

keyed, the current drain may jump to 20, 30, or more amps, depending upon the transmitter power rating. Since the voltage should remain constant under all these conditions, a regulated power supply, or more specifically a voltage-regulated power supply, must be used.

The power supply shown in Figure 3-37 has all the necessary and/or desirable features for testing and servicing mobile transceivers, etc. This is a Ratelco, Inc., model PS-8 regulated power supply. The voltage is adjustable from 1 to 40 V dc. The power capacity is rated at 20 A continuous and 40 A intermittent at 1 to 15 V dc and 300 W continuous/600 W intermittent duty above 15 V. The power supply has a current limiter that can be set to any value between 100 mA and 40 A. RFI circuitry has been provided to avoid erratic operation while testing two-way radio transmitters. Two small output jacks are provided on the power supply. When they are used, they automatically cause the ammeter to indicate 0 to 5 A full scale and reduce the current limiter range to 4 A maximum. The detachable remote control head can be mounted on the workbench, requiring minimum space. The larger part of the supply can then be placed at a distance of up to fifteen feet away from the control unit.

MULTIFUNCTION TEST INSTRUMENTS

There are dozens of professional-quality multifunction test instruments on the market for servicing and testing communications transceivers. Many of these instruments contain almost everything necessary for radio tests and measurements. One of these multifunction instruments is briefly described here.

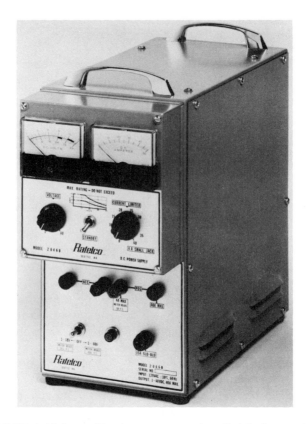

FIGURE 3-37. A high-quality dc power supply suitable for use in servicing and testing mobile transceivers. Courtesy of Ratelco, Inc.

Figure 3-38 shows a Cushman Electronics' model CE-50A FM/AM
Communications Monitor. Tests and measurement functions of the CE-50A
include the following:

1. Synthesized AM/FM signal generator to 1,000 MHz

2. A receiver with 2-μV sensitivity for off-the-air monitoring and measurement

3. Frequency-error meter (discriminator type)

4. AM/FM meter

5. RF power meter

6. Sinad meter (more fully discussed in a following chapter)

7. Oscilloscope

8. Audio tone generator

9. Synthesized offset generator

10. Single-sideband zero capability
11. Cable fault locator (a function of tracking generator and spectrum analyzer)
12. Spectrum analyzer
13. Tracking generator

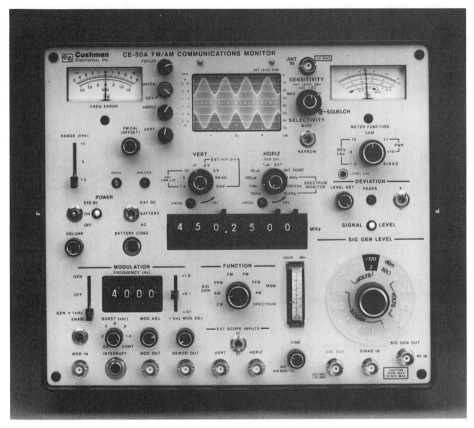

FIGURE 3-38. The Cushman Electronics' Model CE-50A FM/AM Communications Monitor. Courtesy of Cushman Electronics Inc.

Most of these functions are covered in one or more of the following chapters. However, the tracking generator warrants special attention here.

The tracking generator operates in conjunction with the spectrum analyzer. As you learned in Chapter 2, the spectrum analyzer is basically a swept-tuned receiver whose output is detected and displayed on a scope screen. A sweep generator is a signal generator that is swept-tuned.

In the CE-50A, a sweep generator is synchronized with the sweep of the spectrum analyzer so that at any instant during the sweep the instantaneous frequency of the sweep generator is the same as the frequency to which the spectrum analyzer is tuned. Thus, the sweep generator tracks along with the spectrum analyzer. Hence the name tracking generator.

The result of this arrangement is an instantaneous frequency versus amplitude graph that appears on the screen. This method is superior to the commonly used sweep generator/detector method of sweep testing because the RF detector is very broadband, thus indicating not only the fundamental frequency of the generator but also any harmonics or spurious signals that might be present. The display of the TG™ represents the response of the circuit under test to *only* the fundamental frequency at any particular instant. Other advantages of the TG™ sweep method are its greater dynamic range and absolute-level measurement capability.

COUPLING DEVICES

The RF wattmeter is the only measuring instrument that requires *full* transmitter power for the measurement. Other transmitter test instruments —that is, frequency counters or meters, service monitors, modulation meters, spectrum analyzers, etc.—require only a small amount of RF signal input to perform their measuring functions properly. Excessive RF input levels to these instruments can cause severe damage. Some are equipped with RF fuses or switching devices for protection against excessive RF levels. Never key a transmitter into a signal generator! The signal is not supposed to flow in this direction! Several methods of coupling a *sample* of the transmitter RF output to the test instruments are presented here.

The simplest coupling method is to use a short antenna on the test instrument itself to pick stray radiation from the equipment to be tested. Sometimes this may be entirely sufficient, but when operated in an environment where high RF levels from other sources may exist, closed coupling should be used to prevent extraneous RF from causing problems.

Figure 3-39 shows a coupler that permits interconnections of signal generator for receiver testing and frequency meter, modulation meter, wattmeter, spectrum analyzer, etc., for transmitter testing. A typical equipment setup using the Power Pad 2™ is shown in Figure 3-40. The wattmeter is connected between the transceiver and the pad. The pad serves as a dummy load for the transmitter, capable of dissipating 80 W continuously or 110 W intermittently. The frequency counter and deviation meter are connected to the "measure" terminal or port. In the transmit mode a small sample of the transmitter RF output appears at the "measure" port to operate the frequency counter and deviation meter. The signal generator is connected to the "generate" port. The signal appearing at the receiver input from the signal generator is attenuated 40 dB by the pad. This must be taken into account when determining the actual signal level applied to the receiver input. In the

FIGURE 3-39. A coupling device which facilitates the interconnection of various transmitter and receiver test instruments. Courtesy of Communication Instruments.

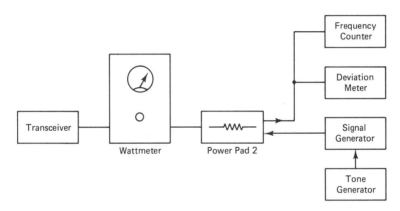

FIGURE 3-40. A typical application of the Power Pad 2™. Courtesy of Communication Instruments.

receive mode, the frequency counter and deviation meter can be used to monitor the signal generator frequency and modulation.

Another method of RF signal sampling is by means of an in-line adjustable "tap-off," such as those shown in Figure 3-41. The difference in these models is the type of connectors used. The sampling-level adjustment range varies with frequency (6 dB per octave). At 100 MHz, the sampling level ranges from approximately −37 dB to −50 dB. At 200 MHz, the sampling level ranges from −31 dB to −44 dB. A chart is included with the sampler to determine sampling-level ranges from 15 to 500 MHz.

At least two wattmeter manufacturers (Bird and GS Dielectric®) manufacture signal samplers in the form of plug-in elements to be used with their

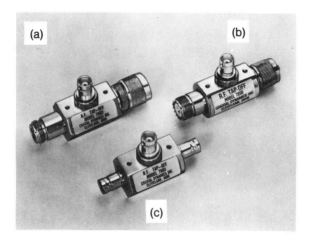

FIGURE 3-41. These "in-line" adjustable "tap-offs" permit adjustable coupling to various test instruments. At *A* is the "N"-connector type. At *B* is the UHF type. At *C* is the BNC type. Courtesy of Coaxial Dynamics, Inc.

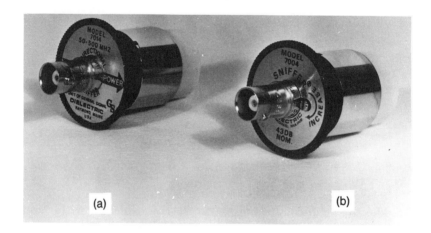

FIGURE 3-42. These coupling devices are designed to be used with Coaxial Dynamic's in-line wattmeter. The one at *A* is a fixed attenuation, directional pickup coupler. The one at *B* is a variable attenuation, nondirectional pickup coupler. Courtesy of Coaxial Dynamics, Inc.

wattmeters. (Examples are shown in Figure 3-42.) The coupler at Figure 3-42A is a fixed directional coupler while the one in Figure 3-42B is an adjustable nondirectional coupler. At least one company (Bird) manufactures a wattmeter that has a special port to supply a sample of the RF passing through the wattmeter. This makes it unnecessary to interrupt the normal use of the wattmeter to insert the special sampling elements. Thus, the RF power can be monitored at the same time other instruments are being operated from the sample port.

A homemade signal sampler can be made by removing the pin from the male portion of a UHF T-connector, as shown in Figure 3-43. The pin is cut and the threaded portion is reinserted to hold the feedthrough in place. As a result, dc continuity will be interrupted but *capacitive* coupling will give sufficient signal transfer to operate most measuring instruments. The T-connector is simply inserted into the transmission line between the transmitter and the load.

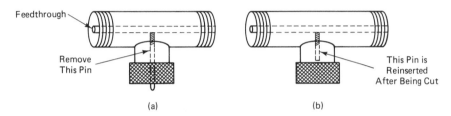

FIGURE 3-43. A method of making a homemade coupling device using readily available T-connectors. *A* shows the T-connector before being modified. *B* shows the T-connector after modification.

BIBLIOGRAPHY

Bird Electronic Corporation, *Instruction Book for the Thruline® Wattmeter Model 43.*

Carr, Joseph. *Elements of Electronic Instrumentation and Measurement.* Reston, VA: Reston Publishing Company, 1979.

Caywood Electronics, Inc., *Instruction Book for the Millen Solid State Dipper No. 90652.*

Heath Company, *Instruction Book for the FM Deviation Meter Model IM-4180.*

Heath Company, *Instruction Book for the Frequency Counter Model IB-1100.*

4
AM TRANSMITTER TESTS AND MEASUREMENTS

RF POWER MEASUREMENT

The measurement of RF power output of a transmitter is usually one of the first steps in making a complete transmitter performance test. In order to make accurate and consistent power measurements, certain criteria must be met. The transmitter *must* be loaded properly. Whenever possible, a dummy load should be used for all transmitter testing. This ensures accuracy and minimizes interference to other radio users on the frequency of the transmitter under test. For mobile transmitters, the dc voltage supply should be set to the proper level and for base stations the voltage-adjust control (if one is provided) should be properly set. Table 4-1 lists the proper voltage levels for various load currents for nominal 12-V equipment.

Operating Current Amps	Test Voltage Volts
Less than 6	13.8
6–16	13.6
16–36	13.4
36–50	13.2
Greater than 50	13.0

TABLE 4-1

CARRIER POWER MEASUREMENT

The setup for measuring the carrier power is shown in Figure 4-1A and 4-1B. In measuring the *carrier* power, the transmitter *must not be modulated*. An *average-reading* type of wattmeter is usually used for measuring the carrier power. An *in-line* or *termination* type of wattmeter may be used. If an in-line wattmeter is used, a separate dummy load will be required, as shown in Figure 4-1A. If a termination wattmeter is used, the dummy load is built into the wattmeter and no separate dummy load is required.

Performing the Test

1. Hook up the equipment as shown in Figure 4-1A or 4-1B.
2. Make sure the supply voltage is correct.
3. Key the transmitter, but *do not modulate it.*
4. Read the power in watts from the proper scale of the wattmeter. Best accuracy is generally obtained by keeping the wattmeter reading above one-half scale. This can be accomplished by using the proper plug-in element or proper switch position.

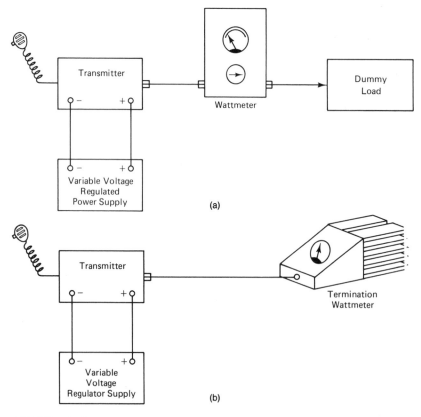

FIGURE 4-1. The setup at *A* shows how an in-line type of wattmeter is used to make power measurements. The setup at *B* shows how the termination-type of wattmeter is used to make the power measurement.

If the power output test is performed with a mobile unit installed in the vehicle, it may be necessary to have the engine running in order to keep the
voltage up to the proper level. The current demands of high power transmitters can drop the battery voltage very quickly unless it is being charged.
It is also possible for the vehicle's battery voltage to exceed the proper voltage level if the engine is running and the voltage regulator is not functioning properly. This can cause excessive RF power output and can damage the
equipment if sustained.

Power Measurement with an RF Voltmeter

If an RF wattmeter is not available, it is possible to make a fairly close
approximation of RF power output through the use of an *accurate* RF voltmeter.

Performing the Test

1. Set up the equipment as shown in Figure 4-2.
2. Key the transmitter, but *do not modulate it.*
3. Measure the *rms* voltage across the dummy load.
4. To determine the RF power, substitute the RF voltage reading into the
   formula: $E^2/50 = P$.

The accuracy of the result will depend upon the accuracy of the RF voltmeter and the accuracy of the computations.

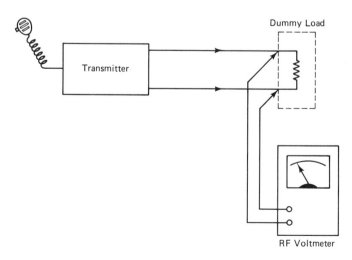

FIGURE 4-2. This setup shows how an RF voltmeter can be used to determine the RF power level across a dummy load.

Peak Power Measurement

The *carrier* power measurement is the most common way of measuring AM transmitter output power. However, there are times when it is desirable or useful to make a measurement of the *peak envelope power* (PEP). When a carrier is modulated 100%, the instantaneous peak power in the modulation envelope will be *four* times the carrier power. This will be explained more fully in the following section on amplitude modulation. In order to measure this peak power, it will be necessary to use a peak-reading type of wattmeter.

Performing the Test

1. Hook the equipment up as shown in Figure 4-3.
2. Key the transmitter and adjust the 1-kHz audio generator for exactly 100% modulation.
3. Read the peak power on the wattmeter.

This peak power indicates whether or not the transmitter is delivering sufficient power on the modulation peaks.

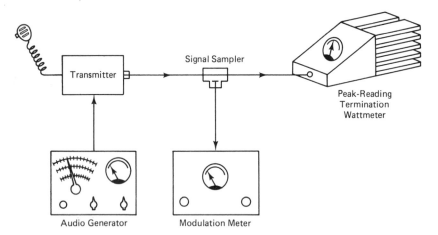

FIGURE 4-3. A typical setup to make a measurement of the peak envelope power at 100% modulation.

Power Output Leveling

In multichannel transmitters, it is important to check the power output or *leveling* over the entire frequency band. Although, in the interest of thoroughness, the power could be checked at every channel, it is usually sufficient to check the upper end of the frequency span, the lower end, and the center. In a properly tuned transmitter, the power output should be about the same across the entire frequency span. However, where a relatively wide band is covered the power at the center of the band may be higher than at the upper and lower limits. The power at the upper and lower limits should be approximately the same amount *lower* than the power at the center frequency. If the transmitter power is higher toward the upper or lower end of the frequency span, this indicates that the transmitter is tuned improperly.

DC Power Input

The power output of a transmitter is a function of the dc power *input* to the final stage and the efficiency of the final stage. The dc input to the final stage is a function of the dc current and voltage applied to the stage. The voltage and current is the plate-to-cathode value for vacuum tubes or the collector-to-emitter value (assuming a common emitter configuration) for transistor stages. The current and voltage measurements should be taken at a point that is not "hot" with RF or else erroneous readings may result. Figure 4-4 shows a typical common emitter transistor output stage. The RF choke, L2, blocks RF from entering the voltmeter, ammeter, and modulation transformer. The bypass capacitors, C2 and C3, offer further protection against RF by placing each end of the meter at RF ground potential. Thus,

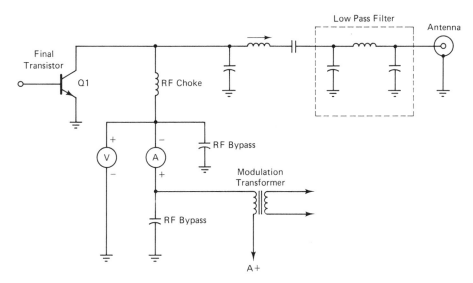

FIGURE 4-4. A typical solid-state transmitter output stage showing where the dc current and dc voltage input to the final stage is made.

the readings obtained by the voltmeter and ammeter are due to dc only. The full dc collector current must flow through the ammeter and the full dc collector voltage must appear across the voltmeter. The RF choke, L2, is wound with heavy-gauge wire so that the dc voltage drop across it is negligible. The product of the voltage and current as indicated on these meters is the dc power input to the final stage.

Measurement Procedure Summarized

1. Connect the voltmeter in the plate or collector circuit at a point that is not "hot" with RF. This point should indicate the full collector/emitter or plate/cathode dc voltage.
2. Connect the ammeter at a point in the collector or plate circuit that will indicate the full collector or plate current but at a point that is not RF "hot."
3. Record the readings; the product of the two readings is the dc power input to the stage.

Efficiency of the Final Stage

Efficiency is basically a measurement of the effectiveness of a device or stage. It can also be defined as how well a device converts energy from one form to another form. In the case of the final output stage of a transmitter, we are concerned with how much RF power output the stage delivers for a given amount of dc power input to the stage.

Calculating Efficiency

1. Measure the RF output power.
2. Determine the dc power input to the stage from the voltage and current measurements.
3. The efficiency is equal to:

$$\frac{\text{RF output power (step 1)}}{\text{dc input power (step 2)}}$$

AMPLITUDE MODULATION MEASUREMENT

Before getting into the actual measurement techniques for amplitude modulation, it is appropriate here to review a few fundamental aspects of amplitude modulation.

Modulation Percentage/Power Relationships

Suppose that a final RF amplifier stage is plate modulated. Assume that the dc power input to the stage is 1,000 W (with no modulation) and assume that the impedance of the stage is 1,000 Ω. The dc *quiescent* plate voltage can then be calculated from the formula:

$$E = \sqrt{PZ}$$

This is derived from

$$P = E^2/Z$$

Substituting, we have:

$$E = \sqrt{1,000 \times 1,000} = \sqrt{1,000,000} = 1,000 \text{ V}$$

Modulator Power   In order to modulate the transmitter, a full 100% the peak value of the audio modulating voltage must equal the quiescent value of the dc plate voltage, which is 1,000 V. Also, the secondary impedance of the modulation transformer must be made equal to the plate circuit impedance for proper operation. Thus, the modulation transformer's secondary impedance must be 1,000 Ω. Since the peak value of the transformer's secondary voltage is 1,000 V, the *rms* voltage will be 707 V. The power delivered by the transformer is: $P = E^2/Z = 707^2/1,000 = 500$ W.

This proves that the modulator must deliver 50% of the dc plate input power in order to 100% modulate the transmitter. For single-tone sine-wave modulation, this extra 50% of power divides equally between the upper and lower sidebands, which are separated from the carrier by an amount equal to the frequency of the modulating tone.

Calculating Sideband Power   In this example, the total sideband power is 500 watts, each sideband at 250 W. To calculate the dB below the carrier for each sideband, use the formula:

$$dB = 10 \log(P1/P2) = 10 \log(1,000/250) = 10 \log(4) = 10(0.602) = 6.02$$
or 6 dB below the carrier

For 50% modulation, the value of the modulating voltage is one-half the value required for 100% modulation. We can calculate the sideband level in dB below the 100% modulation value by using the dB formula for voltage ratios: $dB = 20 \log(E1/E2)$. Simply substitute the modulation percentages for E1 and E2. The 100% reference level is E1. Thus, the formula becomes: $dB = 20 \log(100/50) = 20 \log(2) = 20(0.301) = 6.02$ or 6 dB. Remember this represents the *new* sideband level as compared to the 100% modulation level, which itself was 6 dB below the carrier. Thus, if we want to reference this new level against the carrier level we must add 6 dB which makes the new sideband level 12 dB below the carrier. This can best be expressed by the formula:

$$dB \text{ (below carrier)} = 20 \log [(100/\text{mod}\%) + 6]$$

where mod% is the modulation level in percentage. The percentage of power in each sideband relative to carrier power can be determined from the formula: $\%Psb = 25M^2$, where $\%Psb$ is the power in one sideband expressed as a percentage of the carrier power and M is the modulation factor. Modulation factor is modulation percentage divided by 100—i.e., 50% modulation equals a modulation factor of 0.5.

A formula that expresses the total sideband power (both sidebands) as a percentage of total power (both sidebands and carrier) is:

$$Psb = 100M^2/(M^2 + 2)$$

where Psb is total sideband power expressed as a percentage of total power and M is the modulation factor.

Figure 4-5 correlates the sideband power as a percentage of total power with modulation percentages from 10% to 100%. Notice how rapidly the sideband power falls as the modulation level decreases, especially between

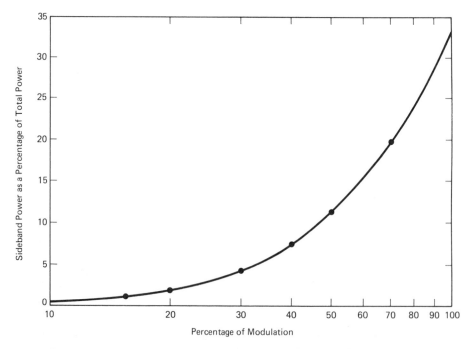

FIGURE 4-5. This graph illustrates the relationship of sideband power to total power for modulation percentages from 10% to 100%.

0% and 40% modulation. Thus, you can see how important it is to keep the modulation level as high as possible.

Instantaneous Peak Power   In our example at the beginning of this section, the quiescent plate voltage (no modulation) of the final amplifier is 1,000 V. At 100% modulation, the positive peak of the modulating voltage is 1,000 V. This *adds* to the dc plate voltage of 1,000 V, resulting in an instantaneous peak voltage of 2,000 V. This is a doubling of the voltage and since power changes as the square of the voltage the instantaneous peak power or peak envelope power (PEP) is *four* times the carrier power. The graph in Figure 4-6 shows the relationship of PEP power to the unmodulated carrier power for modulation percentages between 0 and 100%.

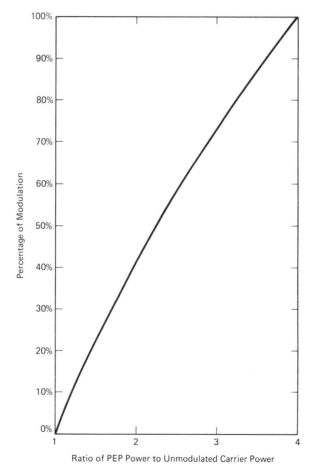

FIGURE 4-6. This graph illustrates the relationship of PEP power to unmodulated carrier power for modulation percentages from 0 to 100%.

Antenna Current Versus Modulation Percentage   Since power (P) varies as the square of the current (I), then current varies as the square-root of power. If the unmodulated carrier power of a transmitter causes 1 A of antenna current, at 100% modulation the power will increase 50%. Thus, the new power is 1.5 times the unmodulated power. The new current will be:

$$\sqrt{1.5} = 1.225 \text{ A}$$

This represents a 22.5% increase in antenna current at the 100% modulation level.

The following formula will yield the percentage of increase in antenna current for any modulation percentage from 0 to 100%.

$$\text{Percent of current change} = 100 \left[\sqrt{1 + (M^2/2)} - 1 \right]$$

where M is the modulation factor. Figure 4-7 is a graph of the percentage of antenna current increase for modulation percentages from 10 to 100%.

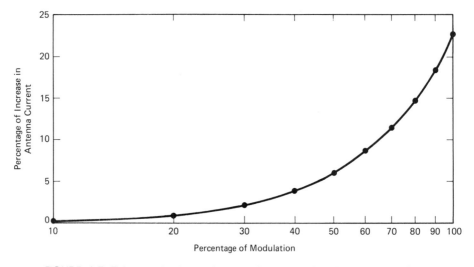

FIGURE 4-7. This graph shows the relationship of antenna current increase with the percentage of modulation.

Modulation Percentage

The modulation percentage of an AM transmitter can best be explained by referring to Figure 4-8. This is an amplitude-modulated signal. The unmodulated carrier is shown at the left. At the right is a modulation envelope. The degree of modulation can be determined from either of three ratios. One ratio is:

$$\frac{E(max) - E}{E}$$

This ratio is multiplied by 100 to give the modulation percentage. This is referred to as the *peak* modulation percentage. Another ratio is:

$$\frac{E - E(min)}{E}$$

Again, this ratio is multiplied by 100 to give the percentage of modulation. This is referred to as the *trough* modulation percentage. The third method uses the *peak and the trough* measurement as shown in Figure 4-8. The formula is:

$$\frac{A - B}{A + B} \times 100$$

where *A* is the *peak-to-peak* measurement and *B* is the *trough-to-trough* measurement.

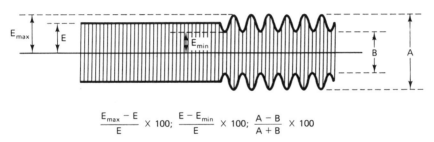

$$\frac{E_{max} - E}{E} \times 100; \quad \frac{E - E_{min}}{E} \times 100; \quad \frac{A - B}{A + B} \times 100$$

FIGURE 4-8. The method measuring the percentage of modulation from the envelope pattern on a scope is illustrated here.

Methods of Amplitude Modulation Measurement

Amplitude modulation measuring instruments were discussed in some detail in Chapter 3. These instruments only require a small *sample* of the RF output signal of the transmitter. Any of the coupling devices discussed in Chapter 3 can be used to couple sufficient RF energy into the modulation meter. Alternately, a small pickup loop can be placed near the coaxial transmission line to get sufficient pickup. Figure 4-9 shows a typical setup for measuring amplitude modulation. When using the microphone itself to modulate the transmitter, a loud sustained whistle is usually sufficient to get a steady indication on the meter. However, better accuracy will result with the use of an audio generator connected to the audio input of the transmitter.

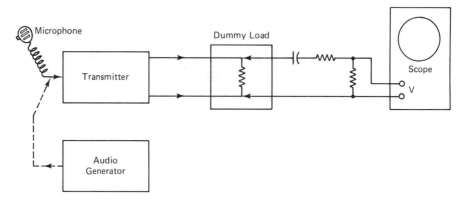

FIGURE 4-9. One method of obtaining an envelope pattern directly across the dummy load resistor is shown here. The vertical amplifier of the scope must be able to handle the frequency of the transmitter when this method is used.

Using the Scope for Determining Modulation Percentage

In some respects, the oscilloscope is superior to the meter-method of checking modulation. The waveform on the scope, when closely examined can reveal problems which won't show up on a meter reading. Furthermore, the scope pattern responds instantly to voice peaks, thus showing overmodulation conditions that a meter might miss. The scope doesn't have to be calibrated in order to determine the 100% modulation point. There are several methods used for checking modulation with a scope. The envelope method, the trapezoid method, and the envelope detector method are presented here.

The Envelope Method The envelope method is quite popular because of the simplicity of the setup. If the signal to be examined is within the passband of the scope's vertical amplifier simply couple a small sample of the signal to the scope's vertical input (see Figure 4-9). Any of the coupling devices described in Chapter 3 can be used to obtain the sampling signal. Alternately, a few turns of wire wrapped around the transmission line will give good coupling as will a small loop held or placed near the transmitter output stage. The method of coupling is not critical as long as sufficient signal is obtained to operate the instrument properly without overdriving it.

If the signal frequency is higher than the scope's vertical amplifier can handle, there are other alternatives.

1. The signal can be fed directly to the vertical plates of the scope, thus bypassing the vertical amplifier. Some scopes provide access to the vertical plates through terminals on the rear of the scope or elsewhere. If your scope doesn't provide access to the vertical plates, you can modify your scope to provide access to them. This simple modification is shown in Figure 4-10. Coupling capacitors C1 and C2 should be of high voltage rating to prevent a shock hazard. Resistors R1 and R2 allow the vertical centering control to operate when using the vertical plates directly. Connections can be brought out to the front panel or rear panel through banana jacks. The direct/normal switch should be installed near the jacks. If long leads are necessary, shielded wire should be used.

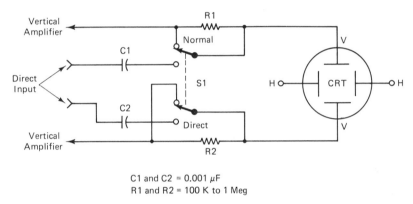

C1 and C2 = 0.001 μF
R1 and R2 = 100 K to 1 Meg

FIGURE 4-10. This simple circuit modification can be made to any scope to allow direct access to the vertical plates for viewing modulation waveforms of transmitters which operate on a frequency far above the reach of the scope's vertical amplifier.

2. If desired, a down converter can be used to convert the high-frequency signal down to a lower frequency that can be handled by the vertical amplifier of the scope. A simplified setup employing the down converter method is shown in Figure 4-11. This is a "poor man's" down converter but will work very well and can be used over a wide range of frequencies. The signal generator should be tuned 455 kHz *above* or below the frequency of the signal to be checked. The two frequencies are mixed in D1. Transformer T1 (which is a simple 455-kHz IF transformer) is tuned to the difference frequency. This difference frequency will have characteristics of both the generator signal and the transmitter signal. For this reason, it is important that the generator signal be as pure as possible.

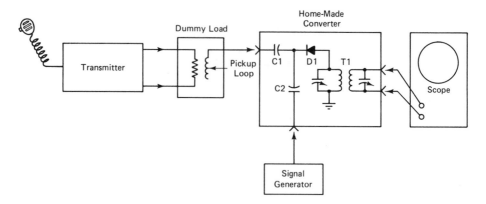

T1   = 455 kHz Standard IF Transformer
D1   = Mixer Diode
C1 and C2 = Small Coupling Capacitors

FIGURE 4-11. This setup, using a homemade converter, can be used to obtain a modulation envelope pattern on the scope when the transmitter frequency is above the vertical amplifier's reach.

3. A popular trick is to use a communications receiver as a down converter. This is done by tapping into the low IF stage of the receiver. The low IF frequency (usually 455 kHz) is coupled to the input of the vertical amplifier of the scope. A very low capacitance (5 to 10 pF) is used to couple the IF signal to the scope. This low capacitance minimizes loading of the IF stage. The communications receiver is then tuned to the frequency of the signal whose modulation envelope is to be checked. Since the receiver has high gain, very little coupling will be necessary; as a matter of fact, the antenna may have to be disconnected from the receiver to prevent overloading the front end if the receiver is placed near the transmitter. The RF gain control (if one is provided) should be operated at as low a setting as possible for good signal pickup. A typical setup using the communications receiver as a down converter is shown in Figure 4-12.

Various modulation envelopes are shown in Figure 4-13. The 100% modulation point is determined from the envelope. It occurs at the point where the upper and lower troughs just touch at the center line. Other percentages of modulation can be determined by using the scope's graticule calibrations to measure the distance between the appropriate points on the envelope (see Figure 4-8). This is somewhat cumbersome and not very accurate.

The Envelope Detector Method   The envelope detector method allows the use of a narrow band scope but the scope must employ *dc coupling* from the vertical amplifier input all the way to the CRT vertical deflection plates. Scopes which feature dc coupling usually have an ac/dc switch on the front

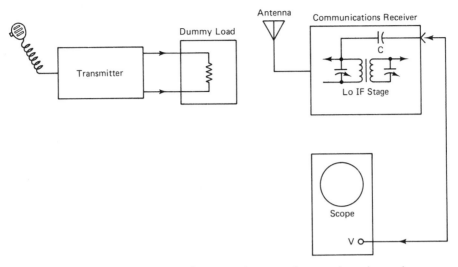

FIGURE 4-12. A communications receiver can be used as shown here to obtain a modulation envelope pattern on the scope.

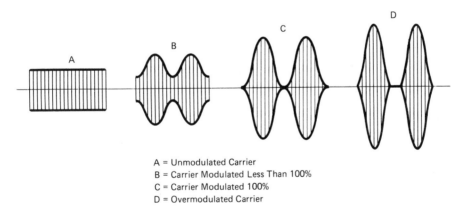

A = Unmodulated Carrier
B = Carrier Modulated Less Than 100%
C = Carrier Modulated 100%
D = Overmodulated Carrier

FIGURE 4-13. Several typical modulation envelope patterns are shown here.

panel near the vertical amplifier input. For this test the switch should be in the dc position.

This test requires an envelope detector which can be constructed in a small box using parts readily available in the service shop. A diagram of an envelope detector is shown in Figure 4-14A. Figure 4-14B shows the proper hookup for checking modulation with the envelope detector.

Measurement Procedure

1. With the transmitter off, set the scope's trace to a couple divisions below center (see Figure 4-15A). This represents "zero."

2. Key the transmitter and adjust the scope's vertical gain control with no modulation applied to position the trace at the center of the screen (see Figure 4-15A).

3. Apply modulation. At 100% modulation, the positive peak of the signal will rise to twice the carrier level and the negative peak will reach the "zero" line (see Figure 4-15B).

4. Other percentages of modulation can be determined by the formula:

$$\text{Percent modulation} = \frac{Ep}{2Ec} \times 100$$

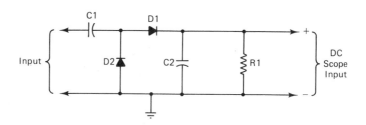

C1   = Use 2–5 pF or "Gimmick" (Twist Two Wires Together for About 1–2 Inches)
C2   = 620 pF
D1, D2 = 1N34 or 1N295

(a)

(b)

FIGURE 4-14. This simple envelope detector can be used to obtain a pattern on the scope for checking modulation. The scope must employ dc coupling in the vertical amplifier for this procedure. Courtesy of Pathcom, Inc. The setup at *B* shows how the envelope detector is connected to get the trace pattern on the scope.

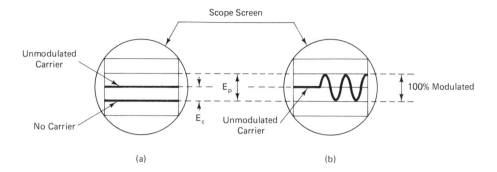

FIGURE 4-15. This illustration shows how the envelope detector pattern is used to determine the percentage of modulation of the transmitter. See text. (Courtesy Pathcom, Inc.)

The Trapezoid Method The trapezoid method is a little more difficult to obtain but can give a more complete analysis of the modulation process after one becomes familiar with it. To obtain the trapezoid pattern, a sample of the RF signal is fed to the vertical input of the scope (or the vertical plates) while a sample of the audio modulating signal is fed to the horizontal input of the scope (or the horizontal plates). Figure 4-16 shows a typical hookup used to obtain a trapezoid pattern from a typical CB set using a narrow bandwidth scope. In this arrangement (which is typical of most CB sets), the audio output transformer becomes the modulation transformer in the transmit mode. The lower winding (L2) is the modulation winding. The upper winding (L1) is the audio winding to the speaker circuit. In the transmit mode, S1 breaks the speaker circuit to prevent feedback from the speaker to the microphone. However, the external speaker jack circuit is not broken in the transmit made. This provides a very convenient source of the modulating signal sample for the horizontal input of the scope.

To understand how the trapezoid pattern is produced, examine Figure 4-17. The vertical dashed lines on either side represent the level which the audio modulating voltage must reach to 100% modulate the carrier. The right side represents the positive half of the audio cycle, while the left side represents the negative half. Notice that at the zero voltage point (vertical centerline) the carrier is represented by a vertical line. Remember that the carrier level increases to maximum level on the positive peak of the audio modulating voltage and decreases to a minimum level on the negative peak

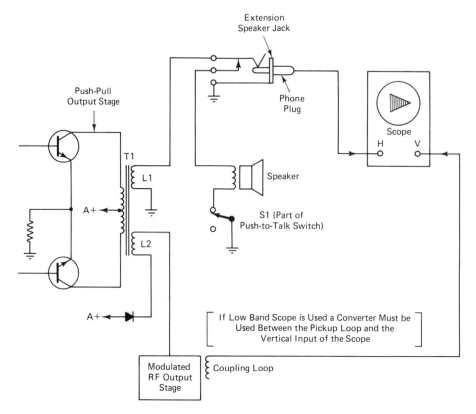

FIGURE 4-16. This setup shows how the trapezoid pattern can be obtained from a typical CB transceiver.

of the audio modulating voltage. At Figure 4-17A, the audio modulating voltage is exactly the level required for 100% modulation. As the audio voltage rises from 0 to the positive peak the vertical line representing the carrier is swept from the center to the right, increasing in height as the amplitude of the modulating voltage increases. When the modulating voltage reaches its peak positive value, the carrier level is at maximum. This is the point of 100% peak modulation, the maximum height of the wedge. As the positive peak is passed, the carrier trace starts moving back to the left, decreasing in size as it moves. As the audio voltage moves through zero toward the negative peak, the carrier trace moves to the left of center still decreasing in size as it moves to the left. At the negative peak of the audio modulating voltage, the carrier reaches zero amplitude. This is the point of 100% trough modulation, the point of the wedge.

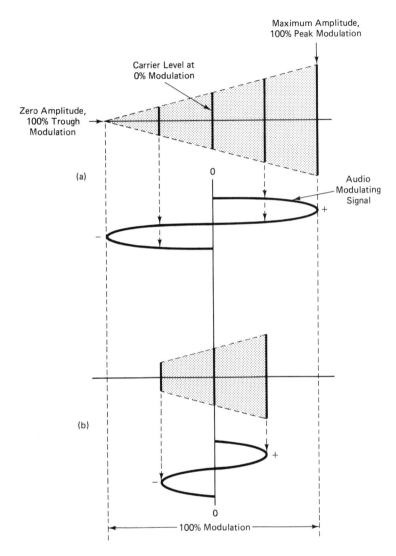

FIGURE 4-17. The relationship of the audio modulating (and horizontal sweep signal) to the trapezoid pattern on the scope is shown here.

After the negative peak has passed, the carrier trace starts moving to the right toward the center, increasing in size as it moves. This cycle is repeated as long as the RF carrier and modulating voltage are applied to the scope. Thus the carrier trace is continuously swept across the CRT increasing in size from left to right, decreasing in size from right to left. This rapid sweeping "paints" a wedge-shaped pattern on the CRT screen.

In Figure 4-17B, the pattern is shown for a modulating signal less than the 100% modulation level. This results in a trapezoid pattern. Several typical patterns are shown in Figure 4-18. In Figure 4-18A, only a dot appears on the screen because there is no horizontal or vertical deflection. This condition should not be allowed to persist or the CRT phosphor may be burned and permanently damaged. In Figure 4-18B, the vertical line is the carrier during the "zero modulation" condition. In Figure 4-11E, the horizontal line at the point of the pattern indicates overmodulation (trough overmodulation). The length of this line indicates the degree of overmodulation. Modulation percentages between 0 and 100% can be determined from the trapezoid pattern as shown in Figure 4-19. A steady sine-wave modulating tone should be used for the measurement. The result is not extremely accurate because of the difficulty of making precise measurements from the scope screen.

Generally, the trapezoid method is preferred over the envelope method. For one reason, with voice modulation the patterns are "laid" directly over each other, rather than appearing at random as in the case of the envelope method. This makes the pattern easier to interpret. Another advantage is that nonlinearity is more apparent from the trapezoid pattern.

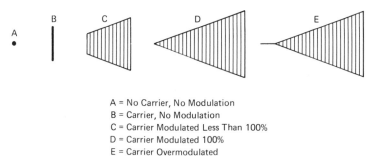

A = No Carrier, No Modulation
B = Carrier, No Modulation
C = Carrier Modulated Less Than 100%
D = Carrier Modulated 100%
E = Carrier Overmodulated

FIGURE 4-18. Various trapezoid patterns for various percentages of modulation are shown here.

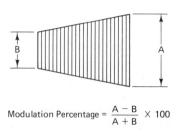

$$\text{Modulation Percentage} = \frac{A - B}{A + B} \times 100$$

FIGURE 4-19. The calculation of the percentage of modulation from the trapezoid pattern is illustrated here.

Measuring Amplitude Modulation
with the Spectrum Analyzer

As previously discussed in this chapter, the level of each sideband varies with the percentage of modulation. The level of each sideband with respect to the carrier can be stated as so many dB below the carrier level. Since the carrier is the reference let's call it the 0-dB reference mark. Then the sideband level can be stated as X dB. With a spectrum analyzer, the sideband components can be measured in relation to the carrier. For 100% modulation, each sideband will be 6 dB below carrier level or simply -6 dB. Figure 4-20 shows a display of a 100% amplitude modulated signal. Notice that both sidebands are 6 dB below the carrier. Other percentages of modulation can be determined from a spectrum analyzer by measuring the difference in the carrier level and sideband level and then using the graph in Figure 4-21 to convert this difference in dB to the modulation percentage.

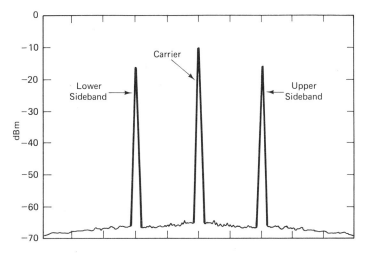

FIGURE 4-20. This spectrum analyzer display is obtained by modulating the AM transmitter 100% with a single tone. Note that the two sidebands are each 6 dB below the carrier level.

Modulation Distortion Check

The modulation waveform superimposed on the carrier should be an exact replica of the audio modulating signal. If a carrier is amplitude modulated by a 1,000-Hz audio tone then when the signal is demodulated the resulting audio signal should be a pure 1,000-Hz tone. However, the modulation process is not perfect and will produce a certain amount of distortion. There are several ways this distortion can be checked.

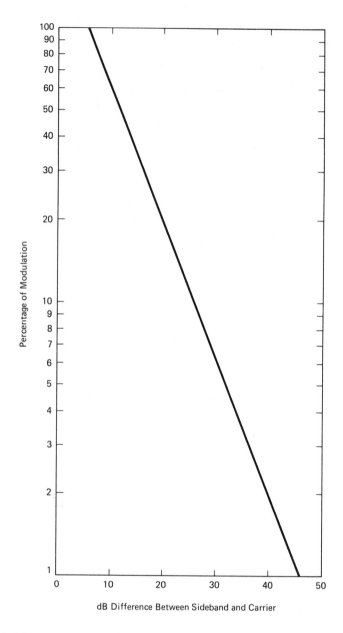

FIGURE 4-21. When single tone modulation is used the level of each sideband (in terms of dB below the carrier level) can be found from this graph for percentages of modulation from 1 to 100%.

The Trapezoid Pattern Method   The trapezoid pattern method is probably the most reliable of all the scope methods for checking modulation distortion. The reason for this is that the sides of the wedge are perfectly straight and come to a well-defined point when the signal is modulated 100%. Any curvature in the sides represents distortion.

Another point to watch is the degree of swing to each side of center. At the beginning of the test the transmitter is keyed and the unmodulated carrier (vertical trace) is set to the center line of the scope screen. Then, with a symmetrical modulating signal (where the positive peaks equal the negative peaks) the horizontal distance from the center of the screen to either side of the pattern should be the same. If it isn't, the modulating signal is not symmetrical.

Figure 4-22 shows a few typical trapezoidal patterns. In Figure 4-22A, the unmodulated carrier is set to the center line. In Figure 4-22B, the trapezoid is linear but the trace extends further to the right side (peak side) than to the left side (trough side) with reference to the center. This indicates unsymmetrical modulating signal. In Figure 4-22C, the sides of the trapezoid are noticeably curved, indicating nonlinearity in the modulated stage. This might escape notice if the envelope method were used. In Figure 4-22D, the peak is flattened, indicating that the modulated stage can't deliver the required peak power, a nonlinear condition. In Figure 4-22E, the ellipses indicate a phase shift, usually caused by improper hookup or sampling the audio for the horizontal input at a point prior to the point at which the modulating signal is applied to the transmitter final stage.

The Envelope/Dual-Trace Scope Method   Defects in the modulating signal are more easily recognized in the envelope pattern while defects in the *modulated* stage show up better in the trapezoid pattern. The modulating signal can be analyzed by simply "scoping" the modulator output. With a *dual-trace* scope, the *modulating* signal can be compared with the *modulated* signal to detect distortion. The procedure is as follows.

Procedure

1. Modulate the transmitter with a steady tone (1 kHz or so). A high percentage of modulation should be used but less than 100%. (Eighty-five percent modulation is fine.)
2. Feed the audio modulating signal at the output of the modulator into one of the scope's vertical channels.
3. Feed a sample of the modulated signal into the other vertical channel using a down converter if necessary.

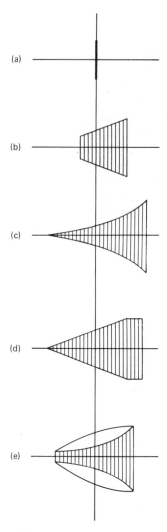

FIGURE 4-22. Various trapezoid patterns and their interpretation are shown here. (See text.)

4. Check the modulated signal against the modulated signal by positioning the modulating signal over the modulated signal (see Figure 4-23). The two signals should perfectly match. If they don't, the modulated signal is distorted. Small amounts of distortion will be difficult but significant amounts of distortion will be apparent.

Modulation Meter/Distortion Analyzer This test permits an absolute measurement of the distortion in the modulation waveform. The test setup is shown in Figure 4-24. The modulation meter used should be a high-quality-type instrument that produces a minimum of distortion of its own.

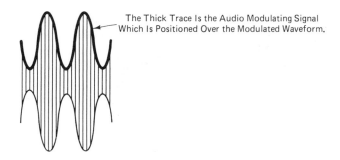

The Thick Trace Is the Audio Modulating Signal
Which Is Positioned Over the Modulated Waveform.

FIGURE 4-23. This illustration shows how a dual trace scope can be used to check the modulation envelope for distortion by comparing it to the original audio modulating signal. Here the original modulating signal is sampled on one of the scope channels while the envelope is obtained on the other channel. The audio modulating waveform is then positioned directly over the modulation envelope for direct comparison.

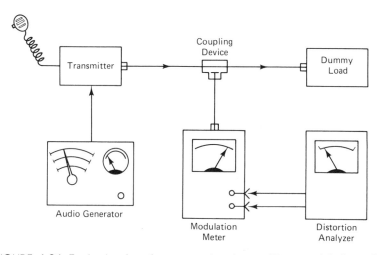

FIGURE 4-24. Typical setup for measuring transmitter modulation distortion.

Measurement Procedure

1. Modulate the transmitter with a 1-kHz tone at approximately 85% modulation.
2. Tune the distortion analyzer to null out the 1-kHz tone.
3. The meter indication remaining is the distortion products.

The Spectrum Analyzer Method The spectrum analyzer (a high-resolution type) can be used to check the distortion of the modulated signal. The setup is shown in Figure 4-25. The transmitter is modulated approximately 85% with a 1-kHz audio tone. A distortion-free signal will appear as in Figure 4-26A. The composite modulated signal will consist of a carrier, an upper sideband, and a lower sideband. The sidebands are separated from the carrier by an amount equal to the frequency of the modulating tone. Figure

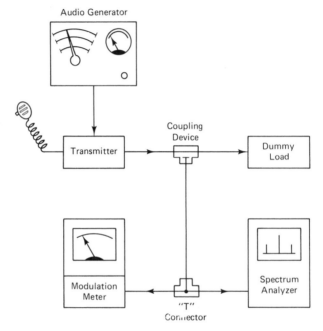

FIGURE 4-25. This setup shows how a spectrum analyzer can be used to check for distortion in the modulated signal.

4-26B shows a heavily distorted modulated signal. The additional sidebands are caused by harmonic distortion. Notice that the sidebands are spaced from the carrier by multiples of the modulating frequency. The relative amplitude of the distortion products are easily seen on the spectrum analyzer.

The Communications Receiver/Distortion Analyzer Method A high-quality communications receiver can be used to check for distortion in the modulated signal by using the hookup shown in Figure 4-27. The inherent distortion produced in the receiver will show up in the result, but if the distortion produced in the receiver is relatively low, the measurement will be valid. The transmitter is modulated approximately 85% by a 1-kHz tone. The audio output of the receiver is then checked with the distortion analyzer

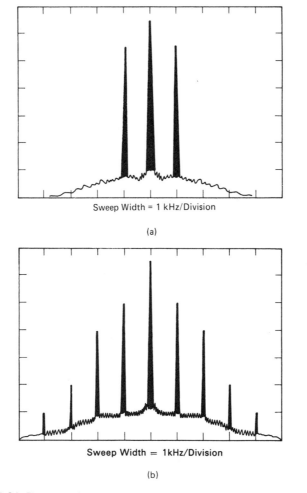

Sweep Width = 1 kHz/Division

(a)

Sweep Width = 1kHz/Division

(b)

FIGURE 4-26. The spectrum analyzer display at *A* shows a clean modulated signal while the display at *B* indicates considerable distortion in the modulated signal.

for distortion content. It is important not to overload the front end of the receiver when performing this test.

Hum and Noise Modulation Measurement

A certain amount of hum and noise modulation is always present on the carrier signal. Basically, any of the setups used for measuring modulation

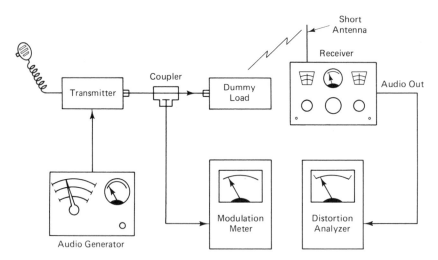

FIGURE 4-27. This setup shows how a communications receiver in conjunction with a distortion analyzer can be used to check for distortion in the modulated signal.

can be used for measuring the hum and noise modulation. However, when measuring the hum and noise modulation produced within the transmitter, certain steps should be taken to prevent the microphone from contributing to the modulation. This can be done by removing the microphone from the transmitter while performing the test. With the microphone disconnected, the transmitter is keyed and the modulation meter or scope is used to check the modulation. The type of modulation (hum or noise) can best be identified by using the scope. Figure 4-28 shows how these hum and noise components appear on the signal.

Noise modulation is difficult to measure exactly because it is usually so erratic in behavior that the meter needle doesn't remain steady long enough to get a usable reading. The scope is superior in such cases because of its faster response. Hum modulation (if steady) can be measured on a modulation meter.

Once this test is completed, the microphone is connected again. The test can be repeated with the microphone in the circuit to determine the amount of hum and/or noise produced by the microphone itself. Of course, a quiet environment must be used for this test to minimize acoustic noise.

Modulation Sensitivity Measurement

Most transmitter service or instruction manuals list the modulation sensitivity as so many millivolts required to produce a given amount of modula-

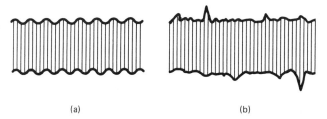

(a)   (b)

FIGURE 4-28. Typical modulation patterns showing hum and noise are shown here. The pattern at *A* indicates significant hum while the pattern at *B* shows noise on the modulation pattern.

tion percentage. Eighty-five percent is the modulation percentage frequently used. If the dc power input to the final modulated stage is changed, the modulation sensitivity will also change. This is because the relationship between the dc input voltage to the final stage and the peak value of the voltage used to modulate the stage; i.e., the peak value of the modulating voltage must be equal to the quiescent value of the dc input voltage. This applies to collector- or plate-modulated final stages. The setup for this test is shown in Figure 4-29.

Measurement Procedure

1. Set the audio generator frequency to 1 kHz and turn the output level all the way down to minimum level.

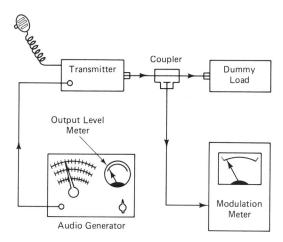

FIGURE 4-29. This setup shows how the modulation sensitivity measurement is made.

2. Key the transmitter.

3. While observing the modulation meter, increase the audio output from the generator until the specified amount of modulation (85%) is reached.

4. Read the output level of the audio generator from the output level meter on the generator or use an audio voltmeter across the input to the transmitter. This audio voltage reading is the modulation sensitivity of the transmitter for that specified amount of modulation.

Modulation Audio Bandwidth

Almost all voice transmitters that are primarily designed for communications (not entertainment) are designed to operate with voice-input signals of 300 to 3,000 Hz. Frequencies outside this range contribute very little to the intelligence of the voice signal. The higher frequencies significantly increase the bandwidth of the transmitted signal, so in order to limit the bandwidth to only that necessary for good communications the upper frequencies are greatly attenuated by properly designed audio filters installed in the audio input circuitry. From time to time, it is desirable to check the operation of the audio filters to make sure that the audio frequencies within the 300 to 3,000 Hz range are being passed by the filters and to make sure that the frequencies above this range are being significantly attenuated so as to hold down the bandwidth of the transmitted signal.

The same setup used for the modulation sensitivity test can be used for this test. The audio generator must be variable over a range of frequencies from below 300 Hz to cover 3,000 Hz in order to check the modulation bandwidth properly. The setup is shown in Figure 4-29.

Test Procedure

1. Set the audio generator to 1 kHz.

2. Key the transmitter and adjust the audio level of the generator to produce a modulation percentage of 85% for the reference level. This audio level will serve as the reference level for the entire test. It should be noted for reference. It is easier if a dB scale is used so that any deviation from this reference level can be measured directly in dB.

3. Tune the audio generator through the range of frequencies from below 300 Hz to well above 3,000 Hz. If the modulation percentage changes, adjust the audio level control to return the modulation percentage to the reference level (85%) and note the change in the audio level required to keep the modulation percentage at 85%. By checking the response at several different frequencies throughout the range and recording the level change, a graph could be plotted to show the frequency response.

Transmitter Modulation Spectrum

The transmitter *modulation spectrum* can be determined best through the use of a spectrum analyzer. Various types of communications services have differing bandwidth requirements which are set forth by the FCC. The modulation spectrum test generally calls for modulating the transmitter to a high degree of modulation using a high-frequency audio signal. Distortion will result in the production of undesirable sidebands. The level of these sidebands must be within prescribed limits depending upon the separation of these sidebands from the carrier frequency. The following test procedure describes how to conduct the modulation spectrum test on an AM transmitter.

Test Procedure[1]

1. Set up the equipment as shown in Figure 4-25.
2a. If the transmitter employs a modulation limiter or clipper, set the AF generator to 2,500 Hz and set the level 16 dB *above* the level required to produce 50% modulation.
2b. If the transmitter does not employ a modulation limiter or clipper, set the AF generator to 2,000 Hz and set the level to produce 85% modulation.
3. Any signal display on the spectrum analyzer should fall within the pattern of Figure 4-30.

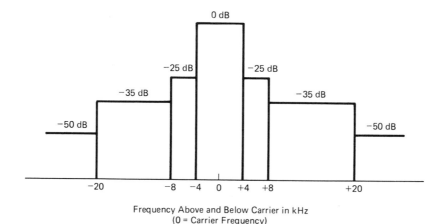

FIGURE 4-30. All transmitter modulation–spectrum modulation components must fall within the pattern shown here.

[1] EIA Standard RS-382: "Minimum Standards-Citizens Radio Service Operating In the 27 MHz Band." pp. 14–15, para. 24.1–24.3. This entire document is available from: Electronic Industries Association, 2001 Eye St. N.W., Washington, DC 20006, Phone (202) 457-4900.

Detection of Incidental FM

Incidental FM caused by the AM process can be detected with a spectrum analyzer. The transmitter should be modulated to a high percentage of modulation but less than 100%. If FM is present, it will show up in the appearance of unequal sidebands above and below the carrier. Figure 4-31 shows a typical pattern produced by the incidental FM in the AM signal.

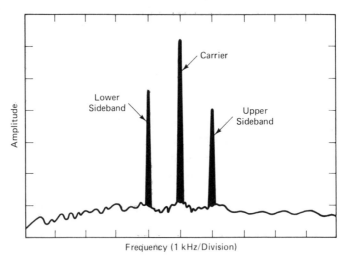

FIGURE 4-31. This spectrum analyzer display shows incidental FM on the single-tone modulated AM signal. The incidental FM shows up as unequal sideband amplitude.

Automatic Modulation Control Test

The function of the automatic modulation control (AMC) is to prevent overmodulation of the transmitter on voice peaks. A typical AMC circuit used in a CB transceiver is shown in Figure 4-32. The level of the audio modulating signal is checked at the primary of the audio modulation transformer. A portion of this signal (as determined by R1) is rectified by D1 and filtered by C1 to obtain a smooth dc control voltage that is then used to forward bias Q1 (through R2). Notice that the emitter-to-collector circuit of Q1 is across the audio input to the audio amplifier. Thus, as the forward bias on Q1 increases the emitter/collector resistance decreases, thus shunting the audio input. Hence, a high output level at the audio output causes Q1 to shunt the input signal, bringing down the output level. Since this output level determines the modulation percentage, the modulation percentage is limited by the setting of R1.

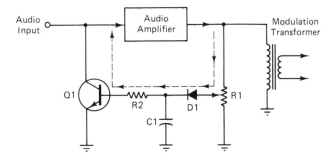

FIGURE 4-32. A simplified schematic showing how an automatic modulation control circuit (AMC) works.

Test Procedure

1. Set up the equipment as shown in Figure 4-29.
2. Set the audio generator to 1 kHz.
3. Initially, set the audio generator to minimum level.
4. Key the transmitter and slowly increase the audio generator level while
   observing the modulation meter. A point will be reached where further
   increases in the audio level will not cause a further increase in the
   modulation percentage. The modulation percentage at this point is the
   modulation *limiting* level. The AMC control can be set by applying a
   high-level audio signal and then adjusting the AMC control for the desired limiting modulation percentage.

Carrier Shift Detection and Measurement

When the positive and negative modulation peaks are unequal or
assymmetrical, a shift in the average carrier level occurs. This is called *carrier
shift.* Figure 4-33 illustrates this factor. In Figure 4-33A is an unmodulated
carrier. Figure 4-33B represents an ideal symmetrical envelope with no carrier shift. In Figure 4-33C, the *peak* (positive) modulation exceeds the *trough*
(negative) modulation, thus the average level shifts in the positive direction.
In Figure 4-33D, the trough (negative) modulation exceeds the peak (positive) modulation, resulting in a negative shift of the average level.

Figure 4-34 is a simple circuit which can be used to detect and measure
carrier shift. A sample of the transmitter signal is loosely coupled to L1. The
diode detects this signal. Capacitors C1 and C2 charge to the peak value of
the carrier. The capacitors bypass the RF to ground. Thus, meter *M* shows a
dc reading determined by *R*. The capacitors offer a high impedance to the
modulated signal, hence they pass through the meter. The meter reading

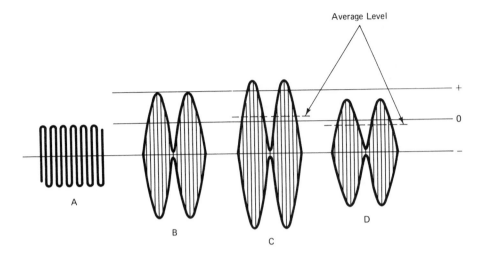

A = Carrier
B = Symmetrical Modulation Envelope; No Carrier Shift
C = Higher Positive Modulation; Positive Carrier Shift
D = Higher Negative Modulation; Negative Carrier Shift

FIGURE 4-33. The effect of carrier shift on the modulation envelope pattern is illustrated here. The waveform at B is free of carrier shift. The waveform at C indicates positive carrier shift. The waveform at D indicates negative carrier shift.

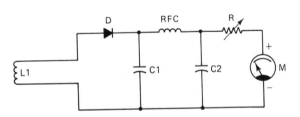

L1 Is Loosely Coupled to the Output Stage of the Transmitter.

FIGURE 4-34. This simple circuit can be used to determine carrier shift.

should be set to the center of the scale with only a carrier transmitted. With modulation applied, an increase in the meter reading indicates positive carrier shift; a decrease in the meter reading indicates negative carrier shift. The carrier shift can be calculated in percent by the following formula:

$$\text{Percent carrier shift} = \frac{\text{Ic} +/- \text{Im}}{\text{Ic}} \times 100$$

where Ic is the current flow with no modulation and Im is the current flow with modulation.

FREQUENCY MEASUREMENT

The frequency of a transmitter must be kept within prescribed limits as set forth by the Federal Communications Commission (FCC). The frequency limit is usually specified as a percentage. For example, the frequency limit of a CB transmitter operating in the 27 MHz band is $+/-0.005\%$. This amounts to a frequency change of $+/-1,350$ Hz. For transmitters operating in the land mobile radio service the frequency tolerance is 0.0005%. For a transmitter operating at 160 MHz, this amounts to a frequency change of $+/-800$ Hz.

Obviously, with such tight frequency tolerances the frequency measuring instruments must be highly accurate. Generally speaking, the accuracy of a measuring instrument should be *at least twice* that required of the equipment which is to be tested. For example, a transmitter which has a frequency allowance of $+/-0.0005\%$ should be checked with an instrument that has a tolerance of no less than $+/-0.00025\%$ at the frequency to be measured.

Various types of frequency measuring instruments were discussed in Chapter 3. The method used to measure the frequency of a transmitter is really very simple. Either the closed coupling method or the off-the-air pickup method can be used. Procedures are described for each of these methods.

Closed Coupling Method   The closed coupling method is normally used when the transmitter to be tested is directly accessible to the technician. Any of the coupling devices described in Chapter 3 can be used to sample the RF signal the frequency of which is to be measured.

The closed coupling method of frequency measurement can be made using a setup similar to that shown in Figure 4-35.

Measurement Procedure

1. Key the transmitter. *Do not modulate it!*
2. Simply read the frequency indication on the frequency measuring instrument.

Off-the-Air Measurements   When the transmitter is not directly accessible to the technician, the frequency can be measured by using a highly sensitive service monitor with its built-in frequency meter. Such instruments can be used to measure the frequency of transmitters that are many miles away by hooking the instrument to an external antenna. Frequency counters are not sufficiently sensitive to measure the frequency of distant transmitters directly. However, there is a method that can be used by which the frequency of a distant transmitter can be determined with a frequency counter. The setup is shown in Figure 4-36.

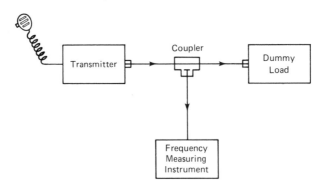

FIGURE 4-35. This is a typical setup used for measuring carrier frequency.

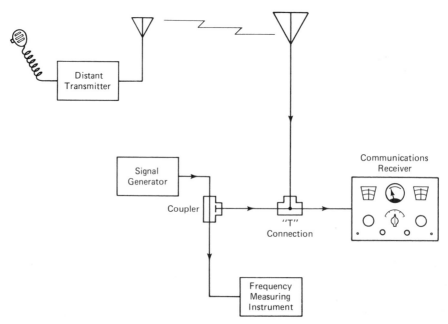

FIGURE 4-36. This setup shows how the frequency of a distant transmitter can be measured.

Measurement Procedure

1. Tune the communications receiver to the signal to be measured.
2. Tune the signal generator to produce a zero beat with the signal to be measured.
3. Measure the frequency of the signal generator with the frequency counter. This is the approximate frequency of the signal to which the receiver is tuned.

Improved accuracy will result if an oscilloscope is used to set the signal generator to zero beat with the received signal. This is because the oscilloscope can determine the zero-beat point much more closely than the ear.

NEUTRALIZATION AND PARASITIC TESTS

Interelectrode capacitance such as *grid-to-plate* or *base-to-collector* can cause an RF amplifier stage to *oscillate* at or near the signal frequency of the transmitter. This is undesirable because of the burden it places on the stage and also because it will cause spurious signals to be generated that might interfere with other communications services. Figure 4-37 shows several types of simple indicating devices that can be used to detect the presence of RF energy in a circuit. The neon bulb requires a comparatively large amount of RF energy to "fire." Its use is therefore limited. An ordinary flashlight bulb is much more sensitive and can therefore detect lower levels of RF energy. The most sensitive of all the devices shown is the diode detector and meter circuit. The flashlight bulb and the diode/meter indicator should be coupled to the tank circuit of the amplifier through a pickup loop attached to the end of a stick such as a small diameter wooden dowel. The pickup loop should be loosely coupled to the tank circuit of the RF amplifier, preferrably near the RF ground side of the coil to minimize detuning. To avoid burning out the indicating device, it should initially be held at a safe distance from the tank circuit and then gradually moved in closer while the indicating device is observed closely.

Testing for Parasitic Oscillation/Improper Neutralization One procedure is simply to remove the drive signal from the RF amplifier under test. CAUTION! With transistors, this is perfectly safe but with tubes be careful. (That's a switch!) Many RF amplifiers using tubes employ a biasing method called "grid-leak" bias. Grid-leak bias is entirely dependent upon the drive signal; no drive signal, no bias! Without proper bias, the plate current will soar, destroying the tube and possibly other circuit components. Before removing the drive signal from a grid-leak biased RF amplifier, you must first supply a fixed-bias to prevent excessive plate current through the tube.

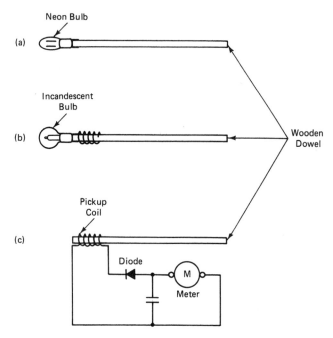

FIGURE 4-37. These simple devices can be used to detect the presence of RF energy in a circuit for neutralization and parasitic tests.

Test Procedure

1. Remove the drive signal. This can be done by pulling the crystal or tube or any other means of disabling the oscillator. *Be sure to observe the proper precautions when removing the drive signal!*
2. Hold the pickup loop of the indicating device near the tank circuit.
3. Observe the indicating device (bulb or meter) for indication of RF energy in the circuit.

An indication on the lamp or meter reveals the presence of RF in the plate circuit. Since the drive signal has been removed, the amplifier has become an *oscillator*, generating its own signal. This may or may not be caused by incomplete neutralization. If the signal is far removed from the normal operating frequency of the transmitter, the cause is not improper neutralization but probably parasitic oscillation. Before adjusting the neutralizing capacitor, you should check the frequency of the signal by coupling the circuit to a frequency counter or other frequency measuring instrument. If the frequency of the oscillation is found to be near the normal operating frequency of the transmitter, improper neutralization is indicated. If the oscillation is found to be far removed from the normal transmitter frequency, parasitic oscillation is indicated.

Trapezoid Pattern Neutralization Test A trapezoid pattern on the scope will indicate improper neutralization in an RF amplifier stage. The pattern in Figure 4-38 indicates improper neutralization. Notice how the slopes are curved inward (concave). The slopes should be perfectly straight, as indicated by the dashed lines.

The Receiver Method for Identifying Improper Neutralization This method employs a communications receiver tuned to the transmitter frequency.

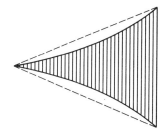

FIGURE 4-38. The trapezoid pattern shown here indicates improper neutralization.

Test Procedure

1. Key the transmitter into a dummy load.
2. Tune the receiver to the transmitter signal.
3. Disable the transmitter drive signal. *CAUTION! Check the biasing method before killing the drive signal.*
4. With the transmitter keyed and the drive signal disabled, tune the receiver around the vicinity of the normal transmitter frequency.

If a signal is found near the normal transmitter frequency, some stage in the transmitter is oscillating. To make sure that the signal is originating within the transmitter, turn the transmitter on and off. If the receiver signal follows the transmitter, you can be sure that the transmitter is the source.

Using the Drive Signal to Test Neutralization One of the indicators in Figure 4-35 can be used for this test.

Test Procedure

1. Remove the plate voltage from the RF amplifier under test.
2. With the plate voltage removed, key the transmitter.
3. Bring the pickup loop of the indicator near the plate (or collector) tank circuit, just close enough to get a useable reading or indication.

Any energy in the tank circuit is due to incomplete neutralization. Since the plate (or collector) voltage is removed, the stage can't oscillate on its own. Therefore, any energy in the tank circuit must be due to the drive signal feeding through the interelectrode capacitance of the tube or transistor. Thus, this test isolates the neutralization problem since parasitics are no longer a factor with the plate or collector voltage removed.

Using a Dip Meter for Neutralization Tests   The dip meter is described in detail in Chapter 3. In this test, the dip meter will be operated as an *absorption-type wavemeter,* which is very sensitive and highly selective. First, the dip meter should be set to the frequency of the transmitter by plugging in the proper coil and then setting the tuning dial.

Test Procedure

1. Disable the transmitter drive signal. *CAUTION!* As always, make sure that the tube (if this is a vacuum-tube stage) is properly biased for safety to prevent exceeding the tube's dissipation rating.
2. With the drive signal disabled, key the transmitter.
3. Couple the dip meter to the plate or collector tank coil.
4. Carefully tune the dip meter around the transmitter frequency while observing the meter for any indication.

If a meter reading is obtained on the dip meter at or near the transmitter operating frequency, improper neutralization is indicated. The absence of an indication means that the stage must be neutralized properly.

Using a Dip Meter for Parasitic Testing   The purpose of this test is to determine the approximate *frequency* of a parasitic. It is assumed that you have concluded (from one of the previous tests) that a parasitic is present in the stage.

The dip meter is set up to be used as an absorption-type wavemeter. In searching for the parasitic, it may be necessary to change plug-in coils several times to tune through several ranges.

Test Procedure

1. Couple the dip meter to the circuit under test.

2. Key the transmitter.

3. Tune the dip meter slowly through the range of interest while closely observing the meter for any indication.

4. When an indication is observed on the meter, the approximate frequency can be determined from the dip-meter scale.

The dip meter can then be used to check for possible sources of the parasitic. For this purpose, the dip meter is set up as a dip oscillator. The power is removed from the transmitter while the dip meter is used to check various circuit wiring and components for resonance at the freqeuncy of the parasitic.

Driven Parasitic Test   Sometimes parasitics occur only at peak power— that is, at the peak of the modulation envelope. These types of parasitics can usually be seen on the modulation envelope pattern or the trapezoid pattern as observed on an oscilloscope. Figure 4-39 shows a *driven* parasitic as displayed on the trapezoid and wave envelope patterns.

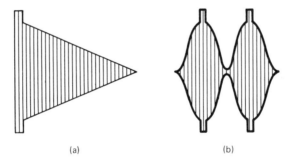

(a)   (b)

FIGURE 4-39. The trapezoid pattern at A and the envelope pattern at B indicate the presence of a driven parasitic.

Test Procedure

1. Modulate the transmitter a full 100% with a single tone (1 kHz or so) but do not overmodulate.

2. Key the transmitter while observing the pattern on the scope.

Determining the Parasitic Frequency with a Receiver   This test procedure should be used only when normal frequency measuring instruments are not available or the parasitic is too weak to measure with normal frequency measuring instruments. This procedure is very slow and tedious, and should only be used as a last resort. It is assumed that the determination has already been made that a parasitic does exist at some unknown frequency.

Test Procedure

1. *Loosely* couple the receiver to the transmitter under test. It may not be necessary to use an antenna on the receiver.

2. Slowly tune through the range of frequencies to be checked while carefully listening for a signal on the receiver.

3. If a signal is found, make sure it is originating within the transmitter by turning the transmitter on and off. If the signal follows the transmitter, you can be sure that the signal is from the transmitter.

4. The frequency can be roughly determined from the receiver dial.

Parasitic Testing with the Spectrum Analyzer   Since parasitics can occur at frequencies far above or below the normal transmitter frequency, pinning down these parasitic frequencies can be very difficult and time-consuming with a *manually tuned* device such as the dip meter or general coverage receiver. However, the spectrum analyzer, with its *electronically swept tuning,* is ideally suited for this application.

Test Procedure

1. Remove the drive signal from the transmitter. *As always, observe the proper precautions concerning bias.*

2. Set the spectrum analyzer for a wide sweep above and below the normal frequency of the transmitter.

3. Determine the frequency of any parasitic that might be found by its position on the horizontal scale of the analyzer.

Driven parasitics can be found with the spectrum analyzer also. As stated before, these usually occur at or near modulation peaks with the drive signal present.

Test Procedure

1. Set the analyzer to scan a wide range of frequencies above and below the normal frequency of the transmitter. (Make certain that the carrier signal does not overload the spectrum analyzer front end.)

2. Key the transmitter, and while observing the spectrum analyzer, slowly increase the modulation of the transmitter using a single-tone modulating signal (1 kHz or so).

3. Study the display carefully. Any signal display that suddenly appears may be a parasitic.

SEARCHING OUT TRANSMITTER SPURIOUS SIGNALS

Spurious signals that are generated by a transmitter are undesirable byproducts that can cause interference to other radio services at frequencies far removed from the assigned frequency of the offending transmitter. These spurious signals must be kept to an absolute minimum. Spurious signals can appear with or without modulation, or under both conditions. Therefore, it is good practice to check for spurious signals with and without modulation applied to the transmitter.

In searching out these spurious signals generated by a transmitter, a very wide range of frequencies must be checked, ranging from the lowest frequency generated within the transmitter to several multiples (harmonics) of the carrier frequency. There are several methods that can be used to search out these spurious signals, some of which can be very tedious and time-consuming. Some of these methods are presented here.

The Spectrum Analyzer Method

When it is necessary to examine a wide range of frequencies, a spectrum analyzer is a "natural" instrument for the task.

Test Procedure

1. Set up the spectrum analyzer for the proper sweep range.
2. Key the transmitter into a dummy load. *Don't* modulate the transmitter.
3. Check the display on the analyzer for any spurious signals. Make certain that the front end of the analyzer is not overloaded by the strong carrier signal.
4. Modulate the transmitter near 100% (but not over 100%) with a single-tone frequency of 2,500 Hz. The modulating tone should be a very low distortion signal.
5. Check the display again for any spurious signals. If a high resolution analyzer is used, only the carrier and the two sidebands should be present, with the sidebands separated from the carrier by the frequency of the modulating tone.

The most frequent cause of spurious signals in AM transmitters is over-modulation.

Spurious Searching with a
Field Strength Meter

A field strength meter can be used for spurious *searching* as long as it is tuneable over the necessary frequency band. Precautions should be taken to guard against overloading the front-end of the field strength meter just as in the use of the spectrum analyzer.

Test Procedure

1. Key the transmitter into a dummy load. (Don't modulate the transmitter.)
2. Tune the field-strength meter through the frequency range to be checked while carefully observing the meter for any indication.
3. The frequency and relative amplitude and any spurious signal can be determined from the field-strength meter dial and meter, respectively.
4. The procedure should be repeated with modulation applied. The transmitter should be modulated near 100% by a low distortion 2,500-Hz, audio-modulating signal.

Spurious Searching with a
Communications Receiver

A communications receiver with a wide tuning range can also be used to check for spurious signals. The procedure is the same as that using a field-strength meter, as described in the previous section. The approximate frequency can be determined from the dial calibrations. A more accurate frequency determination can be made by coupling a signal generator to the receiver and tuning it to zero-beat with the spurious signal. Then the frequency of the signal generator can be measured with a frequency counter or other frequency-measuring instrument.

There are a couple of ways to measure the relative amplitude of the spurious signal with respect to the carrier. Figure 4-40 shows the typical set-up for using a signal generator to determine the relative levels of the spurious signals as referenced against the carrier signal. To reiterate, it is very important to prevent overloading the front end of the receiver during the test.

Test Procedure

1. Turn off the receiver AVC.
2. Key the transmitter and tune the receiver to the carrier frequency.

3. Adjust the receiver RF gain control for full-scale reading on the S-meter.

4. Unkey the transmitter.

5. Tune the signal generator to the receiver frequency (same as carrier) and adjust the signal generator output level for full-scale. Note the signal generator output level in dBm.

6. Key the transmitter again and tune the receiver to the *spurious* signal.

7. Note the reading on the S-meter.

8. Unkey the transmitter and tune the signal generator to the receiver frequency (same as spurious).

9. Adjust the signal generator output level for the same indication as that obtained in step 7.

10. Note the signal generator output level in dBm.

11. The difference in the signal generator levels for steps 5 and 10 is equal to the difference in the levels of the carrier and spurious signals.

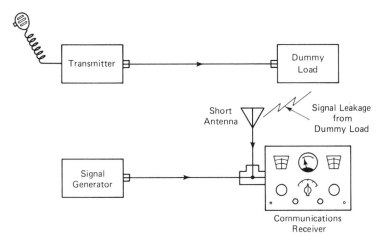

FIGURE 4-40. The setup here can be used to determine the relative level of the spurious signal.

An alternate method is to use a calibrated adjustable attenuator to determine the relative level of the spurious signal as compared to the carrier. The setup is shown in Figure 4-41.

Test Procedure

1. Turn the receiver AVC off.

2. Key the transmitter and tune the receiver to the spurious signal.

3. Set the attenuator to "0" and adjust the RF gain control for a full-scale reading on the S-meter.

4. Tune the receiver to the *carrier* signal.

5. Increase the attenuator setting until the S-meter reads full-scale as it did in step 3.

6. The attenuator setting is equal to the difference in the levels of the carrier and the spurious signal.

In order that this measurement be valid, it is important that all of the signal that is entering the receiver passes through the attenuator. If any signal is bypassing the attenuator by entering the front end of the receiver directly, the measurement will not be accurate. It is important that the receiver front end be well shielded and that a low leakage coax cable be used between the output of the attenuator and the input of the receiver.

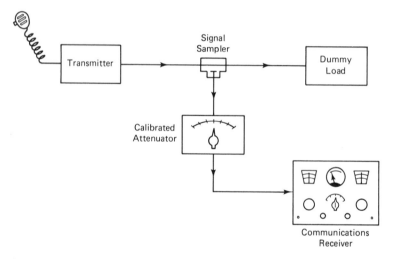

FIGURE 4-41. This setup can be used to determine the relative level of the spurious signal if a calibrated attenuator is available.

BIBLIOGRAPHY

Electronic Industries Association, *Minimum Standards, Citizens Radio Service, AM Transceivers Operating in the 27 MHz Band*, Standard # RS-382. Washington, D.C., 1971.

Miller, Gary M., *Handbook of Electronic Communication*. Englewood Cliffs, N.J.: Prentice-Hall, Inc., 1979.

National Radio Institute, *RF Power Amplifiers*. Washington, D.C., 1968.

Oliver, Bernard M., and Cage, John M., *Electronic Measurements and Instrumentation*. New York: McGraw-Hill, Inc., 1971.

Philco Corporation, *Radio Communication System Measurements*. ©1952, Philco Corporation.

Sands, Leo G., *CB Radio Servicing Guide*. Indianapolis: Howard W. Sams & Company, 1974.

Wolf, Stanley, *Guide to Electronic Measurements and Laboratory Practice*. Englewood Cliffs, N.J.: Prentice-Hall, Inc., 1973.

5
AM RECEIVER TESTS
AND MEASUREMENTS

SENSITIVITY TESTS

Basically, the sensitivity of a receiver can be defined as the minimum signal input to the receiver which will produce a usable output from the speaker. Various methods of sensitivity testing have evolved over the years; some methods produce more meaningful results than others. Several methods of measuring the sensitivity of AM receivers are presented here. These sensitivity tests should be conducted in an environment which is as noise-free as possible; that is an environment which is relatively low in electrical noise disturbances. Such noise sources are fluorescent lights, electric motors, automobile ignition systems, power-line noise, etc.

Standard Output Sensitivity Test

The basic equipment setup for this test is shown in Figure 5-1. The standard output level is usually much less than the full audio power rating of the receiver. Usually, 1/2 W is used as the standard output level for receivers which are capable of producing substantially more audio power than 1/2 W. On low audio power receivers, 50 mW is often used as the standard.

Test Procedure

1. Set the signal generator for 30% modulation at 1 kHz and tune it to the receiver frequency.
2. Set the receiver audio level to maximum. If an RF gain control is used set it for maximum gain. If a squelch control is used, set it to unsquelch the receiver audio circuits fully.

3. While observing the audio output level, increase the signal generator output level until the standard power level is reached. If an audio voltmeter is used, the power level can be determined from the formula: $P = E^2/Z$. The graph in Appendix D can be used to correlate voltage with power across 3.2-Ω and 8-Ω loads.

4. When the audio meter indicates standard output level, note the signal generator output level in microvolts or dBm. This is the standard output level sensitivity of the receiver.

In this test, the standard output measurement consists of the 1-kHz signal component, various distortion components, and noise components. Depending upon the receiver design, some receivers produce a relatively large amount of noise and distortion components. If the output consists of a relatively large amount of noise and distortion, the signal component may be completely masked by these noise and distortion components. At the same time, the sensitivity may appear to be good. The usefulness of this test can be roughly gauged by the ear. If the audio output consists of a high level of noise that tends to mask the 1-kHz audio tone, the test and measurement is not reliable as an indicator of sensitivity. On the other hand, if the 1-kHz signal stands "head and shoulders" above the noise and distortion components, the test and measurement is a reliable indicator of sensitivity.

Full Power Output Sensitivity

This test is performed in the same manner and with the same equipment setup as used in the previous test for standard output sensitivity (see Figure 5-1).

Test Procedure

1. Modulate the signal generator 30% at 1 kHz.

2. Increase the generator output level until the audio output power equals the specified maximum rated power.

3. The generator level setting at this point is the *full-power sensitivity* of the receiver.

The 10-dB $S+N/N$ Sensitivity Test

This test takes into account the signal to noise ratio and is therefore a more reliable indicator of useful sensitivity than the standard output-level sensitivity test. First, a measurement of signal plus noise ($S+N$) is made and then a measurement of the noise (N) alone is measured. The point at which the signal plus noise ($S+N$) is 10 dB greater than the noise (N) alone is the

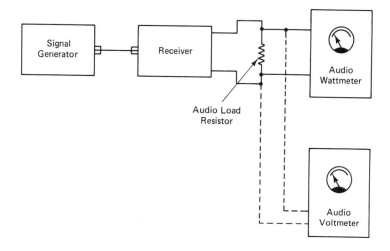

FIGURE 5-1. Basic setup for many of the sensitivity and selectivity measurements described in this chapter.

10-dB $S+N/N$ sensitivity of the receiver. The same basic setup shown in Figure 5-1 can also be used for this test. The audio meter should have a dB scale so that the difference in $(S+N)$ and (N) levels can be determined directly from the scale in dB units.

The audio output power must be at least 1/2 W for receivers that are designed to produce in excess of 1 W audio power. For lower audio power receivers, the audio power must be at least 50 mW. The audio power can be determined from the graph in Appendix D or from the formula $P = E^2/Z$. If an audio power meter is available, so much the better. The power can be read directly in watts rather than converting the audio voltage to watts.

In this test, the composite $(S+N)$ signal is produced by a signal generator modulated by a 1-kHz tone at 30% modulation. The noise signal (N) is measured by turning off the modulation of the signal generator, leaving only the carrier feeding the receiver input.

Test Procedure

1. Unsquelch the receiver and turn the receiver volume control to maximum. Set the RF gain control (if used) to maximum.
2. Set the signal generator modulation to 30% at 1 kHz.
3. While observing the audio power meter (or voltmeter), adjust the signal generator level to produce an output power of 1/2 W (50 mW for receivers designed for less than 1-W audio power). This is the signal plus noise $(S+N)$ level.

4. Turn off modulation. If the audio output level drops 10 dB or more, the signal generator level setting is the 10-dB $S+N/N$ sensitivity of the receiver. If the level drops less than 10 dB, increase the signal generator level setting a small amount, turn on the modulation, and adjust the volume control for 1/2 W audio power (or 50 mW for low-power receivers). Again, turn the modulation off and note the drop in audio level. It may be necessary to repeat the procedure several times before the correct signal generator level is reached that produces the 10-dB $S+N/N$ ratio.

This 10-dB $S+N/N$ test has been widely adopted as a standard method of testing AM communications receivers. In most cases, this method works very well. However, this method doesn't work very well in cases where an *audio* AGC amplifier is used between the receiver audio output and the point in the system where the actual $S+N/N$ measurement is taken. This is often the case in aircraft communications systems (see Figure 5-2). An AGC amplifier is shown connected between the receiver output and the headset. The purpose of this AGC amplifier is to maintain a fairly uniform audio level from the widely varying levels of input signals. If the 10-dB $S+N/N$ measurement is taken at point A in Figure 5-2, the result will be the 10-dB $S+N/N$ of the receiver only. But a truly valid test must take into consideration the system as a whole from the signal input to the receiver to the point at which the audio is applied to the speaker, or in this case the headset. Therefore, in this case the audio AGC amplifier must be included in the system test.

Now let's see what happens at point B (the output of the audio AGC amplifier) when the standard 10-dB $S+N/N$ test is attempted. With a 30% modulated signal at 1 kHz fed to the receiver input, the audio AGC amplifier will set the output level at point B to a specified level (predetermined by the adjustment of the AGC amplifier). When the modulation is removed

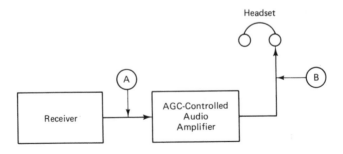

FIGURE 5-2. A situation that might be encountered in an aircraft communications system. The audio AGC amplifier is used to hold the level nearly constant at the output.

from the input signal, the audio level at the output of the receiver will drop by an amount that depends on the receiver sensitivity and the input signal level. However, in an attempt to hold its output level constant, the audio AGC amplifier will give *extra* amplification to the lower level noise-only signal. Hence, the $S+N/N$ ratio measured at point B will be much less than at point A. For example, a 10-dB $S+N/N$ ratio at point A might become 1 or 2 dB $S+N/N$ at point B. This makes the system performance look much worse than it actually is.

A method of sensitivity testing that can be used to overcome this problem is the *SINAD* test (Sinad is an acronym for *si*gnal, *n*oise, and *d*istortion). This method of sensitivity testing is described next.

The Sinad Sensitivity Test

In the 10-dB $S+N/N$ sensitivity test, the audio output measurement is made under two different receiver input signal conditions. First, a modulated signal is fed to the receiver input and an audio output measurement is taken; then, the modulation is removed and a second audio output measurement made. Actually, in the first measurement (with modulation) the receiver is being tested under dynamic conditions while the second measurement might be considered a static condition.

Although the Sinad method has not yet been standardized for AM receiver sensitivity tests, it is nevertheless very useful in AM receiver work. In the Sinad sensitivity measurement, the test is performed under only one condition—that is, with a modulated signal at the receiver input. This is why the Sinad method works even with audio AGC amplifiers incorporated in the system, such as in the aircraft system discussed previously. A common distortion analyzer or a specially designed *Sinad meter* can be used to measure the audio output. A Sinad meter is basically a distortion analyzer designed to meet certain EIA criteria. A full discussion of a typical Sinad meter along with principles of the Sinad measurement are included in Chapter 9 under "FM Receiver Sensitivity Tests."

Basically, the composite signal (signal + noise + distortion) is used to set the reference level on a distortion analyzer. Then the distortion analyzer is used to separate the noise and distortion components from the composite signal and to measure them. The ratio of the composite signal to the noise and distortion components is expressed as:

$$S + N + D/N + D$$

This ratio is usually expressed in dB, and that's the way the Sinad meter or distortion analyzer measures it. As the signal input level to the receiver increases, the $(N + D)$ components decrease while the signal, (S), increases. The signal input level that produces a Sinad ratio, $S + N + D/N + D$, of 10 dB is called the 10-dB Sinad sensitivity of the receiver. The test procedures for both the Sinad meter and the typical distortion analyzer are described below.

The Distortion Analyzer Method Test Procedure (See Figure 5-3.)

1. Set the signal generator modulation to 30% with a 1-kHz tone.
2. Set the squelch control to unsquelch the receiver completely. Set the RF gain control to maximum.
3. Set the volume control to maximum.
4. Adjust the signal generator output level for 1/2 W audio output power.
5. Adjust the "set-level" control on the distortion analyzer for 0 dB on the dB scale.
6. Switch the distortion analyzer to "measure distortion" and adjust the null control for minimum indication on the meter.
7. If the noise and distortion measurement in step 6 is at least 10 dB below the reading in step 5, the signal generator level setting is the 10-dB Sinad sensitivity at 1/2 W audio output.
8. If the noise and distortion measurement in step 6 is less than 10 dB below the reading in step 5, increase the signal generator output level and repeat steps 5 and 6. If necessary, adjust the volume control to maintain the 1/2 W audio output level. It may be necessary to try several different level settings on the signal generator before the 10-dB Sinad point is reached. The signal generator level setting at the 10-dB point is the 10-dB Sinad sensitivity of the receiver.

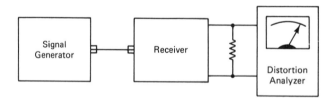

FIGURE 5-3. A test setup that can be used to perform the Sinad sensitivity measurement on a receiver using a standard distortion analyzer.

The Sinad Meter Method Test Procedure

The Sinad meter is simply a distortion analyzer with automatic level setting and a fixed null frequency at 1 kHz. More information on the Sinad meter is presented in Chapter 9 under "FM Receiver Sensitivity Tests."

These Sinad meters were designed primarily for FM receiver work but can be used for AM receiver work also. The procedure is a little different, however.

Test Procedure

1. Set up the equipment as shown in Figure 5-4.

2. Modulate the signal generator 30% at 1 kHz. This 1-kHz modulating signal must fall within the 1-kHz reject notch filter so as to be properly filtered out. If it is supplied from the Sinad meter itself, it will probably be at exactly the correct frequency. If it is from another source, such as within the signal generator itself, the notch filter in the Sinad meter may have to be retuned slightly to ensure maximum rejection of the 1-kHz tone frequency. If the 1-kHz tone frequency is not adequately rejected, proper results cannot be obtained. If an external variable frequency audio generator is used to modulate the signal generator, the frequency of the audio generator should be carefully tuned for minimum reading (maximum null) on the Sinad meter.

3. Set the RF gain and volume controls to maximum and set the squelch control to unsquelch the receiver fully.

4. Set the signal generator on frequency and set the signal generator output level to produce 1/2 W audio output power.

5. If the Sinad meter indicates − 10 dB or more, use this generator level setting as the 10-dB Sinad sensitivity at 1/2 W audio power.

6. If the Sinad meter reading is between − 10 dB and 0 dB, increase the signal generator level setting until the Sinad meter indicates − 10 dB. At this point, the signal generator level setting is the 10-dB Sinad sensitivity.

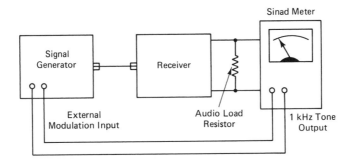

FIGURE 5-4. A test setup in which a special Sinad meter is used to perform the Sinad sensitivity measurement on a receiver.

"Effective" Sensitivity Test

The purpose of this test is to determine just how much the receiver's sensitivity is degraded by interference. The test is made under the actual operating conditions of the receiver. Since the noise level varies from one location to another, it is necessary to perform this test "on-site"—that is, at the location where the radio is to be used. There are many different sources that can contribute to the noise level at a given location. Some such noises are: power line noise, electric motor noise, automobile ignition noise, transmitter noise, and intermod products from nearby transmitters.

Transmitter noise and intermod products will cause intermittent interference depending upon the keying of the transmitter and/or the keying of certain combinations of transmitters. The other types of interference may also be intermittent.

The proper equipment setup for the test is shown in Figure 5-5. The resistor network consisting of R1, R2, and R3 is used to maintain a proper impedance match. Looking from points *A*, *B*, and *C*, the impedance will be 50 Ω. The 40-dB attenuator can be used to minimize the strong signals picked up by the antenna, possibly producing intermod products within the signal generator itself. Thus, the attenuator serves as an *isolator* between the antenna and signal generator. The attenuator can be omitted as shown by the dashed lines when generator-produced intermod is not a problem. The test procedure is the same with or without the attenuator in the circuit.

The resistor matching network itself provides 6 dB of attenuation between any two ports; that is, *A–B*, *B–C*, and *C–A*.

Test Procedure

1. Measure the sensitivity of the receiver in a normal manner; that is, with the signal generator connected directly to the receiver input. Use the 10-dB *S + N/N* or the 10-dB Sinad method, but use the *same* method throughout the test procedure. Note the signal generator level in dBm.

2. Connect the resistor network and attenuator (if used) as shown and connect the 50-Ω termination to part *A* of the resistor matching network.

3. Measure the sensitivity of the receiver and make a note of the generator level setting (in dBm).

4. Remove the 50-Ω termination from point *A* and connect the antenna to point *A*.

5. Measure the sensitivity again and make a note of the signal generator level setting (in dBm).

6. Subtract the generator level in step 5 from the generator level in step 3. The difference is the amount of degradation caused by noise and/or interference present at that locality.

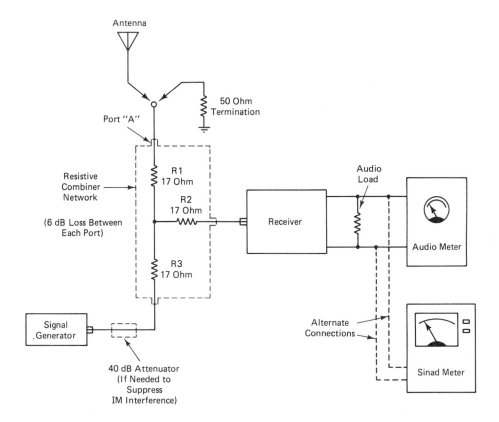

FIGURE 5-5. This test setup is used to determine the effective sensitivity of a receiver. Note the special resistive combiner network used to connect the antenna and signal generator simultaneously to the receiver under test. Alternate connections are shown by the dashed lines for the use of a Sinad meter if the Sinad method is to be used.

7. This difference figure must now be subtracted from the sensitivity measurement level obtained in step 1 to get the effective sensitivity of the receiver at that location.

The following example should help to clarify this test procedure. Suppose the sensitivity measurement in step 1 was −120 dBm. Further suppose that the step 3 measurement (without antenna) was −74 dBm and that the step 5 measurement (with antenna) was −69 dBm. Subtracting step 5 from step 3, we have:

$$-74 - (-69) = -74 + 69 = -5 \text{ dBm}$$

This -5 dBm is the degradation caused by noise and interference. This -5 dBm must now be subtracted from the original measurement in step 1. Thus,

$$-120 \text{ dBm} - (-5 \text{ dBm}) = -120 \text{ dBm} + 5 \text{ dBm} = -115 \text{ dBm}$$

Thus, -115 dBm is the effective sensitivity.

An alternate method of performing this test is to use a homemade co-axial T-coupler, which is described in detail in Chapter 3. Other types of in-line sampling devices are equally suitable. The setup is shown in Figure 5-6. The 50-Ω termination connected to the output of the signal generator through the T-connector provides a proper match for the signal generator. A resistor-matching network is not used. The T-coupler provides a high degree of isolation between the signal generator and the line section. For this reason, the generator has practically no loading effect on the line section and vice versa. The procedure for using this setup to measure effective sensitivity is the same as the procedure just described.

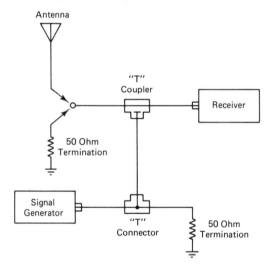

FIGURE 5-6. An alternate method of connecting the signal generator and antenna to the receiver is shown here. The T-coupler is used to provide isolation for the signal generator.

Sensitivity Bandwidth

When radio receivers are designed to operate over a relatively wide range of frequencies, the receiver sensitivity may vary over the range. For this reason, sensitivity checks for such broadband receivers should be conducted at several points throughout the band. It is usually sufficient to spot check rather than to test at every channel in the band. To spot check, simply perform the sensitivity test at the center, upper end, and lower end of the band. Generally, the sensitivity will be slightly better at the center of the band (providing the radio is properly aligned). However, the sensitivity should not drop off significantly at the extreme ends in a properly designed receiver.

SQUELCH TESTS

Squelch circuits are built into receivers to keep the speaker quiet in the absence of a signal at the receiver input. This minimizes operator fatigue, which would result from the constant "noise blowing" without squelch. The amount of RF signal input required to open the receiver is determined by the setting of the squelch control. Thus, the squelch can be set to respond only to strong signals if only these signals are desired. For maximum sensitivity, the squelch control must be set near the *threshold*, or the point at which noise blowing occurs. Squelch tests usually are performed at two squelch settings, the threshold point and the maximum squelch point. The setup is shown in Figure 5-7.

FIGURE 5-7. A simple test setup that is used to test the squelch operation of a receiver.

Critical Squelch Test Procedure

1. Set the signal generator for 30% modulation at 1 kHz.
2. Set the signal generator off to one side of the receiver frequency.
3. Set the squelch control to the critical point, making sure that the receiver is completely quiet.
4. Set the signal generator to the receiver frequency.
5. Slowly increase the signal generator output level until the receiver opens and the signal is heard in the speaker.
6. The signal generator level at this point is the critical squelch sensitivity.

Tight Squelch Test Procedure

The test procedure is the same as above, except that the squelch control is set to maximum or tight squelch in step 3. The signal generator level in step 6 will then be the tight squelch sensitivity.

Squelch Range

The squelch range is the threshold-to-tight squelch sensitivity. For example, if the threshold squelch sensitivity is 0.3 μV and the tight squelch sensitivity is 50 μV, the squelch range is 0.3 to 50 μV.

SELECTIVITY TESTS

This test takes into account all stages of the receiver from the RF input to the audio output of the receiver. A simplified method of checking the selectivity of a conventional superheterodyne AM receiver is shown in Figure 5-8. The AGC or AVC is disabled temporarily by either grounding the AVC line or applying a source of fixed bias to it. This prevents the signal level from affecting the gain of the RF/IF stages.

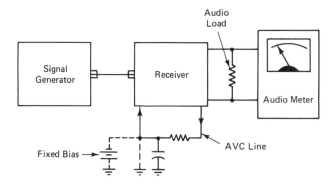

FIGURE 5-8. A test setup that is used to perform certain selectivity tests.

Test Procedure

1. Modulate the signal generator 30% at 1 kHz.
2. Set the receiver RF gain at maximum, the squelch control to fully unsquelch the receiver, and the volume control to maximum.
3. Tune the signal generator to the receiver frequency.
4. Adjust the signal generator to produce the standard output from the receiver (1/2 W or 50 mW). Note the generator level.
5. Tune the signal generator away from the receiver frequency (by 10 kHz or so) and increase the signal generator level until the output level of the signal generator in dBm is the amount of attenuation of the off-resonance signal. Repeat this step at several different frequencies on each side of the receiver frequency.

6. Using linear graph paper, plot the results of the measurements to give the response or selectivity curve. Let the generator level in step 4 serve as the 0-dB reference mark. The other level settings can then be represented as so many dB below the 0-dB mark.

Blocking/Desensitization Test

When a strong signal is present at a frequency near that to which the receiver is tuned, the strong off-channel signal may degrade the response of the receiver to the weaker on-channel signals. Yet, the interfering signal itself may never be heard. Hence, it is possible that the operator may be totally unaware that an interference problem even exists. Such interference is called blocking or desensitization, or simply *desense*.

The test setup for checking desense is shown in Figure 5-9. The two generators are connected to the receiver through a resistive combining network. The combining network provides 6 dB of attenuation between any two of its terminals. Additional attenuators may be inserted between the signal generator and the combining network to provide further isolation between the signal generators. The generator must have sufficient output capability to overcome the attenuation. Signal generator #1 is the desired signal while signal generator #2 is the undesired signal.

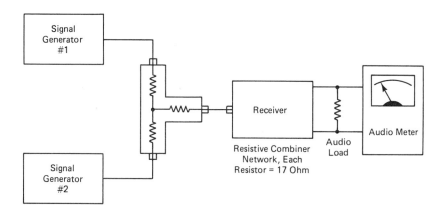

FIGURE 5-9. A test setup for certain blocking, adjacent channel rejection and desensitization tests.

Test Procedure

1. Set up the equipment as shown in Figure 5-9.
2. Modulate generator #1 30% with a 1-kHz tone and set the output lev-

el to the 10-dB $S + N/N$ sensitivity level of the receiver. Then set the volume control to produce the standard output level (1/2 W or 50 mW).

3. Set generator #2 to a frequency off-resonance, usually the adjacent channel above or below the desired channel. Do not modulate generator #2.

4. While observing the audio output level on the meter, gradually increase the output level of generator #2. At some point, the audio output level will drop rapidly with small increases in the output level of generator #2. The point at which the audio level drops 3 dB is the blocking or desense point. The difference in the two generator levels at this point is the desense figure in dB.

Adjacent Channel Selectivity and Desense Test

The main difference in this test and the blocking/desense test just described is that in this test the interfering signal is heard in the speaker and measured on the audio output level meter. While we are primarily interested in the amount of rejection of the adjacent channel, it is necessary to include the *desense factor* because the desense factor of the receiver may actually make the adjacent channel selectivity or rejection appear much better than it really is. This is because at the point at which desense occurs the response of the receiver to any signal is greatly impaired. Thus, the reduced response to the adjacent channel (in the presence of desense) will make it appear that the receiver's adjacent channel rejection is quite good, while in fact the receiver can hardly function at all.

There are several approaches to measuring the adjacent channel rejection figure and desense. Three possible methods are presented here. Two of these methods require the use of two signal generators; the third method requires only one signal generator but represents a less realistic test than the other two methods. Still, it is of some value in evaluating the adjacent channel rejection performance of the receiver.

Test Procedure—Method 1

1. Set up the equipment as shown in Figure 5-9.

2. Set generator #1 to the desired channel (with generator #2 temporarily set to standby) and set the output level of generator #1 to the 10-dB $S + N/N$ sensitivity level of the receiver with the modulation set to 30% at 1 kHz.

3. Set the audio output level of the receiver to the standard output level.

4. Set generator #2 to the adjacent channel, modulated 30% at 400 Hz and increase its output level until the 10-dB $S + N/N$ ratio is reduced or degraded to 6-dB $S + N/N$.

5. The difference in the two signal generator output levels is the adjacent channel rejection figure of the receiver.

6. The same procedure should be repeated with the generator #2 tuned to the opposite adjacent channel and the minimum figure taken as the true adjacent channel rejection figure of the receiver.

The previous test is a measure of the amount of adjacent channel signal strength required to *degrade* the desired signal (which is just strong enough to produce the minimum 10-dB $S+N/N$ ratio) by 4 dB.

Many manufacturers recommend a procedure for testing the adjacent channel rejection and desense with the use of a single signal generator. It is described as follows.

Test Procedure—Method 2

1. Set up the equipment as shown in Figure 5-1, as required for the normal sensitivity measurement.

2. Perform the standard 10-dB $S+N/N$ sensitivity test.

3. Leave the modulation on (30% at 1 kHz) and tune the receiver to an adjacent channel. Do not change any other receiver control settings.

4. Increase the signal generator level to obtain the standard output as obtained in step 1.

5. The difference in signal generator level settings (in dB) is the adjacent channel rejection and desense figure. If the difference in the two signal generator level settings is extreme (on the order of 100 dB or more), acute desensitization is indicated.

6. The test should be repeated for the other adjacent channel on the opposite side and the minimum figure used for the final test result.

In the adjacent channel selectivity test, we are basically interested in determining just how well a desired signal can be received in the presence of a strong adjacent channel signal. If the desired signal is adjusted for 10-dB Sinad level (30% modulation at 1 kHz) at 1/2 W audio output, the interfering signal can be adjusted to degrade the desired signal by a specified amount, 6 dB or so. The 1-kHz signal from one of the signal generators will represent the desired signal. The undesired signal is then set up on another signal generator. The undesired signal is modulated by 400-Hz tone at 30% modulation. The desired 1-kHz tone is filtered out by the distortion analyzer or Sinad meter. When the adjacent channel interfering signal is increased to a sufficient level to cause interference to the desired signal, the distortion analyzer or Sinad meter reading will increase showing the amount of degradation of the desired signal caused by the undesired signal. When the desired signal is degraded 6 dB, the difference in the two signal generator levels is the adjacent channel rejection figure.

Test Procedure—Method 3

1. Set up the equipment for the test as shown in Figure 5-9.
2. Perform the 10-dB Sinad measurement as described previously in this chapter. Use signal generator #1 as the desired signal for this test. At the end of the test, the audio output level from the receiver should be at least 1/2 W, with the Sinad meter or distortion analyzer meter indicating at least 10-dB Sinad.
3. Set generator #2 to the undesired frequency (one of the adjacent channels) and set the modulation to 30% at 400 Hz.
4. Increase the level of generator #2 until the Sinad meter or distortion analyzer meter reading increases 6 dB.
5. The difference in the two signal generator level settings is the adjacent channel rejection figure of the receiver.

Since this test is basically a measurement of the amount of degradation of the desired signal by the undesired signal, it would seem to give a more realistic picture of actual performance of the receiver under real-life operating conditions. It is important to keep in mind that this particular test has not been standardized for AM receiver testing. It is presented here in the interest of showing different approaches that may be used in conducting such tests. If desired, new equipment that is known to be in excellent condition could be tested in such manner in order to establish reference figures for future use. Periodic testing can reveal deterioration in the performance level of the equipment, as long as the tests are repeated in exactly the same manner each time. This is the whole purpose of standards of testing, to ensure repeatability (and, of course, validity).

Image Rejection

The image frequency of a receiver is the frequency that differs from the desired frequency by an amount equal to twice the first IF frequency. The image frequency can be either lower or higher than the desired frequency, depending upon the operating frequency of the first IF frequency. If the first LO frequency is above the desired receive frequency, this is called high side injection. If the first LO operates below the desired receive frequency, this is called low side injection.

If high side injection is employed, the image frequency will be above the desired receive frequency by an amount equal to twice the first IF—that is, $IMAGE = F + 2IF$. If low side injection is employed, the image frequency will be below the desired receive frequency by an amount equal to twice the first IF—that is, $IMAGE = F - 2IF$.

The two primary factors governing the degree of image rejection possible are front-end selectivity and the first IF frequency. For a given amount of front-end selectivity, the higher the first IF frequency, the greater the image rejection. It is important to note that the selectivity of the stages following the first mixer do not and cannot discriminate between the desired frequency and the image frequency. This is because the image and the desired frequencies are both converted to the same IF frequency by the first mixer. Hence, they are inseparable at this point.

For all practical purposes, any of the methods used in testing the adjacent channel selectivity or rejection could also be applied to the image rejection test. The accepted method of determining the amount of image rejection is by using a single signal generator tuned to the image frequency. The setup of the equipment is the same as shown in Figure 5-1 for sensitivity testing.

Test Procedure

1. Perform the standard 10-dB $S+N/N$ sensitivity test and note the signal generator level setting. Note the reference audio output from the receiver.
2. Tune the signal generator to the image frequency and adjust the output level to produce the reference audio output obtained in step 1.
3. Subtract the signal generator level in step 1 (in dBm) from the signal generator level in step 2. The result is the image rejection in dB.

IF Rejection

If a signal at (or very close to) the IF frequency of the receiver is present and of sufficient strength, it may be able to get through (or around) the front-end stages of the receiver and enter the IF amplifier directly. Once it gets into the IF amplifier, it is treated like a normal on-channel signal from that point on to the output. For that reason, it is important that the receiver IF stages be shielded properly and it is equally important that the front end be designed properly to minimize this problem.

The IF rejection figure of a receiver can be determined in basically the same manner that the adjacent channel selectivity is determined. A common procedure is to use one signal generator to perform the test. The basic equipment setup shown in Figure 5-1 can be used for this test.

Test Procedure

1. Perform the standard 10-dB $S+N/N$ sensitivity test at the normal receiver frequency. After performing this test, don't change the setting of the volume control. Note the signal generator level and the audio output reference level.

2. Tune the signal generator to the first IF frequency. (If the receiver is a single-conversion type, there will be only one IF.) Leave the modulation at 30% at 1 kHz.

3. Increase the signal generator output level until the receiver audio output reaches the reference level that was established in step 1.

4. The difference in the generator level settings in step 1 and step 3 (in dB) is the IF rejection figure of the receiver.

In *dual* conversion or *triple* conversion receivers, the IF rejection test can be performed at the second and/or third IF, if desired. Usually the IF rejection figure is given for the first IF.

Cross Modulation

Suppose a receiver is tuned to a relatively weak on-channel signal, while on the next frequency (adjacent channel) there exists a very strong signal. If this strong off-channel signal is sufficient to *overdrive* the receiver's front end, the overdriven stage can become a *nonlinear mixer*. When the desired signal and the undesired signal are combined in the nonlinear mixer, the result is that the original modulation of the undesired signal can appear as modulation on the desired signal. This is called *cross modulation.*

The cross modulation effect is most noticeable when the normal modulation of the desired signal is interrupted as in pauses between sentences, etc. A measurement of a receiver's *immunity* to the effects of cross modulation can be performed with the same basic equipment setup shown in Figure 5-9. Signal generator #1 represents the desired signal, while signal generator #2 represents the undesired signal.

Test Procedure

1. Set generator #2 to standby. Set signal generator #1 to the 10-dB sensitivity level of the receiver. With the modulation set to 30% at 1-kHz adjust, the receiver's volume control to produce the standard reference audio output (1/2 W or 50 mW).

2. Turn off the modulation of generator #1.

3. Tune generator #2 to one of the adjacent channels. Set the modulation to 30% at 1 kHz.

4. Increase the output level of generator #2 until the audio output is at the standard reference level again. To make sure that the response is caused by cross modulation, temporarily turn off signal generator #1 or put it in the standby mode. If the output disappears, cross modulation is occurring. If the output is not affected, cross modulation is not the problem.

5. The difference in the level settings of the two generators in dB is the cross-modulation rejection figure of the receiver.

It is worth noting here that certain of the test procedures used for checking the adjacent channel rejection figure include the cross-modulation effect as part of the overall adjacent channel interference test. Methods 1 and 3 described under adjacent channel rejection tests in this chapter will include the effects of cross modulation; however, method 2 does not include the effects of cross modulation since only one signal is used in the test procedure.

Intermodulation Rejection

Intermodulation interference and cross modulation interference are often confused or considered the same. They are not the same. In its truest form, cross-modulation interference only exists in the presence of the desired signal, since the interference shows up as undesired modulation of the desired carrier. On the other hand, intermodulation is usually most noticeable in the absence of the desired signal. Intermodulation interference results from the mixing or combining of two or more signals (none of which are on the receiver frequency) to produce a resultant signal, which is on the receiver frequency. This mixing process can occur in the receiver itself or at some point external to the receiver. If the mixing process is within the receiver itself, it is *receiver-produced intermodulation.* This is what we are concerned with here. If a receiver's front end is overloaded by a strong off-channel signal, intermodulation can result. This intermodulation can cause nuisance interference, or if severe enough can cause a serious impairment of communications.

As a rule, odd-order intermods fall in-band, while even-order intermods fall out-of-band. This holds true regardless of the channel spacing. Probably the most troublesome is the third-order intermod. These can occur when the frequency spacings of two signals are 2-to-1 as referenced to the receiver frequency. As an example, suppose that two transmitters are operating simultaneously, one at 27.085 MHz and the other at 27.005 MHz. If a nearby receiver is tuned to 27.165 MHz and the other two signals are strong enough to overload the receiver, the two signals could mix together there to produce a 27.165-MHz intermod signal. The second harmonic of the 27.085-MHz signal mixes with the 27.005-MHz signal in the following manner:

$$2(27.085 \text{ MHz}) - 27.005 \text{ MHz} = 27.165 \text{ MHz}$$

This is a third-order intermod signal. Note that the frequency spacing of the two combining signals is 2 to 1 (27.005 MHz is 160 kHz away from the receiver frequency, while 27.085 MHz is 80 kHz away).

There are several possible approaches to testing a receiver for intermod rejection performance. Three possible methods are presented here.

Test Procedure—Method 1

1. Set up the equipment as shown in Figure 5-10. Generator #1 will be used to produce the desired signal while generators #2 and #3 will be used to produce the intermod signal.

2. Set the receiver RF gain to maximum, the squelch control to unsquelch the receiver fully, and the volume control to maximum.

3. Set generators #2 and #3 to the standby mode.

4. Tune generator #1 to the desired receiver and set the modulation to 30% at 1 kHz.

5. Set the output level of generator #1 to produce 1/2 W audio output power. If the Sinad meter or distortion analyzer indicates less than 10-dB Sinad, increase the output of generator #1 to produce 10-dB Sinad.

6. Turn generator #2 on and tune it to one of the adjacent channels. *Do not modulate it.*

7. Turn generator #3 on and tune it to the second channel away from the receiver channel, on the same side of the receiver channel as generator #2. For example, if generator #2 is tuned to the first channel

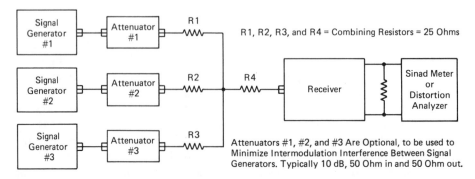

FIGURE 5-10. A test setup for intermodulation testing.

above the receiver frequency, generator #3 should be tuned to the second channel above the receiver channel. Modulate generator #3 30% at 400 Hz.

8. Increase the output level of generator #2 and #3 together, being careful to keep both outputs the same. Continue increasing the generator levels in this manner until the Sinad meter drops 6 dB (-10 to -4).

9. The difference in the output levels of generators #1 and #3 (or #2) is the receiver's intermodulation rejection figure as referenced to a 6-dB degradation of the Sinad ratio.

Test Procedure—Method 2

1. Set up the equipment as shown in Figure 5-9. Set the receiver controls the same as in method 1, step 2.

2. With generator #2 off or in the standby mode, set generator #1 to the receiver frequency and perform the standard 10-dB $S + N/N$ sensitivity test on the receiver. Note the signal generator output level required (in dBm) and the reference audio output from the receiver.

3. Without disturbing any of the receiver's operating controls, tune generator #1 to the first channel above or below the receiver channel. Turn off the modulation.

4. Turn on generator #2 and tune it to the second channel above or below the receiver channel (on the same side as generator #1) and set the modulation to 30% at 1 kHz.

5. Increase the output of both generators together, being careful to keep them both at the same level. Continue increasing the output level of both generators in this manner until the audio output reaches the reference level.

6. Note the generator level setting (either generator) and compare this with the generator level noted in step 2. The difference in dB is the intermodulation rejection figure of the receiver as referenced to the standard output level at 10-dB $S + N/N$ ratio.

Identifying Receiver-Produced Intermod

It is often desirable to determine whether an intermod interfering signal is produced within the receiver itself or if the intermod signal is produced at some point external to the receiver. When frequency multiplication is involved in the intermod-producing process (and it usually is), a simple attenuator test can be used to determine whether or not the intermod is receiver-produced or externally produced intermod. The following examples and illustrations will serve to clarify this for you.

Figure 5-11A shows a simple frequency multiplier. The input frequency
(*A*) is multiplied by a factor *N* in the multiplier, the output then becomes
(*N*)*A*. In Figure 5-11B, a 1-dB attenuator is inserted in the input line. The
actual input to the multiplier then is *A* − 1dB. The output from the multiplier is then (*N*)(*A*-1dB) or (*N*)*A* − (*N*)dB. Thus the output level change (in
dB) is *N* times as much as the input level change. Figure 5-11C shows a
tripler. The input frequency (*A*) becomes 3*A* at the output. In Figure 5-11D
of Figure 5-11, a 1-dB attenuator has been inserted in the input line. The input to the tripler then becomes *A* − 1 dB. The output from the tripler is

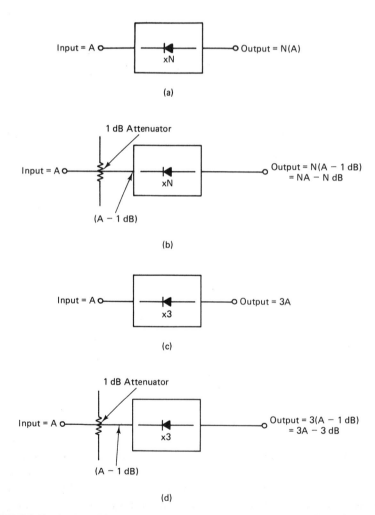

FIGURE 5-11. An illustration of the nonlinear mixing action that can
occur when the front end of a receiver is overloaded by strong signals.

then $3(A - 1 \text{ dB}) = 3A - 3 \text{ dB}$. Hence, the tripler not only triples the frequency of the input but also triples the *change* in input level (in dB) so that the output level ($3A$) changes three times as much as the input level (in dB).

Now let's apply this to a practical situation. Figure 5-12A shows two transmitter signals, A and B, present at the antenna of a receiver. Suppose that the receiver is tuned to frequency C and that $C = 2A - B$. If signals A and B are strong enough to overload the front end of the receiver, nonlinear mixing (and multiplying) will occur causing the generation of many intermod products. One of these intermod products will be $2A - B$. Since the receiver frequency (C) is equal to $2A - B$, this intermod will pass through the receiver, causing interference to desired signals and/or nuisance interference. Figure 5-12B shows an attenuator inserted in the input line to the receiver. This attenuator has an attenuation factor of 3 dB. Both signals A and B will be attenuated 3 dB so that the signals at the receiver input becomes $A - 3$ dB and $B - 3$ dB. The intermod product ($2A - B$) then becomes:

$$2(A - 3 \text{ dB}) - B - 3 \text{ dB} = 2A - 6 \text{ dB} - B - 3 \text{ dB} = 2A - B - 9 \text{ dB}$$

Thus, the 3-dB change in the input signal level becomes a 9-dB change in the intermod signal level. This factor gives us a decided advantage in dealing with receiver-produced intermod. However, in order to realize this advantage the attenuation must *precede* the mixing stage. Any attenuation inserted after the mixing (and multiplying) will give a 1-dB change for each dB of attenuation, so there is no advantage there. Hence, if the intermod product ($2A - B$) in our example were produced externally and then entered the antenna as an *intermod* signal, there would be no advantage in inserting an attenuator in the input line.

A simple test procedure to determine whether an intermod signal is receiver-produced or externally produced is described below.

Test Procedure

1. Set up the equipment as shown in Figure 5-13A.
2. Disable the AGC line by applying fixed bias or grounding it.
3. With the antenna connected and the intermod signal present, set the voltmeter to a scale that gives a good reference reading.
4. Switch the receiver to the signal generator and apply an on-channel signal to the receiver (see Figure 5-13B).
5. Set the RF output level of the generator to produce the reference reading again on the voltmeter. Note the generator level setting.
6. Insert a 3-dB pad into the line and reconnect the antenna (see Figure 5-13C).
7. With the intermod signal present, note the new reading on the voltmeter.

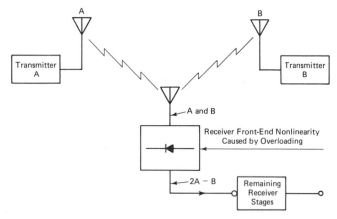

FIGURE 5-12A. An intermod signal caused by nonlinear mixing of signals A and B causes interference to the receiver, which is tuned to frequency C, where frequency C equals 2A − B.

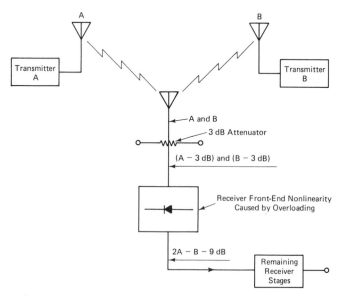

FIGURE 5-12B. The effect of an attenuator before the nonlinear mixer is multiplied, therefore giving an advantage in dealing with receiver-produced intermodulation products.

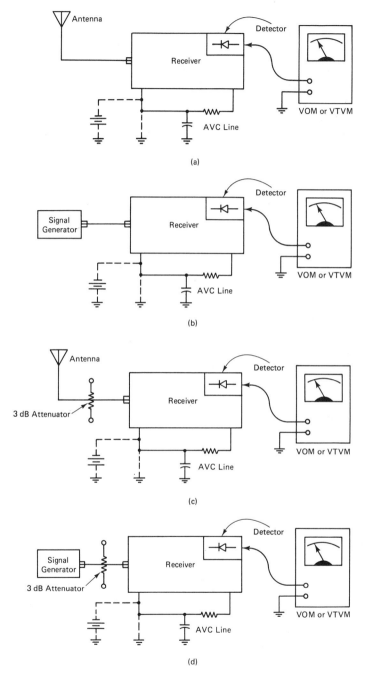

FIGURE 5-13. A test setup to determine whether an intermod signal is produced within the receiver or is produced externally.

183

8. Reconnect the signal generator and leave the attenuator in the line (see Figure 5-13D).

9. Set the signal generator output level to produce the same voltmeter reading as obtained in step 7. Note the signal generator setting.

10. Determine the *difference* in the two signal generator settings in step 5 and step 9. If this difference is greater than 3 dB (say, 9 dB to 15 dB), the intermod is *receiver-produced*. If the difference is approximately 3 dB or so, the intermod signal is produced at some point *external* to the receiver.

NOISE LIMITER/NOISE BLANKER EFFECTIVENESS TEST

Many receivers have built-in noise *limiters* to reduce the level of impulse noise. Some receivers utilize noise *blankers*, which are usually more effective in suppressing impulse noise. In order to test the effectiveness of noise blankers or noise limiters, it is necessary to use a noise generator device that simulates impulse noise. In order to establish a proper test, the EIA has specified that the noise generator produce noise pulses that are (1) at a repetition rate of 100 pps; (2) a pulse width of 1 μs; and (3) at a level of 1 V.[1] The rise time characteristic is also specified. Some signal generators have a built-in noise generator that meets the EIA's specifications. This simplifies impulse noise testing of receivers.

There are several possible approaches that can be used in evaluating the effectiveness of noise-limiting or noise-blanking circuits. The basic approach is outlined in the following procedures.

Test Procedure—Method 1

1. Set up the equipment as shown in Figure 5-14.

2. With the noise generator turned off, perform the standard 10-dB $S+N/N$ sensitivity test. Note the signal generator level in dBm required to produce the 10-dB $S+N/N$ ratio.

3. Turn on the noise generator and repeat the 10-dB $S+N/N$ test. Again, note the signal generator level in dBm required to produce the 10-dB $S+N/N$ ratio.

4. The difference in the signal generator levels in steps 2 and 3 in dB is a measure of the receiver's immunity to the effects of impulse noise.

---

[1] EIA STANDARD #RS-382, "Minimum Standards, Citizens Radio Service-AM Transceivers Operating in the 27 MHz Band," Electronic Industries Association, 1971, p. 8, para. 12.2. This complete standard is available from: Electronic Industries Association, 2001 Eye St. N.W., Washington, D.C. 20006, Phone: (202) 457-4900.

Lower dB figures indicate better receiver impulse noise rejection. Another approach to determine the effectiveness of the noise blanker or noise limiter is to perform the test in the following manner if the receiver has a switchable noise limiter or noise blanker.

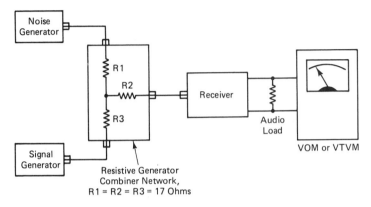

FIGURE 5-14. Test setup for measuring the effectiveness of noise limiters and noise blankers.

Test Procedure—Method 2

1. Set up the equipment as shown in Figure 5-14.
2. With the noise generator turned on and the receiver noise limiter or noise blanker turned off, perform the standard 10-dB $S+N/N$ sensitivity test. Note the level in dBm.
3. Turn on the noise limiter or blanker and repeat step 2. Note the level in dBm.
4. The difference in the two generator levels (steps 2 and 3) is the amount of improvement produced by the noise limiter or noise blanker.

Noise Figure

The noise figure of a receiver is a measure of the amount of noise produced within the receiver itself. The noise figure has a great effect on the weak-signal performance of the receiver. The noise figure can be defined by:

$$\text{Noise figure } (Nf) = \frac{\text{Input } S/N \text{ ratio}}{\text{Output } S/N \text{ ratio}}$$

In an ideal receiver, the output S/N ratio is the same as the input S/N ratio. Therefore, an ideal receiver has a noise figure of 1. Such a receiver doesn't exist, but noise figures of actual receivers are referenced to the ideal receiver.

If the signal at the input to a receiver has an S/N ratio of 20:1 and the S/N ratio at the output is 5:1, the noise figure of the receiver is:

$$Nf = \frac{\text{Input } S/N}{\text{Output } S/N} = \frac{20}{5} = 4$$

This reduction in the S/N ratio at the output is caused by the noise *added* to the signal as it passes through the receiver. Most of the noise originates in the receiver's front end, since any noise generated there is amplified by all the following stages of the receiver.

Noise figure is frequently expressed in dB. In the example above, the dB equivalent of the noise factor (4) can be found from the following formula:

$$Nf(\text{dB}) = 10 \log 4 = 10(.6) = 6 \text{ dB}$$

If a signal at the input to a receiver has a S/N ratio of 10 dB and the noise figure of the receiver is 6 dB, the output signal will have a S/N ratio of 4 dB (input S/N in dB minus noise figure in dB).

A close approximation of the actual noise figure of a receiver can be made with the use of a simple noise generator, such as the one in Figure 5-15. Such a noise generator can be constructed in a small metal box. Resistor R1 should equal the input impedance of the receiver with which it is to be used. The diode should be a silicon crystal diode (not germanium). The reversed-biased diode becomes a broadband noise generator, with the level of the noise increasing as the reverse current through the diode increases.

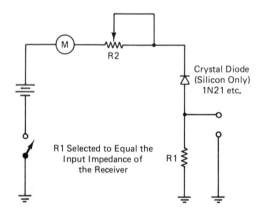

FIGURE 5-15. A simple noise generator that can be constructed from readily available parts can be used in noise tests.

The noise generator can be used to determine the approximate noise figure of a receiver by using the following procedure.

Test Procedure

1. Set up the equipment as shown in Figure 5-16.
2. Set the squelch control to fully unsquelch the receiver and set the RF gain (if used) to maximum.
3. With the noise generator turned off and the 3-dB attenuator out, set the volume for a good reference level on the meter.
4. Insert the 3-dB attenuator between the receiver output and the meter. Don't change the volume setting.
5. Turn on the noise generator and increase the noise amplitude until the receiver output meter again indicates the reference reading.
6. Note the current level on the noise generator meter. The approximate noise figure in dB can then be determined from the formula:

$$Nf(\text{dB}) = 10 \log 20IR$$

where I is the diode current and R is the source impedance of the generator (same as the impedance of the receiver).

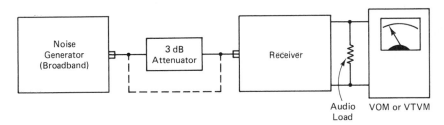

FIGURE 5-16. The test setup in which a simple noise generator is used to determine the approximate noise figure of a receiver.

AUTOMATIC GAIN CONTROL

Automatic gain control (AGC) or automatic volume control (AVC), as it is sometimes called, serves to keep the audio output of the receiver from varying too widely as the input signal strength varies.

In order to measure the effectiveness of a receiver's AGC circuit, certain test procedures have been devised. Generally, the test involves varying the RF signal input level to the receiver through a wide range of input levels while measuring the audio output level of the receiver. The AGC figure may be specified in different ways by different manufacturers, and depending upon the test procedure used to perform the AGC test. For example, one manufacturer might list the AGC specification as: $+/-$ 6 dB from 250,000 μV to 5 μV with a 15-dB roll-off ($+/-$ 4 dB) from 5 to 0.5 μV. This specification is perfectly clear.

Another AGC specification might simply state: "AGC figure = 60 dB." Unless we know just how this 60-dB figure was derived, this specification is meaningless. A common method used to obtain such a figure is to state the ratio of the RF input level ratio to the audio output level ratio expressed in dB. This can be expressed by the following formula:

$$\text{AGC figure (dB)} = 20 \log \frac{\text{RF input maximum/minimum}}{\text{AUDIO output maximum/minimum}}$$

Example: The RF input level to a receiver is varied from 50,000 μV to 1 μV. During this range of input signal variance, the audio output level reaches a maximum of 5 V and a minimum of 2.5 V. Using the formula above, calculate the AGC figure in dB.

Solution: Substituting into the formula, we have:

$$\text{AGC figure} = 20 \log \frac{50,000/1}{5/2.5} = 20 \log \frac{50,000}{2} = 20 \log (25,000) =$$

$$20(4.4) = 88 \text{ dB}$$

If the RF input level change and the audio output level change are both expressed in dB, the AGC figure can be determined by simply *subtracting* the audio level change in dB from the RF input level change in dB. This is expressed by the following formula:

$$\text{AGC figure (dB)} = \text{RF input level change (in dB)} - $$
$$\text{Audio output level change (in dB)}$$

Example: In the previous example the signal generator level was changed from -13 dBm (50,000 μV) to -107 dBm (1 μV), representing a change of $107 - 13 = 94$ dB. The audio output level change was: 20 log (5/2.5) = 20 log 2 = 6 dB. Thus, the AGC figure = 94 dB $-$ 6 dB = 88 dB.

The following test procedure describes the basic principles of AGC testing.

Test Procedure

1. Set up the equipment as shown in Figure 5-17.
2. Set the signal generator to 50,000 μV (-13 dBm) or higher if desired.
3. Modulate the signal to a level of 30% at 1 kHz.
4. Set the receiver's RF gain (if used) to maximum and the squelch control to unsquelch the receiver fully.

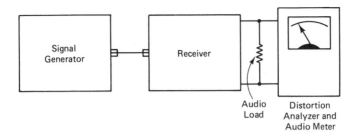

FIGURE 5-17. A test setup by which the AVC or AGC of a receiver is tested. Also used to measure the modulation distortion.

5. Adjust the audio output power of the receiver to produce the desired reference level (typically 50% rated power).

6. The signal generator level is slowly adjusted from 50,000 μV (−13 dBm) to 1 μV (−107 dBm).

7. As the signal generator level is reduced, any significant increases in the audio level *above* the established reference level should be examined carefully by making a check in the distortion level at these points. Those points that show high distortion should be noted. These will generally occur at the higher input signal levels and are caused by nonlinearity in any of the tuned amplifier stages, usually the last IF stage.

8. The audio output versus input signal level is plotted. The resulting curve should be smooth and free of abrupt changes.

The smaller the change in audio output level during this test, the better the AGC performance. If the specification is listed as simply "AGC figure" as described earlier, then the higher the figure, the better the AGC performance.

An ideal way in which to run the AGC test would be to set the volume control to maximum and leave it there while the input signal level is changed through the range from minimum usable sensitivity (10-dB $S+N/N$) to maximum signal generator level (not to exceed 1 V). However, if the test were performed in such a manner, the audio amplifier would overload and the output could not increase above the maximum power capability of the audio amplifier. The following method simulates such a test.

Test Procedure

1. Set up the equipment the same as in Figure 5-16.

2. Set the receiver squelch fully unsquelched, the RF gain to maximum, and the volume to maximum.

3. Modulate the signal generator 30% at 1 kHz and set the output level to the 10-dB $S+N/N$ sensitivity level of the receiver. Note and record the signal generator level and the audio output power.

4. Increase the generator level in small increments (recording the signal generator level and audio output level at each setting) until the audio power level reaches 50% of the maximum rated audio power. At this point, reduce the volume control setting to one-tenth of this level.

5. Continue as in step 4, but multiply the audio power readings by 10.

6. When the audio power meter again reaches the 50% rated power level, reduce the volume control setting until the audio meter indicates one-tenth this level.

7. Continue as in step 4, but now multiply the audio power meter reading by 100. The test is continued in this manner until the input signal level reaches the maximum specified level. The data obtained is then used to plot a graph of signal level input versus audio power output. The graph should be plotted on semilog graph paper with the signal generator level in microvolts on the log scale and the audio power in watts on the linear scale. From the graph, it can be determined how much input signal is needed to produce a specified amount of audio output power, or vice versa. The residual noise level can also be clearly seen from the graph.

HUM MEASUREMENTS

When making hum measurements, it is desirable to distinguish between *residual* hum and *modulation* hum. Residual hum is the hum that exists at the receiver output even when no signal is present at the receiver input. Modulation hum is the hum that appears only in the presence of an input signal. The hum signal actually modulates the input signal. This hum modulation usually occurs in the mixer stage.

Test Procedure—Residual Hum

1. Set up the equipment as shown in Figure 5-18 and disable the last IF stage.

2. Set the volume control to maximum and the squelch control to unsquelch the receiver fully.

3. Measure and record the hum voltage at the output. This is the residual hum level of the receiver.

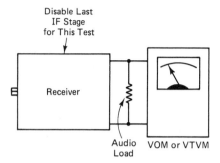

FIGURE 5-18. Test procedure for measuring the residual hum of a receiver.

Test Procedure—Modulation Hum

1. Set up the equipment as shown in Figure 5-19.
2. Set the squelch control to unsquelch the receiver fully and the volume control for maximum undistorted output.
3. Apply a moderate to strong signal (modulated 30% at 1 kHz) to the receiver input.
4. Measure and record the audio output level.
5. Turn off the generator's modulation and measure and record the hum level at the output.
6. Turn off the signal generator and, without disturbing anything else, measure and record the hum level at the output. If the hum level doesn't change from step 5, modulation hum is well below the residual hum and is therefore not a problem.

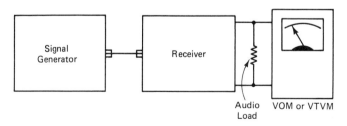

FIGURE 5-19. Test setup for measuring the modulation hum of a receiver.

7. If there is a significant difference in the hum level in steps 5 and 6, modulation hum is significant. The modulation-hum level can be expressed as so many dB below the signal level in the following manner. Subtract the measurement in step 6 from the measurement in step 5. This difference is the modulation-hum level. The dB relationship of the modulation-hum level to the signal level can be determined from the formula:

$$\text{Modulation hum (dB)} = 20 \log \frac{Es - Eth}{Emh}$$

where Es is the output voltage level with the modulation on, Eth is the total hum level (modulation hum and residual hum), and Emh is the modulation hum level (the difference in the total hum level and the residual hum level). In a similar manner, the total hum (residual hum and modulation hum) may be expressed in dB through the formula:

$$\text{Total hum (and noise) in dB} = 20 \log \frac{Es - Eth}{Eth}$$

S-Meter Calibration

In order to calibrate an S-meter properly, the manufacturer usually specifies the input signal strength necessary to produce an S-9 reading. Typical levels of input signal required to produce an S-9 reading are 50 and 100 μV; however, other levels are also used.

Test Procedure

1. Set up the equipment as shown in Figure 5-7.
2. Set the receiver's RF gain to maximum.
3. While observing the S-meter, slowly adjust the signal generator level to produce the S-9 reading.
4. Note the generator level required to produce the S-9 reading. If it is very much different from the level specified by the manufacturer, the S-meter should be adjusted to produce an S-9 reading with the specified amount of input signal level.

MODULATION DISTORTION

Modulation distortion is distortion produced in the stages ahead of the audio amplifier and which changes as the modulation percentage changes.

Test Procedure

1. Set up the equipment as shown in Figure 5-17.
2. If an RF GAIN control is used, set it for maximum gain.
3. Set the RF signal generator output level to 1,000 μV.
4. Set the generator's modulation to 10% at 1 kHz and set the volume control to produce approximately one-quarter the rated output of the receiver.
5. Measure and note the distortion at this modulation percentage.
6. Increase the modulation percentage in increments of 10% from 10% to 100% modulation. At each increment, adjust the volume control to maintain the same audio output level (one-quarter rated audio output). At each increment, measure and note the distortion percentage.
7. If desired, a graph could be plotted to show the affect of modulation percentage on distortion level.

BIBLIOGRAPHY

Carr, Joseph, *Elements of Electronic Instrumentation and Measurement.* Reston, Virginia: Reston Publishing Company, 1979.

Electronic Industries Association, Standard # RS-382. *Minimum Standards, Citizens Radio Service, AM Transceivers Operating in the 27 MHz Band.*

Miller, Gary M., *Handbook of Electronic Communication.* Englewood Cliffs, N.J.: Prentice-Hall, Inc., 1979.

National Radio Institute, *Operation and Maintenance of Communications Receivers.* Washington, D.C., 1970.

Oliver, Bernard M., and Cage, John M., *Electronic Measurements and Instrumentation.* New York: McGraw-Hill, 1971.

Philco Corporation, *Radio Communication System Measurements.* © 1952, Philco Corporation.

Terman, Frederick E., and Pettit, Joseph M., *Electronic Measurements,* 2nd ed. New York: McGraw-Hill, 1952.

6

SINGLE SIDEBAND TRANSMITTER TESTS AND MEASUREMENTS

This chapter deals with the basic tests and measurements that can be used to evaluate the performance of single sideband suppressed carrier transmitters. Hereafter, the abbreviation *SSB* will be used, suppressed carrier being understood. Although this book assumes that the reader possesses a knowledge of the basic working principles of SSB equipment, it is worthwhile to review the operation of an SSB transmitter briefly.

A BRIEF REVIEW OF SSB
SIGNAL GENERATION AND TRANSMISSION

A simplified block diagram of a typical SSB transmitter is shown in Figure 6-1. The balanced modulator receives two input signals: the audio modulating signal and the RF carrier signal. When no audio signal is applied to the balanced modulator, there is no signal at the output (providing the balance controls are properly adjusted). When a single audio tone is applied to the balanced modulator, two signals appear at the output: an upper sideband signal and a lower sideband signal. The upper sideband signal is equal to the RF carrier frequency plus the audio-modulating frequency—that is, $Fusb = Fc + Fa$; where $Fusb$ is the upper sideband frequency, Fc is the carrier frequency, and Fa is the frequency of the audio-modulating signal. The lower sideband signal is equal to the RF carrier frequency *minus* the audio-modulating frequency—that is, $Flsb = Fc - Fa$, where $Flsb$ is the lower sideband frequency, Fc is carrier frequency, and Fa is the frequency of the audio-modulating signal.

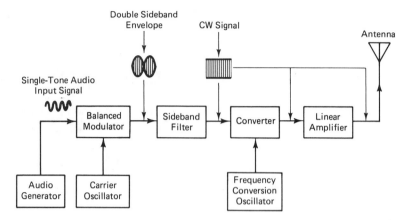

FIGURE 6-1. A simplified block diagram of a single-sideband suppressed carrier transmitter of the filter type, showing the development of the signal at the various points in the transmitter.

These two signals can be represented by phasors, as shown in Figure 6-2. The opposite sidebands are represented by phasors that rotate in opposite directions. Each phasor makes one complete (360°) revolution in the time period equal to one complete audio cycle. In the example in Figure 6-2, the audio-modulating frequency is 1,000 Hz. The time period required for 1 cycle of a 1,000-Hz signal is 1/1,000 of one second or 1 millisecond (ms). Hence, each phasor makes one complete revolution in 1 ms. As the phasors rotate, they alternately add and subtract, thus producing an envelope pattern as shown at Figure 6-2F. At time 0, the two phasors are 180° apart, causing complete cancellation (assuming equal amplitude sidebands). Thus, the envelope at Figure 6-2F is at zero amplitude at time 0. At time 1 (Figure 6-2B), the two phasors are exactly in phase, thus adding to produce maximum amplitude in the envelope. At time 2 (Figure 6-2C), the phasors are again 180° apart, thus cancelling to produce zero amplitude in the envelope waveform. At time 3 (Figure 6-2D), the phasors are again exactly in phase, thus producing maximum amplitude in the envelope. At time 4 (Figure 6-2E), both phasors are back to their starting points and 180° apart, causing the envelope to return to zero amplitude. This process is repeated as long as the audio signal is applied to the balanced modulator. This double-sideband single-tone signal is fed to a filter, which removes one of the sidebands. The output from the sideband filter is a single-sideband signal in the form of a single-frequency *CW signal*, which can be represented by a single phasor as shown in Figure 6-3A. The oscilloscope would show a CW signal as shown at Figure 6-3B. Hence, a single-tone audio input produces a CW RF output from the sideband filter. This SSB signal from the output of

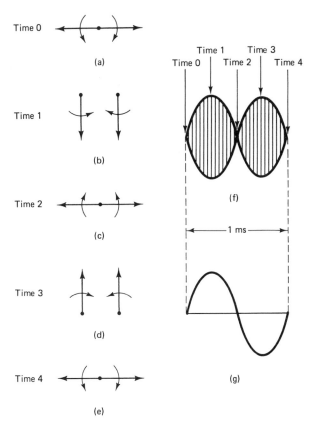

FIGURE 6-2. Phasors *A* through *E* illustrate how a single-tone audio input produces a double-sideband, two-tone envelope at the output of the balanced modulator. The waveform at *G* is the audio modulating signal at the transmitter intput.

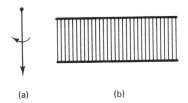

FIGURE 6-3. The phasor at *A* represents a single-tone SSB signal at the output of the sideband filter with a single-tone audio modulating signal at the transmitter input. The illustration at *B* shows the way the envelope appears at the output of the sideband filter.

the filter is fed to the input of a mixer or converter stage where the signal is increased to a higher frequency at which it is to be transmitted. From the mixer, the SSB signal goes to the *linear* amplifier, which increases the power level of the SSB signal sufficiently for transmission. The linear amplifier must amplify the SSB signal without distortion; that is, the envelope at the output of the linear amplifier must be the same as the envelope at the input to the linear amplifier. Easier said than done!

Various Power Relationships in the SSB Signal

The accepted method of testing SSB transmitters is to use a *two-tone* audio generator to modulate the transmitter. The two audio tones should be of equal amplitude and *nonharmonically* related. Figure 6-4 shows a block diagram of a two-tone generator.[1]

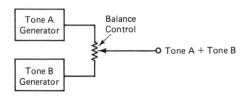

FIGURE 6-4. Block diagram of a two-tone generator used to test SSB transmitters.

Assume that a two-tone audio signal is applied to an SSB transmitter and the transmitter is switched to the USB (upper sideband) mode. Let audio tone $A = 500$ Hz and audio tone $B = 2,400$ Hz. If we let Fc represent the carrier frequency of the transmitter (for reference purposes), the two-tone input will produce two discrete frequencies at the transmitter output, one at $Fc + 500$ Hz and the other at $Fc + 2,400$ Hz. The phasor representation of these two SSB frequencies is shown in Figure 6-5. If we let the lower frequency signal ($Fc + 500$ Hz) serve as the *reference* phasor, the higher frequency signal ($Fc + 2,400$ Hz) will rotate around the reference phasor at a rate equal to the frequency *difference* between the two signals. In this case,

---

[1] Complete construction plans for a two-tone audio generator for SSB testing is described in the August 1981 issue of *QST Magazine.*

the frequency difference is 2,400 Hz − 500 Hz = 1,900 Hz. Thus, the phasor representing Fc + 2400 Hz revolves around the reference phasor (which represents Fc + 500 Hz) at a rate of 1900 revolutions per second. In terms of time period this would be 1/1900 of one second or 526 μs.

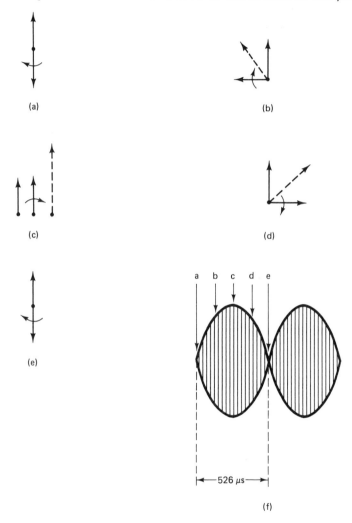

FIGURE 6-5. Phasors *A* through *E* represent two single sideband signals which are produced by two audio tones that are 1,900 Hz apart (500 Hz and 2,400 Hz). The resultant signal which is produced by the two phasors will form the two-tone envelope at *F*. The time period of each envelope is equal to the time period of the frequency difference of the two audio tones, in this case 1,900 Hz. The time period of 1,900 Hz is 1/1,900 = 526 μs.

As the phasor revolves, it produces an envelope in the same manner as previously described for the double sideband signal. Figure 6-5 illustrates how the phasors produce the envelope. Notice that the time period for one complete envelope pattern (526 μs) is the same time period required for one cycle of the difference frequency, 1,900 Hz (tone B — tone A). Stated another way: The repetition rate of the envelope pattern is equal to the difference between the two-tone frequencies. In this case, the repetition rate of the envelope is 1,900 times per second.

Single sideband transmitters are rated in terms of peak power in the envelope or *peak-envelope power. For the single-tone SSB signal the peak-envelope power is the same as the average power* (see Figure 6-6). The power will read the same on an average or peak-reading wattmeter. When a complex waveform (consisting of more than one frequency) is used to modulate an SSB transmitter, the relationship of peak-to-average power changes drastically. The more frequencies present in the complex wave, the higher the peak-to-average power ratio becomes. Let's further analyze the two-tone SSB envelope and phasors to determine the power relationship that exists. The two-tone envelope is shown in Figure 6-7. This envelope waveform is a plot of voltage versus time. Assume that this waveform appears across a 50-Ω dummy load connected to an SSB transmitter. The peak voltage is given as 10V. The *instantaneous peak power* can be computed as follows: $IPP = 10^2/50 = 100/50 = 2$ W. It is very important to note that *instantaneous peak power* and *peak envelope power* are not the same. The PEP power is based on the *rms* voltage in the two-tone envelope. Referring again to Figure 6-7A, the rms voltage in the envelope is 7.07 V. Hence, the PEP power can be computed as follows: PEP $= 7.07^2/50 = 50/50 = 1$ W. Thus, PEP power in a two-tone envelope is equal to 1/2 the instantaneous peak power.

Wattmeters used to measure PEP power in the two-tone envelope are *peak-reading* and *rms-calibrated.* They are generally referred to as peak-reading wattmeters. Wattmeters that are used in AM/FM transmitter measurements are usually the average-reading, rms-calibrated type. If an *average-reading, rms-calibrated* wattmeter were used to measure the power in the two-tone envelope of Figure 6-7A, the meter reading would not indicate the true PEP power in the envelope. The meter indication can be computed as follows: $(10 \times .636 \times .707)^2/50 = 20.22/50 = 0.4$ W. Thus, an average-reading, rms-calibrated wattmeter will indicate approximately 40% of the true PEP power in a two-tone envelope. This relationship holds true only in the two-tone envelope.

In the example above, it is important to note that the power indicated by the average-reading wattmeter (0.4 W) is not the *true* average power in the two-tone envelope of Figure 6-7A. The true average power in the two-tone (equal amplitude) envelope is equal to the sum of the average power in each tone or twice the average power of one tone (since they are of equal amplitude). The average power is based on the rms voltage of the tone. In Figure 6-7A the peak voltage of each tone is 5 V (since the two tones combine in phase to produce a peak-envelope-voltage of 10 V). The rms voltage of each tone is $0.707 \times 5 = 3.535$ V. The average power in each tone is then $3.535^2/50 = 0.25$ W. The total *average* power in the two-tone envelope is then $2 \times 0.25 = 0.5$ W. An average reading wattmeter can be used to determine the PEP and true average power in a two-tone envelope if the following relationship is observed:

Watts on average-reading meter $= 40.5\%$ PEP $= 81\%$ True average power

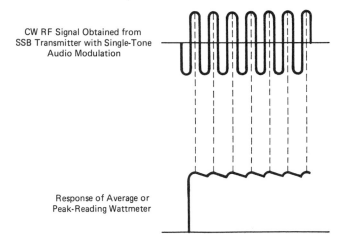

CW RF Signal Obtained from
SSB Transmitter with Single-Tone
Audio Modulation

Response of Average or
Peak-Reading Wattmeter

FIGURE 6-6. The CW waveform resulting from single-tone modulation of an SSB transmitter is shown at the top while the response of an average-reading or peak-reading wattmeter is shown at the bottom.

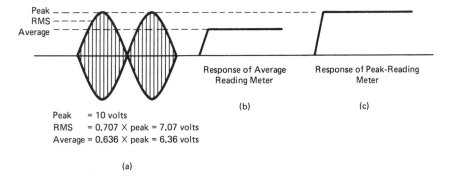

Peak
RMS
Average

Response of Average
Reading Meter

Response of Peak-Reading
Meter

(b)

(c)

Peak   = 10 volts
RMS   = 0.707 X peak = 7.07 volts
Average = 0.636 X peak = 6.36 volts

(a)

FIGURE 6-7. The waveform at *A* is a two-tone envelope produced by modulating an SSB transmitter with two equal-amplitude tones. The illustration at *B* shows how an average-reading wattmeter would respond to this waveform. The illustration at *C* shows how a peak-reading wattmeter would respond to this waveform.

The chart in Table 6-1 summarizes the relationship of power in the two-tone envelope.

PEP (Reference) Power	True Average Power	Average Power in One Tone
1 Watt (Reference)	1/2 watt	1/4 watt
0 dB (Reference)	−3 dB	−6 dB

TABLE 6-1

Now consider the case of three equal amplitude tones (see Figure 6-8). The three phasors that are stacked vertically indicate that the three tone components are in phase. At this instant, the peak of the envelope will be formed. If the three phasors are each 10 V (rms), the PEP power is:

$$30^2/50 = 18 \text{ W PEP}$$

The power in each tone is:

$$10^2/50 = 100/50 = 2 \text{ W}$$

This is one-ninth PEP power. The average power is the sum of the power in each tone. Since the average power in each tone is 2 W, the total average power is:

$$3 \times 2 = 6 \text{ W}$$

This is one-third PEP power. Thus, it is shown that as the number of equal tones increases, the ratio of PEP-to- average power increases.

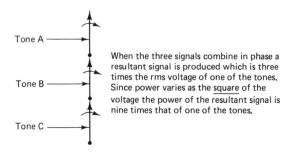

FIGURE 6-8. Phasor representation of an SSB signal produced by modulating it with three equal-amplitude tone signals.

USING THE OSCILLOSCOPE TO ANALYZE THE SSB SIGNAL

By far the best way to analyze an SSB signal is with a spectrum analyzer. However, spectrum analyzers are not always readily available to the technician. On the other hand, an oscilloscope is available in almost every shop. While a complete *quantitative* analysis of an SSB signal can't be made with an oscilloscope, a fairly good *qualitative* analysis can be made by closely examining the various waveforms.

Obtaining the Vertical Deflection   If the upper frequency limit of the scope's vertical amplifier is greater than the frequency of the SSB signal, the SSB signal can be applied directly into the vertical input of the scope.

If the bandwidth of the vertical amplifier isn't sufficient to accommodate the SSB frequency directly, there are two alternatives: (1) the SSB signal can be converted to a lower frequency, which can be handled by the scope's vertical amplifier; or (2) the SSB signal can be applied directly to the vertical deflection plates of the scope. For more information on these methods, refer to Chapter 4 under the heading, "Using the Scope for Determining Modulation Percentage."

A scope that has been especially designed for monitoring transmitter signals is shown in Figure 6-9. This instrument can operate over a frequency range of 3.5 MHz to 54 MHz, with power levels of 10 to 1,000 W at the antenna input and 10 to 300 W at the demodulator input labeled "exciter input."

FIGURE 6-9. This commercially available monitor is excellent for monitoring and/or examining SSB envelopes. (Courtesy of Heath Company.)

Carrier Suppression Test

In a properly operating SSB suppressed carrier transmitter, the carrier component should not appear in the output signal. A simple test with a single audio-modulating tone applied to the transmitter input can be used to determine the *relative* degree of carrier suppression. The basic test setup is shown in Figure 6-10.

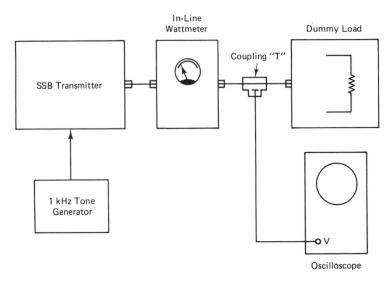

FIGURE 6-10. Basic test setup for carrier suppression test and the unwanted sideband suppression test.

Test Procedure

1. Key the transmitter and adjust the level of the 1-kHz audio-modulating tone to produce a power output of approximately one-half the rated PEP output.

2. Closely examine the envelope waveform on the scope. With good carrier suppression, the envelope should be a CW waveform as shown in Figure 6-11A. If a ripple appears in the waveform as in Figure 6-11B, it may be due to insufficient carrier suppression. Use the scope to compare the ripple waveform with the audio modulating tone in the following manner: (1) using the scope's graticule, measure the distance between the peaks in the ripple waveform as shown in Figure 6-12A; (2) *without changing the horizontal sweep rate*, remove the SSB signal and apply the audio-modulating signal to the scope as shown in Figure

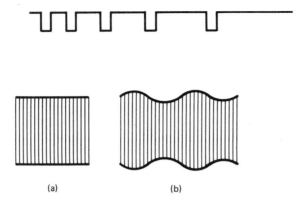

(a)   (b)

FIGURE 6-11. A single-tone SSB envelope with good carrier suppression is shown at *A*. There is no evidence of the carrier in the envelope. At *B* a single-tone SSB envelope with insufficient carrier suppression. The carrier shows up as a ripple in the envelope.

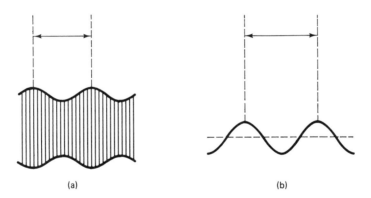

(a)   (b)

FIGURE 6-12. If the ripple in the single-tone envelope is caused by carrier leak-through, the peaks of the ripple will be spaced the same distance apart as the audio modulating signal used to produce the single-tone output, providing the same horizontal sweep rate is used.

6-12B. If the distance between the peaks of the audio-modulating signal is the same as the distance measured in step 1, the ripple is due to insufficient carrier suppression. The relative degree of carrier suppression (or lack of it) is indicated by the relative degree of ripple. Figure 6-13A shows a mild case of insufficient carrier suppression, while the illustration at Figure 6-13B shows a more severe case.

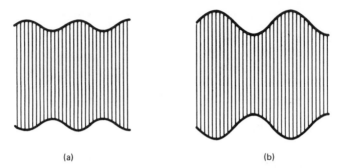

(a)   (b)

FIGURE 6-13. A mild case of insufficient carrier suppression is shown at
A while a more severe case is shown at *B.*

If the timebase of the scope is accurately calibrated, it is not necessary to
compare the two waveforms. Instead, a simple time measurement can be
made. For example, if the modulating signal is exactly 1 kHz, the time period required for 1 cycle is 1 ms. If the scope's timebase is set for 0.2 ms per
division and the distance between the peaks is 5 divisions, the ripple is
caused by insufficient carrier suppression. The ripple frequency caused by
the carrier is equal to the audio-modulating frequency.

The approximate carrier suppression in terms of dB below the single
tone can be determined by measuring the ripple component and envelope
height (see Figure 6-14) and then substituting the two values into the formula:

$$\text{Carrier suppression (in dB)} = 20 \log \frac{\text{Envelope height}}{\text{Ripple height}}$$

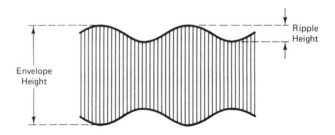

FIGURE 6-14. To determine the approximate suppression in dB, measure the ripple height and the envelope height and substitute the
two measurements into the formula.

Figure 6-15 correlates dB suppression to envelope-to-ripple ratios of 10 to 100. The main limiting factor is the relatively small amounts of ripple. The following example illustrates the use of the graph. A single tone envelope shows a ripple component with a peak-to-peak height of 0.5 divisions on the scope graticule. The peak-to-peak height of the envelope is 9 divisions. This makes the envelope-to-ripple ratio 9/0.5 = 18. A ratio of 18 corresponds to a suppression of 25 dB.

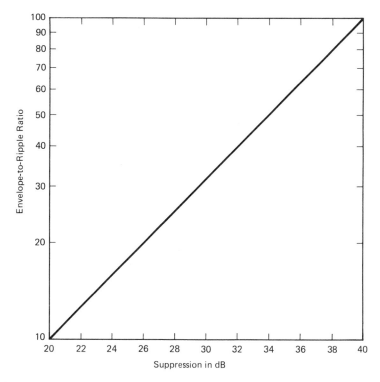

FIGURE 6-15. This graph correlates envelope-to-ripple height ratios with the suppression in dB for ratios of 10 to 100.

Unwanted-Sideband Suppression Test

The test setup and procedure for this test is the same as the test just described for carrier suppression. If a ripple occurs in the waveform, it may be caused by insufficient suppression of the undesired sideband. A comparison of the audio-modulating tone and the ripple should be made, as described in the previous test. If the distance between the ripple peaks is one-half the distance between the peaks of the audio waveform, the ripple is caused by insufficient suppression of the unwanted sideband (see Figure 6-16).

If the scope permits time measurement, the time between peaks of the ripple will be one-half the time of 1 cycle of audio. For a 1-kHz audio frequency, the time between ripple peaks would be 1/2 ms. The ripple frequency caused by insufficient suppression of the unwanted sideband will be twice the frequency of the audio modulating signal.

The formula used in the previous section for determining the carrier suppression in dB also applies to the unwanted sideband suppression. Figure 6-15 also applies here.

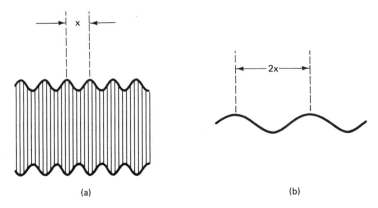

(a)   (b)

FIGURE 6-16. The relationship of ripple caused by insufficient sideband suppression and the audio modulating tone is shown here. The distance between the peaks of the ripple will be one-half the distance between the peaks of the audio modulating waveform when viewed on a scope at the same horizontal sweep rate.

Two-Tone Linearity Tests—Envelope Pattern

The two-tone test can be used to evaluate the performance of an SSB transmitter with a fair degree of accuracy, depending to a large extent upon the ability of the technician to recognize faults in the pattern. Some faults are recognized easily while others are not so apparent, depending upon the type of fault and the severity of the fault.

Usually, a special two-tone generator is used to obtain the two-tone SSB pattern, but if a two-tone generator is not available there is a trick that can be used to obtain a two-tone SSB pattern by using only one audio-modulating tone. Another alternative is to use two separate audio generators with a special combining network.

The One-Generator Method If a two-tone audio generator is not available, don't despair—you can still get a two-tone SSB envelope test signal by using the following procedure.

1. Hook up the test equipment as shown in Figure 6-10.

2. With the transmitter keyed and the 1-kHz tone generator turned off, deliberately misadjust the carrier balance control in the balanced modulator. Adjust the carrier balance control to produce approximately one-quarter the rated PEP output from the transmitter. If sufficient carrier output can't be obtained by adjusting the carrier balance control, it may be necessary also to make a slight change in the carrier frequency to move the signal into the passband of the filter. The response curve of the filter is very steep at this point, so a slight adjustment of the carrier frequency should make a large change in the signal output.

3. Once the proper carrier output is obtained, turn on the audio tone generator and adjust its level to produce the two-tone SSB envelope, as shown in Figure 6-18A. The envelope won't go to zero until the two tones are exactly equal in amplitude. Figure 6-18F shows a two-tone pattern produced by two tones of unequal amplitude. When the proper two-tone display is obtained, the peak-reading wattmeter should show approximately four times the power produced by the carrier alone.

The Two-Tone Generator Method With a two-tone audio generator, the procedure is simplified.

1. Use the same basic equipment setup as shown in Figure 6-10, except that a two-tone audio generator is used.

2. Key the transmitter and adjust the audio level to produce the rated PEP from the transmitter.

3. The two-tone SSB envelope should look like the one in Figure 6-18A. If the two tones are not equal, the pattern will resemble Figure 6-18F.

The Two-Generator Method The outputs from two separate audio generators can be combined by using a combining network such as the ones shown in Figure 6-17. The combining network at Figure 6-17A is a resistive network that offers approximately 20-dB attenuation between each generator and the transmitter input point. This sounds like a lot of loss, but it only takes a very small signal at the microphone input to produce an output from the transmitter. An alternate method of combining two generators is shown at Figure 6-17B. This network offers little attenuation.

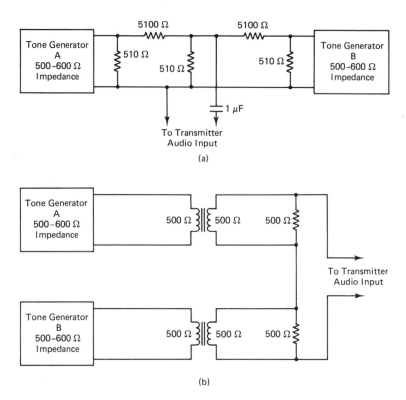

FIGURE 6-17. Two methods of connecting two separate audio generators for use in two-tone testing of SSB transmitters are shown here. At *A* a resistive matching network is used to combine the two signal generators while at *B* transformer coupling is used to combine the two audio signals.

Procedure

1. Set the two tone-generators approximately 1 kHz apart, but one should not be an exact multiple of the other. For example, the frequencies 800 Hz and 1,800 Hz are good choices.

2. With one tone generator off, key the transmitter and adjust the other tone generator level to produce one-quarter the rated PEP output.

3. Turn on the other generator and adjust its level to produce the proper two-tone pattern (where the trough goes to zero). If the transmitter is performing properly, the peak-reading wattmeter should indicate approximately four times the power obtained with only one tone, which should be approximately equal to the rated PEP output.

Evaluation of the Two-Tone SSB Envelope   The two-tone pattern shown in Figure 6-18A represents the ideal SSB two-tone envelope. At first glance, this two-tone pattern might be mistaken for a 100% amplitude-modulated signal envelope. However, a proper two-tone SSB envelope can be distinguished from a 100% AM envelope by careful examination of the *trough*. In the SSB two-tone envelope, the trough comes to a sharp point resulting in a

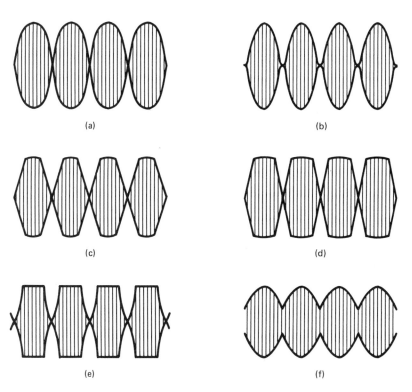

FIGURE 6-18. Several examples of typical two-tone signals are shown here. At *A* the two-tone envelope is typical of a proper envelope which can be obtained when the distortion produced by the amplifier is low. At *B* the cross-over point is not sharp, indicating improper bias of the amplifier. At *C* mild flat-topping is shown, caused by distortion in the amplifier. At *D* a more severe case of flat-topping is shown. The two-tone envelope at *E* is produced by an amplifier which is nonlinear both at the peaks and near the zero level. The cause is insufficient loading and/or excessive drive level combined with improper bias of the amplifier. The envelope at *F* is caused by the two audio tones not of equal-amplitude. When the audio tones are of equal-amplitude the envelope will go to zero.

well-defined X at the *crossover point*. In the AM envelope, the trough is rounded like a sine wave. The peaks of the SSB two-tone envelope and the 100% AM envelope are both shaped like a sine wave. The dinstinction is in the trough or crossover point.

When evaluating a two-tone SSB envelope, carefully study the crossover point and the peaks. The crossover point must be sharp and the sides of the X should be straight up to a point just below the peak. The peaks should be rounded like a sine wave. Figure 6-18B through E shows several examples of faulty SSB two-tone envelopes. A properly adjusted transmitter should produce a good quality two-tone envelope at or near the rated PEP output.

Figure 6-18B shows a pattern obtained when the linear is incorrectly biased. Notice that the crossover becomes rounded resembling a 100% AM envelope. An envelope such as this will result in sharp increases in the level of the third-order intermodulation products (see "Intermodulation Distortion" in this chapter). Figure 6-18C shows a mild case of "flat-topping." Notice the slight flattening of the peaks. This causes an increase in the level of the fifth and higher-order intermodulation distortion products. Figure 6-18D shows a more severe case of flat-topping. This would result in excessive intermodulation distortion products and would cause "splatter" interference to the adjacent channels. Flat-topping can be avoided by proper loading and/or not overdriving the amplifier.

Linearity Tests with a Trapezoid Pattern

Figure 6-19 shows a method of obtaining a trapezoid pattern to analyze the SSB transmitter performance. The demodulator probe demodulates the envelope. This demodulated signal is then used to sweep the scope beam. A sample of the transmitter output signal is then fed to the vertical input of the scope. Figure 6-20 shows how the trapezoid pattern is formed in this manner.

The demodulator probe can be moved to points A, B, or C to check the various stages between these points and the output. Although the two-tone signal is usually used for testing, this setup will also give a stable trapezoid display even with speech signals.

Figure 6-21 shows several typical trapezoid patterns which may be obtained. The pattern at Figure 6-21A is obtained when the two-tones are of equal amplitude. Good linearity is indicated by the straight sides and the sharp, well-defined point. The pattern at Figure 6-21B indicates that the two tones are not of equal amplitudes since the trapezoid doesn't come to a point. The pattern at Figure 6-21C indicates a moderate amount of flat-topping. At Figure 6-21E, nonlinearity near the zero level is indicated by the curving near the point.

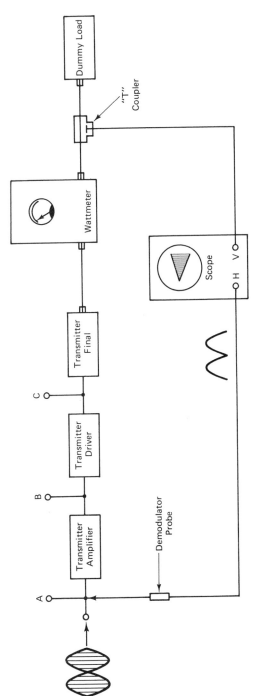

FIGURE 6-19. The trapezoid pattern can be used to test several stages of an SSB transmitter. The horizontal sweep of the scope is obtained by demodulating the envelope and using the demodulated signal to sweep the scope beam. The vertical input to the scope is obtained by sampling the signal at the point where the linearity is to be tested. If necessary, a down-converter can be used between the sampling point and the vertical input of the scope. Another alternative is to apply the signal directly to the vertical plates of the scope if the signal level is sufficient to drive the CRT plates directly.

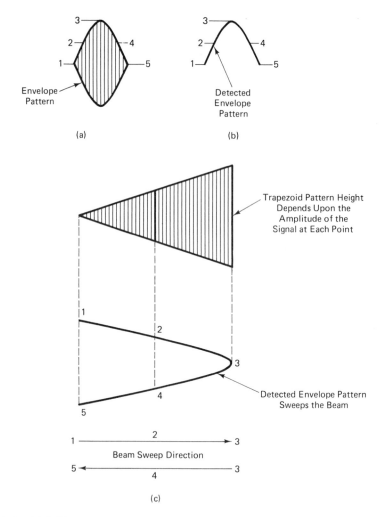

FIGURE 6-20. The illustration at *A* shows one complete envelope of a two-tone SSB envelope. The waveform at *B* shows the waveform produced by demodulating the envelope. The trapezoid pattern at *C* is formed by using the demodulated signal to sweep the scope while the undetected two-tone pattern is fed to the vertical input.

The point to remember when using this test is that if the envelope is already distorted at the point where the demodulator probe is connected, this distortion won't show up in the trapezoid pattern. Only the distortion that occurs between the two sampling points will show up in the trapezoid.

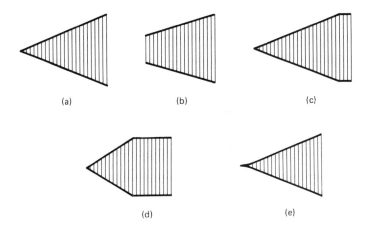

FIGURE 6-21. This illustration shows several trapezoid patterns which may be obtained. The pattern at *A* is the ideal pattern produced by low-distortion amplifiers. The envelope at *B* is also good but the envelope does not go to zero as indicated by the trapezoid not coming to a point. At *C* the trapezoid pattern indicates flat-topping distortion in the amplifier. At *E* distortion caused by improper bias is indicated. Notice that the distortion is more pronounced near the point.

Linearity Tests Using Two Envelope Detectors

Figure 6-22 shows the setup in which two envelope detectors are used to make linearity checks. In this case, both the horizontal and the vertical inputs to the scope are obtained from an envelope detector. The two envelope detectors should be identical. If they are, the distortion caused by them will cancel out. The schematic of an envelope detector is shown in Figure 6-22. When using this test setup, the envelope detector for the vertical input is connected to the point nearest the transmitter output.

With the two envelope detectors connected to the desired test points, the horizontal and vertical gain of the scope is adjusted to produce an angle of 45° between the trace and the baseline.

Figure 6-23 shows typical traces which might be observed with this setup. At Figure 6-23A, the trace indicates perfect linearity between the two test points. The trace at Figure 6-23B shows nonlinearity caused by overdriving or underloading. This is the same as flat-topping distortion. The trace at Figure 6-23C indicates nonlinearity near the "zero" point. This is the same as crossover distortion.

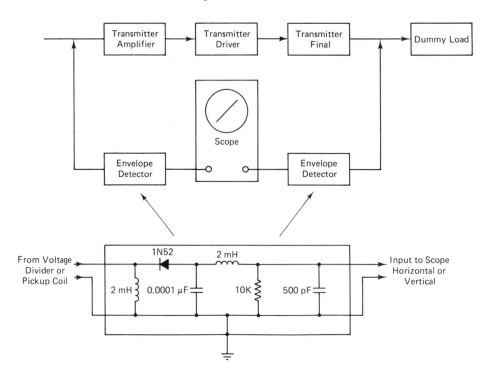

FIGURE 6-22. A method of connecting envelope detectors to obtain a scope trace which can be used to detect nonlinearity in the amplifier. The make-up of the envelope detectors is shown in the diagram in the inset.

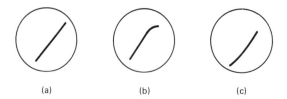

(a)   (b)   (c)

FIGURE 6-23. These traces are typical of those obtained by using the envelope detectors. At *A* the pattern is produced by a well-adjusted and properly operating amplifier. The pattern at *B* is caused by nonlinearity in the amplifier and the curving of the line at the top indicates flat-topping distortion in the amplifier. The curve at *C* indicates nonlinearity caused by improper biasing of the amplifier.

Typical Speech Waveforms

A characteristic of speech waveforms is a high peak-to-average signal level. Almost invariably, SSB transmitters employ some means to increase the average power in the speech envelope. Even with these built-in circuits to increase average power, a typical SSB speech envelope will still show a fairly high peak-to-average power ratio. Figure 6-24 shows examples of speech envelopes that might occur under various conditions. A speech envelope containing little compression is shown at Figure 6-24A. Notice the very high peak in the envelope. Although the peak is high, the average level in the envelope is relatively low. In Figure 6-24B, the envelope shows a higher average level. In Figure 6-24C, too much drive increases the average level dramatically but causes severe flat-topping, and splatter will result. While it is difficult to illustrate just how a "typical" speech waveform envelope should appear on the scope, Figure 6-24A gives a fairly good idea of what one should expect to see from a good signal. Relatively high peaks tapering off in the shape of a Christmas tree are an indication of a fairly good, clean signal.

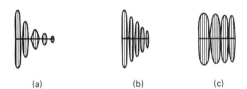

(a)　　　　(b)　　　　(c)

FIGURE 6-24. Several speech patterns as observed on a scope are shown here. The pattern at *A* is typical of a good pattern produced by a normally operating SSB transmitter driven by a low level speech input. The pattern at *B* is produced by a higher drive level at the transmitter input causing the higher peaks to be compressed slightly. The pattern at *C* is caused by excessively overdriving the transmitter to the point at which excessive distortion is produced. Notice the severe flat-topping.

Checking Noise and Hum with the Scope

A simple method that can be used to check for excessive hum and/or noise in the SSB signal is to reinsert the carrier and check for hum ripple and noise components on the carrier.

Test Procedure

1. Hook up the equipment as shown in Figure 6-25. The resistor serves as a dummy microphone to prevent acoustic pickup from causing interference.

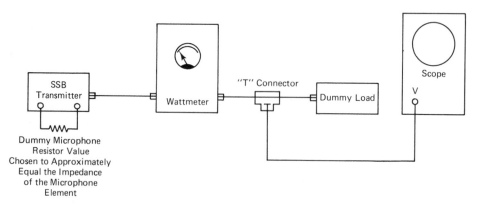

FIGURE 6-25. Typical setup used for checking the hum and noise produced by an SSB transmitter.

2. Key the transmitter and unbalance the balanced modulator by adjusting the carrier balance control. Adjust the carrier balance control to produce approximately one-quarter the rated PEP output from the transmitter. If sufficient carrier level can't be obtained by adjusting the carrier balance control alone, adjust the carrier oscillator frequency slightly to increase the carrier level.

3. Once the proper carrier level is obtained, a clean, smooth CW envelope should appear on the scope as shown at Figure 6-26A. Figure 6-26B shows a hum component present while Figure 6-26C shows noise on the envelope.

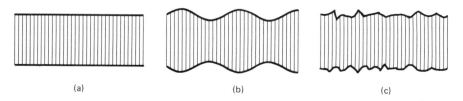

FIGURE 6-26. The pattern at *A* is free of any hum and noise components. The pattern at *B* contains a hum component. The pattern at *C* contains noise.

USING THE SPECTRUM ANALYZER FOR SSB SIGNAL TESTS

The spectrum analyzer can provide more information about the quality of an SSB signal than any other single instrument. However, in order to do the job properly the spectrum analyzer must be able to resolve signal frequencies that are only a few hundred cycles apart. Before getting into the applications of the spectrum analyzer, a brief description of intermodulation is in order.

Intermodulation Distortion

Before being transmitted, the SSB signal must be amplified. This amplification must be linear so that the SSB signal appearing at the output is a true reproduction of the SSB signal at the input. Since an SSB signal can contain many different frequencies, any nonlinearity in the amplifier can cause nonlinear mixing of these various frequencies. Intermodulation distortion is a direct result of nonlinearity in the amplifier. As an example, let's suppose that a two-tone audio test signal is applied to an SSB transmitter and that the frequencies of the two tones are 800 Hz and 1,800 Hz. Assume further that the transmitter is in the upper sideband mode and that the carrier frequency (which won't appear in the output if properly suppressed) is 7 MHz. Under perfect conditions (which can never be achieved in practice), only two frequencies should appear in the output: 7.000800 MHz and 7.001800 MHz. Since a practical amplifier always has an inherent amount of nonlinearity, a certain amount of intermodulation distortion will be produced. This results in the generation of even-order and odd-order sum and difference frequencies.

All even-order sum and difference IM products will fall so far out of band that they will not be any problem. Also, the odd-order sum products will fall far out of band. However, the odd-order difference IM products fall *in-band* and thus can cause problems. Third-order difference IM products are $2F1-F2$ and $2F2-F1$. Fifth-order difference IM products are $3F1-2F2$ and $3F2-2F1$. Higher order IM products can also be significant when the degree of nonlinearity is excessive. In our two-tone test described above, the frequency of the third-order difference IM products will be 2(7.0018 MHz) -7.0008 MHz and 2(7.0008 MHz) -7.0018 MHz, or 6.9998 MHz and 7.0028 MHz.

A display of a two-tone SSB signal on a spectrum analyzer is shown in Figure 6-27. The vertical scale is calibrated in dB (10 dB/division), while the sweep width is set for 1 kHz/division on the horizontal scale. The center frequency of the spectrum analyzer is set to 7.000 MHz. As pointed out in Chapter 2, it is important to prevent overloading the front end of the spectrum analyzer, because this would cause spurious signals to appear on the display which might be mistaken for transmitter spurious. Refer to Chapter 2 for methods of minimizing this problem.

Signal-to-Distortion Ratio   In the display shown in Figure 6-27, the two tones are represented by A and B while the various intermod products are presented in terms of A and B. The third-order intermod products are $2A — B$ and $2B — A$. The fifth-order intermod products are $3A — 2B$ and $3B — 2A$; the seventh-order intermods are $4A — 3B$ and $4B — 3A$. A measurement figure used to indicate the level of intermodulation distortion is called the *signal-to-distortion ratio* and is commonly abbreviated as the S/D ratio. The S/D ratio is the ratio of the amplitude of one of the test tones (in a two-tone signal) to the amplitude of one of the third-order IM products. This ratio is determined in the following test procedure.

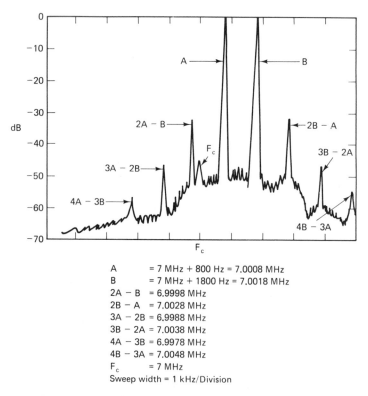

A	= 7 MHz + 800 Hz = 7.0008 MHz
B	= 7 MHz + 1800 Hz = 7.0018 MHz
2A − B	= 6.9998 MHz
2B − A	= 7.0028 MHz
3A − 2B	= 6.9988 MHz
3B − 2A	= 7.0038 MHz
4A − 3B	= 6.9978 MHz
4B − 3A	= 7.0048 MHz
F_c	= 7 MHz

Sweep width = 1 kHz/Division

FIGURE 6-27. A typical spectrum analyzer display produced by an SSB signal produced by a transmitter modulated by a two-tone audio input signal. The intermod products and the residual carrier component are clearly indicated.

Test Procedure

1. Set up the equipment as shown in Figure 6-28.
2. Key the transmitter and apply a two-tone (equal amplitude) audio test signal to the transmitter input.
3. Increase the amplitude from the two-tone generator until the peak-reading wattmeter indicates the full rated PEP output power.
4. Adjust the spectrum analyzer attenuator control so that the top of one of the tones just touches the zero dB reference level.
5. Next, determine the level of the highest amplitude distortion product (usually the third-order IM product). This level in dB is the S/D ratio of the SSB signal.

In Figure 6-27, the level of the third-order IM product is −33 dB. Since the level of each of the tones is set to 0 dB, the S/D ratio of this SSB signal is 33 dB. The S/D ratios of SSB transmitters are sometimes referenced to the PEP level rather than to the level of one of the tones. This results in an S/D ratio that is 6 dB greater since the PEP level in a two-tone SSB signal is 6 dB above the level of one of the tones.

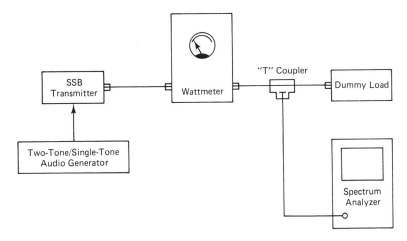

FIGURE 6-28. Typical setup for testing an SSB transmitter using a spectrum analyzer.

Carrier Suppression Measurement

The carrier suppression measurement can be made with either a single-tone or a two-tone signal. However, the display will be less cluttered with a single-tone signal.

Test Procedure

1. Hook up the test equipment as shown in Figure 6-28. In this case, the audio generator is a single-tone generator.

2. Set the audio generator to 1 kHz. Tune the spectrum analyzer's center frequency to the carrier frequency of the SSB transmitter and set the sweep width to 1 kHz division.

3. Key the transmitter and set the audio generator level to produce approximately one-half the rated PEP output of the transmitter.

4. Adjust the spectrum analyzer attenuator so that the top of the single tone is at the 0-dB reference mark on the display (see Figure 6-29).

5. With the spectrum analyzer's center frequency set to the carrier frequency, the carrier will show up as a display at the center of the horizontal scale (see Figure 6-29).

6. The carrier suppression (in dB below single tone) can now be read directly from the scale. In Figure 6-29, the carrier suppression is 40 dB below single tone.

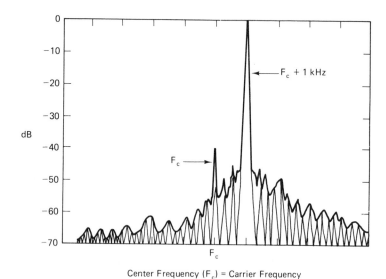

Center Frequency (F_c) = Carrier Frequency
Single-Tone Frequency = F_c + 1 kHz
Sweep Width = 1 kHz/Division

FIGURE 6-29. Typical spectrum analyzer display when a single-tone audio signal is used to test an SSB transmitter for carrier suppression. The residual carrier component appears at *Fc* when the spectrum analyzer is tuned to the SSB transmitter carrier frequency.

Unwanted Sideband Suppression Measurement

This measurement procedure is very similar to the one just described for the carrier suppression test. However, in this test the choice of the frequency of the single-tone modulating signal will have a greater influence on the results. For this reason, it is desirable to perform this test at several different single-tone audio frequencies. The reason for this is that the unwanted sideband signals generated by the lower audio frequencies will be closer to the passband of the sideband filter and thus will be attenuated less than the signals generated by the higher audio frequencies.

Test Procedure

1. Hook up the equipment as shown in Figure 6-28. In this case, the audio generator is a variable-frequency single-tone generator.
2. Set the audio generator to 500 Hz, tune the spectrum analyzer's center frequency to the carrier frequency of the SSB transmitter, and set the sweep width to 1 kHz/division.
3. Key the transmitter and set the audio generator level to produce approximately one-quarter the rated PEP output.
4. Adjust the spectrum analyzer attenuator so that the single-tone level in the desired sideband is at the 0-dB reference line.
5. The undesired sideband will show up on the opposite side of the carrier frequency (Fc) and be separated by the same amount as the desired SSB single-tone signal (see Figure 6-30).
6. The suppression of the undesired sideband as referenced to the single-tone desired sideband now can be read directly from the scale. In the display in Figure 6-30, the suppression is 45 dB. The procedure is then repeated at another audio frequency. Usually only three or four audio frequencies need be used. For example, 500 Hz, 1,000 Hz, 1,500 Hz, and 2,000 Hz should give a pretty good indication of the suppression response.

SSB POWER MEASUREMENTS

There are several possible ways to measure the PEP output from an SSB transmitter. However, it is not only desirable to know the level of PEP that the transmitter can produce, but also it is important to know the quality of the SSB signal at the particular PEP power level indicated on the wattmeter. It is therefore desirable to monitor the SSB signal on a scope or spectrum analyzer while measuring the PEP power.

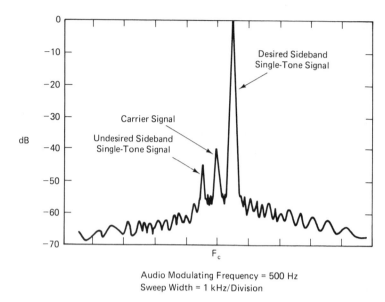

Audio Modulating Frequency = 500 Hz
Sweep Width = 1 kHz/Division

FIGURE 6-30. Typical spectrum analyzer display used to test for unwanted sideband suppression.

The Two-Tone PEP Measurement—With Scope

Test Procedure

1. Set up the equipment as shown in Figure 6-31.
2. Key the transmitter and adjust the level of the two-tone generator until the peaks of the envelope start to flatten (see Figure 6-32A).
3. Decrease the level of the two-tone generator until the peaks of the envelope are no longer flattened (see Figure 6-32B).
4. At this point, the PEP power is read on the peak-reading wattmeter.

The Two-Tone PEP Measurement—Without Scope

While not as accurate, the following method can be used in the absence of a scope.

Test Procedure

1. Hook up the equipment as in Figure 6-31 but omit the scope.
2. Key the transmitter and slowly increase the level from the two-tone audio generator while watching the peak-reading wattmeter carefully.

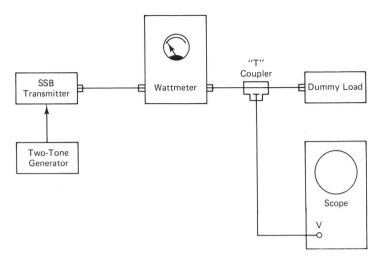

FIGURE 6-31. Typical setup that can be used to check the SSB transmitter output PEP properly.

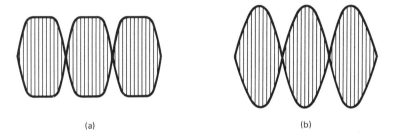

(a)   (b)

FIGURE 6-32. If the two-tone pattern at *A* is obtained, reduce the audio input level until the flat-topping disappears and the envelope looks like the pattern at *B*. This is the point at which the PEP power should be measured.

3. The wattmeter reading should increase rapidly at first and then slow down abruptly. Set the two-tone generator level to the point where the wattmeter reading slows down abruptly.

4. Read the PEP power from the peak-reading wattmeter at this point.

Using the Average-Reading Wattmeter for PEP Measurement

An average-reading wattmeter can be used in either of these two-tone PEP measurements. However, most average-reading wattmeters will indicate approximately 40% PEP in a two-tone SSB envelope. Therefore, the reading must be multiplied by 2.5 to get the PEP power. This relationship only holds true in a properly shaped two-tone envelope.

Single-Tone PEP Measurement

In a single-tone SSB signal, the average-reading and peak-reading watt-meters will both indicate the same power. This is because the average-power in a single-tone SSB signal is the same as the PEP power. However, sustained testing with a single-tone signal should be avoided unless you are sure the transmitter can handle this. Otherwise, you run the risk of blowing out tubes, transistors, etc.

Test Procedure

1. Set up the equipment as shown in Figure 6-33.
2. Key the transmitter and slowly increase the single-tone audio level while carefully watching the wattmeter.
3. The wattmeter reading will increase rapidly up to a point at which it will abruptly slow down.
4. Set the audio generator level to the point at which the wattmeter reading levels off and read the PEP power at this point.

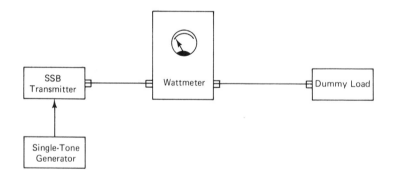

FIGURE 6-33. By using a single-tone audio generator, the PEP output can be measured by using either an average-reading or a peak-reading wattmeter.

FREQUENCY MEASUREMENT

There are two ways in which the frequency of an SSB transmitter can be measured. One way is to use a single audio tone to modulate the transmitter and then measure the frequency of the SSB single-tone signal. If the SSB transmitter is in the upper sideband mode, the carrier frequency can be determined by subtracting the audio-modulating frequency from the single-tone SSB signal frequency. If the transmitter is in the lower sideband, the carrier frequency can be determined by adding the audio-modulating frequency to the frequency of the single-tone SSB signal.

An alternate method is to unbalance the balanced modulator to allow sufficient carrier level to reach the output where the carrier frequency is measured directly.

Test Procedure—Method 1

1. Hook up the test equipment as shown in Figure 6-34A.

2. Apply a single-tone audio-modulating signal (1 kHz nominal) to the transmitter input. The frequency of this modulating signal must be determined by measuring it with a frequency counter.

3. Key the transmitter and increase the level of the single-tone audio signal until the transmitter output signal reaches a level at which the frequency counter gives a stable display of the frequency.

4a. If the transmitter is in the upper sideband mode, the carrier frequency is determined by subtracting the audio-modulating frequency (step 2) from the SSB single-tone signal frequency (step 3).

 b. If the transmitter is in the lower sideband mode, the carrier frequency is determined by adding the frequency of the modulating signal (step 2) to the frequency of the SSB single-tone signal (step 3).

Test Procedure—Method 2

1. Set up the equipment as in the first procedure but omit the audio generator (see Figure 6-34B).

2. Key the transmitter and misadjust the carrier balance control until sufficient carrier appears in the output to give a stable display on the frequency counter.

FREQUENCY STABILITY

It is imperative that an SSB transmitter maintain a highly stable carrier frequency. An AM or FM transmitter frequency can drift considerably without degradation at the receiving end. However, a small change in the frequency of the SSB transmitter can make the signal totally unintelligible at the receiving end. If the carrier frequency in an SSB transmitter is correct, the modulating frequencies will be correct also, since they are referenced to the carrier frequency.

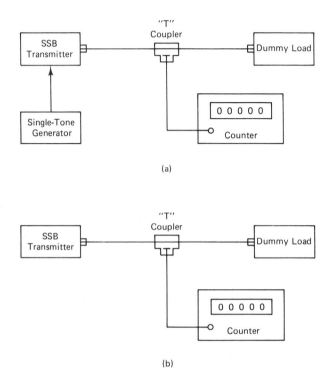

(a)

(b)

FIGURE 6-34. Two methods of measuring the frequency of an SSB transmitter are illustrated here. At *A* a single-tone audio generator is used to produce a single-tone output signal. The single-tone output signal is then measured with a frequency counter. The actual carrier frequency is determined by adding or subtracting the audio modulating frequency from the frequency measured at the output. At *B* no audio modulating signal is used; instead the carrier balance control in the balanced modulator is misadjusted to produce a small amount of carrier output from the transmitter. The frequency counter then measures the carrier directly.

Test Procedure

1. Hook up the equipment as shown in Figure 6-34B.

2. Key the transmitter and misadjust the carrier balance control until sufficient carrier appears in the output to give a stable indication on the frequency counter.

3. Vary the supply voltage 10 or 15%. The frequency should not change more than a few Hz.

4. Leave the transmitter keyed for three to four minutes (at low carrier power level only) and note the frequency drift.

AUDIO FREQUENCY RESPONSE TEST

The purpose of this test is to determine the transmitter's audio response characteristic. In order to get an overall picture of the audio response, the test is repeated at several different frequencies.

Test Procedure

1. Hook up the equipment as shown in Figure 6-35.
2. The audio generator is first tuned to 1 kHz.
3. Increase the audio level until the wattmeter indicates one-quarter the rated PEP.
4. Measure the audio voltage from the generator, preferably using an audio voltmeter with a dB scale. This will serve as the 0-dB reference mark.
5. The test is then repeated at several frequencies, such as 500 Hz, 1,500 Hz, 2,000 Hz, 2,500 Hz, etc. The audio generator level is adjusted each time to produce the one-quarter PEP output level. The audio voltage is then measured and compared with the reference level in step 4.

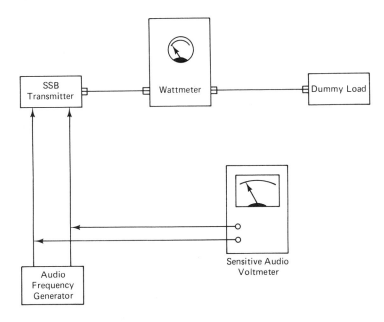

FIGURE 6-35. Setup used to test the audio frequency response of an SSB transmitter. A sensitive audio meter is used here.

Since the audio voltage level required to produce one-quarter the rated PEP will be quite low (on the order of a few millivolts), a very sensitive meter will be required to make the measurement. If a sensitive voltmeter is not available, a voltage divider can be used between the output of the audio generator and the transmitter input (see Figure 6-36). Resistor R1 is chosen to match the output impedance of the audio generator, R3 is chosen to match the input impedance of the transmitter, and R2 is approximately 100 times the resistance of R3. Then the audio voltage appearing across R3 at the transmitter input will be a small fraction of the voltage from the audio generator. Then a typical VOM with a dB scale can be used to monitor the voltage level at the output of the audio generator (across R1). We are not especially interested in the exact level of the voltage; rather, we are interested in knowing how much the voltage level must change to keep the PEP output from the transmitter at the same level at the different input frequencies. The change in the voltage level (in terms of dB) will be the same across the transmitter input as it is across the generator output. For example, if the voltage across R1 drops 3 dB as read on the VOM, the voltage across R3 will also drop 3 dB.

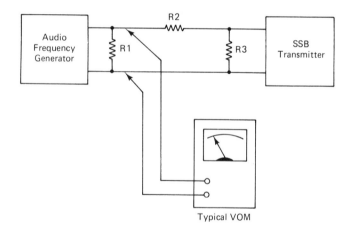

R3 = Microphone Input Impedance of the Transmitter
R2 = 100 Times R3 (R3 × 100)
R1 = Output Impedance of the Audio Generator

FIGURE 6-36. If a voltage divider such as this one is used between the audio generator and the transmitter input, a typical VOM can be used to measure the output of the audio generator at various frequencies to determine the audio response of the transmitter.

HARMONIC DISTORTION

Harmonic distortion only occurs in the audio stages of an SSB transmitter. This is usually not a significant problem since SSB transmitters do not use audio power amplifiers. The relatively low audio voltage needed at the input to the balanced modulator can be obtained from straight class A audio voltage amplifiers, and these can be designed for low distortion. However, some harmonic distortion may still occur in the audio stages and in the balanced modulator. The sideband filter will attenuate frequencies that lie outside its passband.

Two procedures will be outlined here for measuring harmonic distortion in SSB transmitters. The first method is with the use of a spectrum analyzer; the second method uses an SSB receiver of known quality along with an audio distortion meter.

Test Procedure—Method 1 (Spectrum Analyzer)

1. Hook up the equipment as shown in Figure 6-28. In this case, the audio generator is a single-tone generator.
2. Tune the center frequency of the spectrum analyzer to the transmitter carrier frequency and set the sweep width to 1 kHz/division.
3. Set the audio generator frequency to 1 kHz.
4. Key the transmitter and adjust the audio generator level to produce the rated PEP output. *Caution*!!! The transmitter should be keyed for only brief periods with full PEP single-tone output.
5. Adjust the spectrum analyzer attenuator to place the top of the single-tone signal at the 0-dB reference mark on the scale. Observe the level of the harmonics as referenced to the single-tone signal. Figure 6-37 shows a display of an upper sideband single-tone signal with the second and third harmonics of the 1 kHz audio signal shown to the right of the single-tone signal and spaced 1 kHz apart. The second harmonic signal is approximately 35 dB below the single-tone signal. The third harmonic signal is approximately 50 dB down.

Test Procedure—Method 2 (SSB Receiver and Distortion Meter)

1. Hook up the equipment as shown in Figure 6-38. The amount of coupling to the receiver will be very small, depending upon the amount of transmitter output power. The amount of coupling should be minimal to prevent overloading the receiver. The receiver should be operated at minimum RF gain.

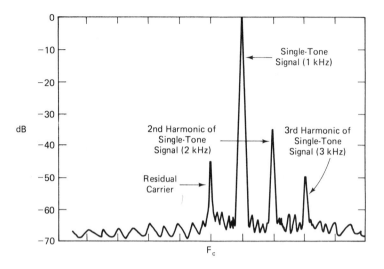

FIGURE 6-37. An SSB transmitter can be tested for audio harmonic distortion by using a spectrum analyzer to check the output signal. A typical spectrum analyzer display is shown in this illustration.

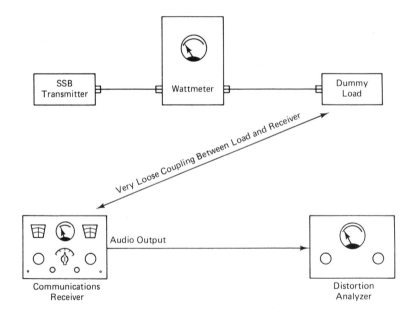

FIGURE 6-38. A setup in which an SSB communications receiver and a distortion analyzer are used to check for audio harmonic distortion. If the inherent distortion of the receiver is quite low compared with the transmitter-produced distortion then the measurement will be valid.

2. Tune the receiver to the transmitter frequency.

3. Key the transmitter and adjust the audio generator level to produce full rated PEP. *Caution*!!! The transmitter should be keyed for only brief periods with a full PEP single-tone signal.

4. Check the distortion at the receiver output with the receiver set for a low audio output level.

The results of this test depend largely upon the quality of the receiver. If the receiver is producing large amounts of distortion of its own, the test results are obviously invalid. A measure of the receiver's quality can be made using a good quality CW signal generator. If the receiver is capable of producing a low distortion output with the signal generator supplying the signal input, it will be satisfactory for this test.

AUTOMATIC LEVEL (OR LOAD) CONTROL (ALC)

In order to minimize flat-topping distortion and the resulting spurious signals caused by it, SSB transmitters generally employ a method to reduce the gain of some stage (or stages) of the transmitter when the output level reaches a certain level. The automatic level control is normally set to come into action when the output reaches the PEP level or just prior to the PEP level. The point at which the ALC is set to take control is called the ALC threshold. Once the ALC has come into action, the output PEP level should remain fairly constant even though the audio input level may increase by many dB. Since typical speech envelopes have a high peak-to-average ratio, the ALC must be fast-acting both on the attack and release. It must attack fast to hold down those sudden peaks, and it must release fast so that low-level signals won't be further reduced in level.

ALC Threshold Test

The typical method of determining the ALC threshold is to increase the audio input to the transmitter while monitoring the output level. If only a wattmeter is used as an indicator, it is difficult to tell whether the levelling off of the output is due to ALC action or flat-topping distortion. To distinguish between the two, a scope is used in conjunction with the wattmeter while a two-tone audio generator is used to furnish the audio-modulating signal for the test.

Test Procedure

1. Hook up the equipment as shown in Figure 6-31.

2. Key the transmitter and increase the audio generator level until the wattmeter reading starts to level off.

3. At this point, observe the two-tone pattern on the scope. If the two-tone envelope has begun to flat-top, the levelling off is due to distortion rather than ALC action. In this case, the ALC threshold needs to be lowered.

4. If the two-tone envelope doesn't flat-top at the point where the wattmeter reading begins to level off, the ALC threshold has been reached. Increasing the audio level further will cause the two-tone envelope to deviate from its normal shape because of the ALC action—but not flat-topping. Rather, the two-tone envelope may become a little cluttered and fuzzy appearing. Still, the two-tone pattern can be seen through the clutter and the peaks should not appear flat.

"Speech-Testing" the ALC Action

The ALC response to a two-tone envelope and an actual speech envelope can be quite different. The ALC response time is very important to the way it performs with an actual speech signal. The previous test for determining the ALC threshold with a two-tone SSB envelope doesn't give any indication of the response time of the ALC circuit. The following method will give a fair indication of the response time of the ALC under actual operating conditions.

Test Procedure

1. First perform the ALC threshold test using the procedure just described.

2. To get a reference, measure the height of the two-tone envelope (on the scope) with an input level that exceeds the ALC threshold (see Figure 6-39A). The envelope pattern may become a little fuzzy and cluttered, but the two-tone envelope should still be very visible.

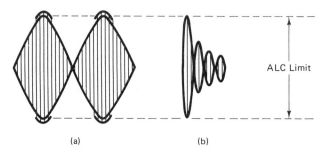

(a)   (b)

FIGURE 6-39. At *A* is a two-tone envelope pattern which is obtained when the input audio level is increased past the point at which the ALC comes into action. Notice the pattern becomes cluttered. At *B* a speech input signal is used and the resulting envelope is compared with the level produced by the two-tone signal.

3. Remove the two-tone generator from the microphone input and connect the microphone.

4. Key the microphone and loudly pronounce the vowel "I." This will produce a speech envelope with a high peak-to-average ratio. The peak should not exceed the reference height produced by the two-tone envelope while under ALC control. If it does significantly exceed this reference height, the ALC is not acting fast enough on speech signals.

BIBLIOGRAPHY

American Radio Relay League, *Single Sideband for the Radio Amateur.* American Radio Relay League, Newington, Conn., 1970.

Collins Radio Company, *Amateur Single Sideband* (1st ed., 2nd printing) Greenville, N.H.: The Ham Radio Publishing Group, 1977.

Hoonton, Harry D., *Single Sideband Theory and Practice.* New Augusta, Ind.: Editors & Engineers, 1967.

Kahn, Leonard R., *Comparison of Linear SSB Transmitters with Envelope Elimination and Restoration SSB Transmitters.* Proceedings of the Institute of Radio Engineers, Volume 44, No. 12 (Dec. 1956).

Miller, Gary M., *Handbook of Electronic Communication.* Englewood Cliffs, N.J.: Prentice-Hall, Inc., 1979.

Pappenfus, E. W.; Bruene, Warren B.; and Schoenike, E. O., *Single Sideband Principles & Circuits.* New York: McGraw-Hill, 1964.

Rusgrove, Jay B. (W1VD) "Spectrum Analysis—"One Picture's Worth." *QST Magazine,* August 1979, pp. 15–21.

7

SINGLE SIDEBAND RECEIVER TESTS AND MEASUREMENTS

It is assumed that the reader already has a basic understanding of the operating principles of SSB receivers. However, for the sake of review, a brief discussion of the manner in which SSB signals are processed is presented here.

Figure 7-1 is a simplified schematic of an SSB receiver that is capable of operating in the USB mode (upper sideband) or the LSB mode (lower sideband). The frequencies chosen for use in this example are strictly arbitrary, not intended to represent any particular radio service. This is a VHF dual-conversion SSB receiver. The high IF is *centered* at 10.7 MHz while the low IF is *centered* at 455 kHz. The low IF bandpass filter is designed to pass a band of frequencies from 453.5 kHz to 456.5 kHz. Those frequencies which fall outside this band are greatly attenuated. The operation of the receiver can be best understood by following a USB and an LSB signal through the receiver from the RF amplifier to the audio amplifier.

First, the USB mode. Assume that an SSB transmitter is being operated in the USB mode and is being modulated by a two-tone audio signal, one of the tones being 500 Hz and the other 2,400 Hz. The frequency of the suppressed carrier is 159.300 MHz. The 500-Hz tone will be represented by a 159.3005-MHz USB signal, while the 2,400-Hz tone will be represented by a 159.3024-MHz USB signal.

Assume further that a second transmitter is being operated at the same frequency (159.300 MHz), but in the LSB mode. The same two-tone frequencies are used to modulate the LSB transmitter (500 Hz and 2,400 Hz). The 500-Hz tone will be represented by a 159.2995-MHz LSB signal, while the 2,400-Hz tone will be represented by a 159.2976-MHz LSB signal. Thus,

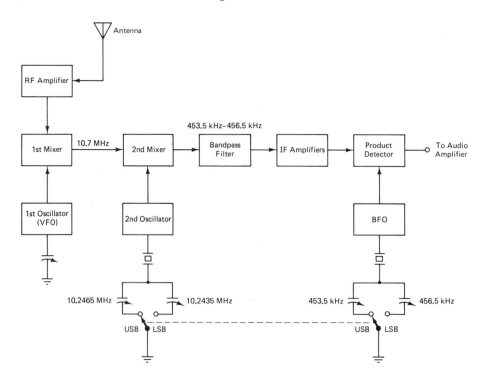

FIGURE 7-1. Simplified block diagram of a dual-conversion SSB receiver.

a USB two-tone signal and an LSB two-tone signal at the same (suppressed carrier) frequency are present at the receiver's front end.

To receive the USB signal, the receiver must first be switched to the USB mode. This sets the second oscillator to a frequency of 10.2465 MHz and the BFO to 453.5 kHz. In order to tune in the 159.300-MHz USB signal, the receiver's first oscillator or *VFO* must be tuned to:

$$159.300 \text{ MHz} - 10.7 \text{ MHz (first IF)} = 148.6 \text{ MHz}$$

With the VFO tuned to 148.6 MHz, the two-tone USB signal is converted to:

$$159.3005 \text{ MHz} - 148.6 \text{ MHz} = 10.7005 \text{ MHz}$$

representing the 500-Hz tone, and

$$159.3024 \text{ MHz} - 148.6 \text{ MHz} = 10.7024 \text{ MHz}$$

representing the 2,400-Hz tone. Next, the first IF USB two-tone signal is mixed with the second oscillator frequency, 10.2465 MHz, to produce a second IF frequency of:

$$10.7005 \text{ MHz} - 10.2465 \text{ MHz} = 454 \text{ kHz}$$

representing the 500-Hz tone, and

$$10.7024 \text{ MHz} - 10.2465 \text{ MHz} = 455.9 \text{ kHz}$$

representing the 2,400-Hz tone.

Notice that this two-tone low IF signal falls within the bandpass of the filter and thus is passed on to the IF amplifiers and then on to the product detector. In the product detector, the 454-kHz signal mixes with the 453.5-kHz BFO signal to produce a 500-Hz difference frequency; hence, the 500-Hz tone is recovered. The 455.9-kHz signal mixes with the 453.5-kHz BFO signal to produce a 2,400-Hz difference frequency; hence, the 2,400-Hz tone is recovered.

Now, let's see what happens to the LSB signal when the receiver is in the USB mode. The 159.2995-MHz LSB signal (representing the 500-Hz tone) is converted to a high IF of:

$$159.2995 \text{ MHz} - 148.6 \text{ MHz} = 10.6995 \text{ MHz}$$

which is then converted to a low IF of:

$$10.6995 \text{ MHz} - 10.2465 \text{ MHz} = 453 \text{ kHz}$$

Since this is out of the passband of the bandpass filter, it will be rejected. The 159.2976-MHz LSB signal (representing the 2,400-Hz tone) is converted to a high IF of:

$$159.2976 \text{ MHz} - 148.6 \text{ MHz} = 10.6976 \text{ MHz}$$

which is then converted to a low IF of:

$$10.6976 \text{ MHz} - 10.2465 \text{ MHz} = 451.1 \text{ kHz}$$

This, too, is out of the passband of the bandpass filter and is therefore rejected.

In the LSB mode, the frequency of the BFO and the frequency of the second oscillator are changed to 456.5 kHz and 10.2435 MHz, respectively. Tables 7-1a and 7-1b show in condensed form how the USB and LSB signals are processed in either mode. Notice that in the LSB mode the BFO frequency operates *above* the low IF frequency, while in the USB mode the BFO frequency operates *below* the low IF frequency. It really makes no difference which is higher to the product detector; just remember that the *difference* between the low IF and the BFO frequency is the resulting audio frequency.

SENSITIVITY TESTS

Three basic test procedures for checking the sensitivity of an SSB receiver are presented here. These are: (1) the 10-dB $S + N/N$ test; (2) the 10-dB Sinad test; and (3) the *effective* sensitivity test. In performing these tests, the signal generator must be *unmodulated* and preferably switched to

USB MODE		
	(500 Hz)	(2,400 Hz)
USB	159.3005 MHz &	159.3024 MHz
1st Osc.	− 148.6000 MHz &	148.6000 MHz
Hi IF	10.7005 MHz &	10.7024 MHz
2nd Osc.	− 10.2465 MHz &	− 10.2465 MHz
Lo IF	0.4540 MHz &	0.4559 MHz
BFO	− 0.4535 MHz &	− 0.4535 MHz
Audio	0.0005 MHz &	0.0024 MHz
	(500 Hz) &	(2,400 Hz)
	(500 Hz)	(2,400 Hz)
LSB	159.2995 MHz &	159.2976 MHz
1st Osc.	− 148.6000 MHz &	− 148.6000 MHz
Hi IF	10.6995 MHz &	10.6976 MHz
2nd Osc.	− 10.2465 MHz &	− 10.2465 MHz
Lo IF	0.4530 MHz &	0.4511 MHz
	(453 kHz)	(451.1 kHz)
	└──── Rejected ────┘	

(a)

LSB MODE		
	(500 Hz)	(2,400 Hz)
LSB	159.2995 MHz &	159.2976 MHz
1st Osc.	− 148.6000 MHz &	− 148.6000 MHz
Hi IF	10.6995 MHz &	10.6976 MHz
2nd Osc.	− 10.2435 MHz &	− 10.2435 MHz
Lo IF	0.4560 MHz &	0.4541 MHz
	(456 kHz) &	(454.1 kHz)
BFO	456.5 kHz &	456.5 kHz
Lo IF	− 456.0 kHz &	− 454.1 kHz
Audio	0.5 kHz &	2.4 kHz
	(500 Hz) &	(2,400 Hz)
USB	159.3005 MHz &	159.3024 MHz
1st Osc.	− 148.6000 MHz &	− 148.6000 MHz
Hi IF	10.7005 MHz &	10.7024 MHz
2nd Osc.	− 10.2435 MHz &	− 10.2435 MHz
Lo IF	0.4570 MHz &	0.4589 MHz
	(457 kHz)	(458.9 kHz)
	└──── Rejected ────┘	

(b)

TABLE 7-1. This table shows how the receiver of Fig. 7-1 processes the SSB signal. The table at (a) shows how a sample USB signal is processed, while the table at (b) shows the processing of a sample LSB signal.

the CW position if one is provided on the generator. The generator must be tuned carefully to the proper sideband frequency and should be stable so as to remain there during the test. In order to simulate a USB signal representing a 1-kHz modulating tone, the signal generator is simply tuned 1 kHz above the suppressed carrier frequency. Conversely, to simulate an LSB signal representing a 1-kHz modulating tone, the signal generator is tuned 1 kHz below the suppressed carrier frequency.

The 10-dB $S+N/N$ Sensitivity Test

The basic equipment setup for this test is shown in Figure 7-2. The test procedure is summarized as follows:

Test Procedure

1. Set up the equipment as shown in Figure 7-2.
2. Tune the signal generator to 1 kHz above the suppressed carrier frequency (for the USB mode) or 1 kHz below the suppressed carrier frequency (for the LSB mode). *Do not modulate the generator!*
3. Set the receiver to the proper mode (USB or LSB), set the RF gain to maximum, the squelch control fully open, and the volume to maximum.
4. Increase the signal generator output level until the audio meter indicates 1/2 W. (This can be determined by the formula: $E = \sqrt{PR}$.)
5. Switch the receiver's mode switch to the opposite sideband. If the audio level drops 10 dB or more, the output level setting of the generator is the 10-dB $S+N/N$ sensitivity at 1/2 W audio output.
6. If the audio output doesn't drop at least 10 dB in step 5, return the mode switch to the original position and increase the signal generator

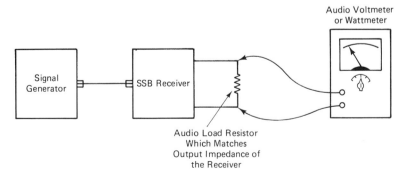

FIGURE 7-2. A basic test equipment setup that is used for many of the tests and measurement procedures described in this chapter.

output level a small amount, readjusting the volume control to maintain the output at the 1/2 W level.

7. Switch the receiver mode switch back to the opposite sideband and again note the drop in the audio output level. It may be necessary to repeat steps 6 and 7 two or three times until the proper signal generator level is found that produces a 10-dB $S+N/N$ ratio.

8. The final signal generator level setting is the 10-dB $S+N/N$ sensitivity of the receiver at a minimum output level of 1/2 W.

If the receiver is designed for low audio output power, the standard reference audio output obviously will be lower. For example, 50 mW is commonly used as a standard audio output for low audio power receivers such as those used in walkie-talkies and other battery operated portable receivers.

The 10-dB Sinad Sensitivity Test

In this test, a distortion analyzer is used. A general-purpose type distortion analyzer can be used or a special instrument called a Sinad meter can be used, such as the Helper Instruments model Sinadder 3. The advantage of this type of instrument is the automatic level-setting feature which simplifies the task. The procedure for using both types of instruments will be described here. It is very important that the signal generator be very stable in order that the beat note produced in the receiver output will remain centered in the rejection notch of the distortion analyzer or the Sinad meter. Very little generator drift can produce very erroneous results in this test. It is equally important that the receiver is stable to maintain a constant beat note.

Test Procedure—General-Purpose Distortion Analyzer

1. Set up the equipment as shown in Figure 7-3.

2. Set the receiver controls as follows: RF gain to maximum, mode switch to desired sideband (USB or LSB) and squelch control fully open, initially set the volume to maximum.

3. Initially set the function switch of the distortion analyzer to the voltmeter position.

4. Set the signal generator frequency 1 kHz above or below the suppressed carrier frequency to which the receiver is tuned; above for the USB mode and below for the LSB mode.

5. Set the signal generator level to produce a voltage across the audio load resistor which corresponds to 1/2 W (or 50 mW for low-power receivers). Use the formula $E = \sqrt{PR}$ to calculate the power or refer to

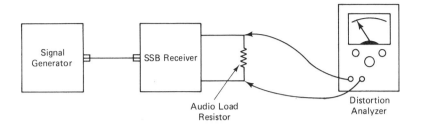

FIGURE 7-3. The test setup used to perform the 10-dB Sinad sensitivity test using a standard distortion analyzer.

the graph in the appendix. For an 8-Ω load the voltage required for 1/2 W is 2 V.

6. Set the distortion analyzer function switch to the set-level position and adjust the set-level control to 0 dB on the dB scale. This is the reference level.

7. Switch the distortion analyzer function switch to "measure distortion" and adjust the null or balance control for maximum null or minimum indication on the meter.

8. If the minimum indication is down 10 dB or more from the 0-dB reference mark, the signal generator level is the 10-dB Sinad sensitivity at 1/2 W audio output (or 50-mW output for low-power receivers).

9. If the null is less than 10 dB down, increase the signal generator level slightly. Switch the distortion analyzer function switch to "set-level" and adjust the receiver volume control to produce the 0-dB indication. This maintains the receiver output at the initial level.

10. Switch the distortion analyzer function switch back to "measure distortion" and retune the null or balance control for minimum indication. It may be necessary to repeat steps 9 and 10 two or three times to find the proper signal generator level required to produce the 10-dB Sinad ratio.

Test Procedure 2—Sinad Meter Method

1. Set up the equipment as shown in Figure 7-4.

2. Set the Sinad meter function switch to measure audio voltage.

3. Set the signal generator frequency exactly 1 kHz above or below the receiver frequency (above for USB mode and below for LSB mode).

4. Set the receiver controls as follows: RF gain to maximum, squelch fully open, and volume to maximum.

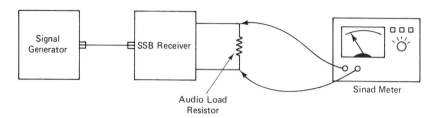

FIGURE 7-4. The test setup used to perform the 10-dB Sinad sensitivity test using a special Sinad meter.

5. Increase the signal generator level to produce 1/2 W output (or other proper reference level). Compute the voltage required across the load resistor from the formula $E = \sqrt{PR}$, or refer to the graph in the appendix.

6. Set the Sinad meter function switch to Sinad and note the reading on the dB scale. Carefully fine-tune the signal generator (or the receiver) for minimum indication on the Sinad meter. If the minimum indication is at least 10 dB down, use this signal generator level as the 10-dB Sinad sensitivity at the established reference output level.

7. If the minimum indication obtained in step 6 is not at least 10 dB down, increase the signal generator level while watching the meter indication. Set the generator level for an indication of -10 dB. This signal generator level is the 10-dB Sinad sensitivity of the receiver.

"Effective" Sensitivity Test

The "effective" sensitivity test procedure is the same as the test procedure described under the "effective" sensitivity test procedure for AM receivers described in Chapter 5. However, it is important to note that the actual sensitivity test procedure noted in steps 1, 3, and 5 of that procedure must be performed by one of the methods just described for SSB receiver sensitivity. Either the 10-dB $S+N/N$ test or the 10-dB Sinad test can be used, as long as the same method is used throughout. Aside from this, the overall procedure is the same.

Sensitivity Bandwidth

For details, refer to the sensitivity bandwidth test described in Chapter 5 for AM receivers. The only difference is that the sensitivity tests must be conducted in accordance with the procedures described in this chapter for SSB receivers.

SQUELCH TESTS

The setup shown in Figure 7-2 is used for making both the squelch tests described here. While the squelch tests can be performed without the audio load resistor and voltmeter—that is, by the ear alone—better accuracy can be obtained from the voltmeter indication.

Critical Squelch Threshold Test Procedure

1. Set the receiver controls as follows: RF gain to maximum and volume to approximately mid-range.
2. With no signal input to the receiver, adjust the squelch control until the receiver is completely quiet. Do not advance the squelch control past the critical point.
3. Tune the signal generator to the receiver at a frequency that will produce a 1-kHz tone in the audio output. (For the USB mode, this will be 1 kHz above the suppressed carrier, and for the LSB mode this will be 1 kHz below the suppressed carrier.)
4. Increase the signal generator output level slowly until the voltmeter indicates an audio output signal. Note the signal generator level.
5. While observing the voltmeter indication, adjust the squelch control toward the fully unsquelched position. The audio voltage should change no more than 1 or 2 dB as the squelch control is moved to the fully unsquelched position. If the voltmeter indicates a larger change than 1 or 2 dB, the test sequence should be repeated and the signal generator should be set to a slightly higher output level in step 4 this time. The sequence should be repeated with increasingly higher signal generator levels until the audio voltage changes no more than 1 or 2 dB as the squelch control is varied in step 5. When this point is reached, the squelch threshold sensitivity is equal to the signal generator level setting.

Maximum Squelch Threshold Test Procedure

The maximum or tight squelch sensitivity test is performed in the same manner as the threshold squelch test except that the squelch control is adjusted to the maximum or tight squelch point.

Squelch Range

The squelch range is defined as the critical-to-maximum squelch threshold levels. For example, if the critical squelch threshold is found to be 0.5 μV and the maximum squelch threshold is 100 μV the squelch *range* is: 0.5 μV − 100 μV.

SELECTIVITY TESTS

While all of the following tests are not necessarily considered to be selectivity tests per se, they are nevertheless related to and/or dependent upon the selectivity of the receiver or certain stages thereof. For this reason, they are all grouped under the general heading of selectivity.

Desensitization/Blocking Test

Test Procedure

1. Set up the equipment as shown in Figure 7-5.
2. Set up generator *A* to produce a 1-kHz beat note on the desired channel (USB or LSB) and set the generator level to produce a 10-dB $S+N/N$ ratio at the receiver output.
3. Adjust the volume control to produce the standard reference output. This will serve as the 0-dB reference.
4. Set generator *B* to a frequency 10 kHz away from the desired channel frequency.
5. Without modulating generator *B*, increase the output level until the receiver's audio output level drops 3 dB.
6. The difference in the output levels (in dB) of generator *A* and generator *B* is the desense or blocking immunity of the receiver to the 10-kHz, off-resonance signal.

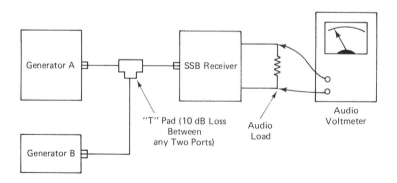

FIGURE 7-5. A test setup used to perform the desensitization or blocking test.

If desired, this test could be performed at several different frequencies above and below the desired frequency in order to determine the receiver's blocking or desense immunity over a range of frequencies.

Opposite Sideband Rejection

When an SSB receiver is tuned to the LSB mode, the USB signals (as referenced to the same suppressed carrier frequency) should be greatly attenuated. Conversely, when the receiver is in the USB mode the LSB signals should be greatly attenuated. The following test procedure can be used to determine the receiver's ability to reject the undesired opposite sideband.

Test Procedure

1. Set up the equipment as shown in Figure 7-2.
2. Set up the signal generator to produce a 1-kHz beat note in the desired sideband.
3. Set the signal generator output level for a 10-dB $S+N/N$ ratio audio output signal.
4. Adjust the receiver's volume control to produce the standard output level. If the standard output level can't be reached, increase the signal generator level until the standard output level is reached. Note the signal generator level.
5. Switch the receiver to the opposite sideband mode *or* tune the signal generator 1 kHz on the *opposite* side of the suppressed carrier.
6. Increase the signal generator output level until the standard output level is obtained from the receiver.
7. Note the signal generator level. The difference in this level and the generator level in step 4 or step 3, in dB, is the opposite sideband rejection figure of the receiver.

In performing this test, be aware of the desense factor. If an unusually high rejection figure is obtained from the opposite sideband rejection test, suspect possible desense. The desense figure obtained from the previous test procedure should be much higher than the opposite sideband rejection figure. An alternate procedure that can be used to determine the opposite sideband rejection figure is to use a Sinad meter and measure the amount of degradation caused to the desirable sideband signal by the undesirable opposite sideband signal. The test procedure is as follows:

Test Procedure

1. Set up the equipment as shown in Figure 7-6.

2. Set generator *A* to the desired sideband at a frequency that will produce a 1-kHz beat note in the audio output.

3. Set the output level of generator *A* to produce a 10-dB Sinad signal at the audio output. The level of the audio output should be at least the standard reference level.

4. It may be necessary to retune signal generator *A* slightly for maximum null on the Sinad meter. The point of maximum null should be the 10-db Sinad point. If not, readjust the level of signal generator *A* to make it so.

5. Tune generator *B* to a frequency 400 Hz to the opposite side of the suppressed carrier frequency and increase the output level until the 10-dB Sinad reading is degraded to 6 dB Sinad.

6. The difference in the output level of the two signal generators is the opposite sideband rejection figure.

This test can also be performed with a distortion analyzer. It is also important to note that the signal generator must be very stable in order to stay in the rejection notch of the Sinad meter or distortion analyzer.

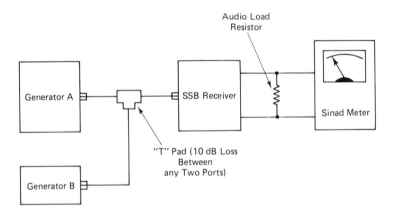

FIGURE 7-6. A basic test setup used to perform many of the interference-type tests described in this chapter.

Image Rejection Figure

Two procedures that can be used to test the image rejection ability of an SSB receiver follow. The first procedure is very similar to the method described in Chapter 5 for testing the image rejection in AM receivers. The second method uses a Sinad meter to indicate a certain amount of degradation caused to the desired signal by the undesired image signal.

Test Procedure 1 (Standard Output Level)

1. Set up the equipment as shown in Figure 7-2.
2. First set the generator to produce a 1-kHz beat note in the audio output by setting the generator to the receiver frequency +1 kHz for the USB mode or to the receiver frequency −1 kHz for the LSB mode.
3. Adjust the generator level for a 10-dB $S+N/N$ ratio in the output signal. Set the volume control to produce the standard reference level. Note the signal generator level.
4. Retune the signal generator to the image frequency *offset* by 1 kHz (−1 kHz for USB and +1 kHz for LSB). In low-side injection receivers, set the generator to:

$$\text{Receiver's tuned frequency} - (2IF +/- 1 \text{ kHz})$$

In high-side-injection receivers, set the generator to:

$$\text{Receiver's tuned frequency} + (2IF +/- 1 \text{ kHz})$$

If the receiver is in the USB mode, offset the generator −1 kHz; for the LSB mode, offset the generator +1 kHz.

5. With the generator now set to the proper image frequency, increase the generator output level until the standard reference output is indicated on the output meter. Note the signal generator level.
6. The difference in the signal generator level setting in steps 3 and 5 is the image rejection figure of the receiver.

Test Procedure 2 (10-dB to 6-dB Sinad Degradation)

1. Set up the equipment as shown in Figure 7-6.
2. Tune generator *A* to the receiver frequency offset by +/− 1 kHz (+1 kHz for USB and −1 kHz for LSB).
3. Adjust generator level for 10-dB Sinad. It may be necessary to fine-tune the signal generator for maximum null on the Sinad meter. Adjust the generator level and fine tuning as necessary to produce the 10-dB Sinad.
4. Turn on generator B and set it to the image frequency offset −400 Hz for the USB mode or +400 Hz for the LSB mode (see step 4 of Test Procedure 1).
5. While observing the Sinad meter, increase the level of generator B to degrade the Sinad reading from 10-dB to 6 dB Sinad.
6. The difference in the output levels (in dB) of the two generators is the image rejection figure of the receiver.

IF Rejection Figure

Test Procedure 1 (Standard Output Level)

1. Set up the equipment as shown in Figure 7-2.
2. Tune the signal generator to the receiver frequency +1 kHz offset for the USB mode or −1 kHz offset for the LSB mode.
3. Set the generator level to produce a 10-dB $S+N/N$ ratio in the audio output and adjust the receiver volume control for the standard output level. Note the signal generator level.
4. Retune the signal generator to the first IF frequency. For high-side-injection receivers, offset the generator frequency +1 kHz for the LSB mode or −1 kHz for the USB mode. For low-side-injection receivers, offset the generator frequency +1 kHz for the USB mode or −1 kHz for the LSB mode.
5. Increase the signal generator output level until the standard reference output is reached. Note the signal generator level.
6. The difference in the signal generator levels in steps 5 and 3 is the IF rejection figure of the receiver.

Test Procedure 2 (10-dB to 6-dB Sinad Degradation)

1. Set up the equipment as shown in Figure 7-6.
2. Tune generator A to the receiver frequency +1 kHz offset for the USB mode or −1 kHz offset for the LSB mode.
3. Set the generator level and fine tuning for 10-dB Sinad on the Sinad meter. Make certain that the signal generator is tuned for maximum null on the Sinad meter. Adjust the generator fine tuning and output level as necessary to produce the 10-dB Sinad level. The audio output should be at the standard output level at least.
4. Tune generator B to the first IF frequency. For high-side-injection receivers, offset the generator frequency +400 Hz for the LSB mode or −400 Hz for the USB mode. For low-side-injection receivers, offset the generator frequency −400 Hz for the LSB mode or +400 Hz for the USB mode.
5. Increase the output level of generator B until the 10-dB Sinad reading is reduced to 6 dB Sinad.
6. The difference in the output level (in dB) of the two signal generators is the IF rejection figure of the receiver.

Cross-Modulation Rejection Tests

Two basic methods of testing the cross-modulation rejection of an SSB receiver will be described here. One method is to use an AM signal as the interfering or undesired signal. Hence, the test result is an indication of the receiver's ability to reject cross modulation caused by AM signals. This test has merit since AM transceivers and SSB transceivers are often operated in the same frequency band. The other method of testing cross-modulation rejection is to use a two-tone signal as the interfering signal. It is necessary to use a two-tone signal as the interfering signal because a single CW tone signal won't produce cross modulation on the desired signal. The frequency that shows up as cross modulation on the desired signal is the *difference* frequency of the two-tone interfering signal, or in the case of the AM signal it is the difference between the carrier and sidebands (which is the same as the modulating frequency).

Test Procedure 1 (AM Interfering Signal)

1. Set up the equipment as shown in Figure 7-6.
2. Set up the desired signal on generator *A*. Tune the generator to the receiver frequency, offset by $+1$ kHz for the USB mode, or -1 kHz for the LSB mode.
3. Adjust the signal generator output level for 10-dB Sinad. It may be necessary to fine-tune the signal generator for maximum null on the Sinad meter.
4. Turn on generator *B* and tune it to 50 kHz above or below the receiver frequency. Modulate generator *B* 100% (don't overmodulate) at 400 Hz.
5. Increase the output level from generator *B* until the 10-dB Sinad reading is reduced to 6 dB Sinad.
6. The difference in the output levels of generators *A* and *B* is the cross-modulation rejection figure of the receiver to an AM interfering signal.

The principle of the test just described is to null out the desired 1-kHz single-tone SSB signal. The cross modulation produced by the interfering AM signal produces sidebands 400 Hz above and below the desired signal. These 400-Hz sidebands will show up as a 600-Hz and 1,400-Hz tone in the audio output. Since these will not be nulled out by the Sinad meter, they show up as distortion resulting in degradation of the 10-dB Sinad reading.

Test Procedure 2 (Two-Tone Interfering Signal)

1. Set up the equipment as shown in Figure 7-7.

2. With generators B and C on standby, set up the desired signal on generator A. Tune generator A to the receiver frequency offset $+1$ kHz for the USB mode or -1 kHz for the LSB mode.

3. Adjust the output level of generator A for 10 dB Sinad. It may be necessary to fine-tune the signal generator for maximum null on the Sinad meter.

4. Turn on generator B and tune it to 50 kHz away from the receiver frequency.

5. Turn on generator C and tune it 400 Hz above or below generator B.

6. Increase the output level of generators B and C coherently until the 10-dB Sinad reading is degraded to 6 dB Sinad.

7. The difference in the output level of generator A and generator B (or C) is the *cross-modulation rejection figure* of the receiver to a *two-tone interfering signal*.

Test procedure 2 is very similar to test procedure 1. The difference frequency between generators B and C (400 Hz) shows up as an undesirable sideband of the desired signal. The undesired sidebands are converted to a 600-Hz tone (1 kHz $-$ 400 Hz) and a 1400-Hz tone (1 kHz $+$ 400 Hz) in the receiver. These will not be nulled out and hence will show up as distortion on the Sinad meter resulting in degradation of the 10-dB Sinad reading.

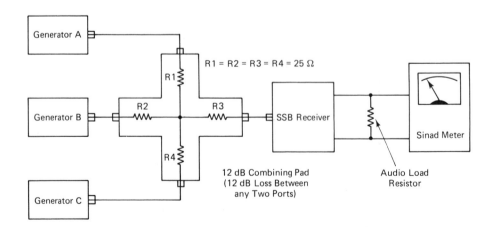

FIGURE 7-7. A test setup used to determine cross-modulation caused by a two-tone interfering signal.

Intermodulation Rejection Tests

Intermodulation can occur between on-channel signals resulting in distortion in the receiver output. Intermodulation also can occur between off-channel signals, resulting in the formation of a third frequency which is equal to either the first IF frequency or the frequency to which the receiver is tuned. This intermod signal will interfere with the reception of the desired signal to a degree dependent upon the strength of the signals that form the intermod and also upon the intermod rejection ability of the receiver.

On-Channel Intermod Probably the simplest way to describe on-channel intermod is through the use of a practical example. Suppose that a CB single sideband receiver is tuned to the lower sideband of channel 12 (27.105 MHz). Further suppose that a transmitter is operating on the lower sideband of channel 12 and is being modulated by a two-tone audio signal, one of the tones being 1,500 Hz and the other at 1,000 Hz. The resulting two-tone LSB signal at the transmitter output and the receiver input will be 27.1035 MHz (the 1,500-Hz tone) and 27.104 MHz (the 1,000-Hz tone). If we let A represent the 27.1035-MHz signal and B represent the 27.104-MHz signal, the intermod relationship can be established as follows:

	(Receiver front end)	*(Receiver output)*
A	27.1035 MHz	1500 Hz
B	27.1040 MHz	1000 Hz
2A-B	27.1030 MHz	2000 Hz
2B-A	27.1045 MHz	500 Hz

These are for the *third-order* intermod products only, usually the most troublesome.

If no intermodulation distortion is produced, only two audio frequencies would be present in the output: 1,500 Hz (representing A) and 1,000 Hz (representing B). The two additional frequencies are the result of third-order intermodulation distortion produced in either the RF or IF section of the receiver or both sections. The third-order intermod products are: 2,000 Hz ($2A - B$) and 500 Hz ($2B - A$).

The following test procedure is found in the EIA publication RS-424, which covers SSB CB transceivers.[1] The purpose of this test is to check the ability of the AGC-controlled RF/IF stages to amplify a strong two-tone SSB signal on the desired channel and desired sideband without producing excessive amplitude intermod signals. The test differs from the conventional

---

[1] EIA Standard RS-424, "Minimum Standards—Citizens Radio Service—SSB Transceivers Operating in the 27 MHz Band" (Electronic Industries Association) pp 4–5, para. no. 5.2, 5.3 This document is available in its entirety from: Electronic Industries Association, 2001 Eye St. N.W., Washington, D.C. 20006, Phone: 202-457-4900.

concept of intermod tests in that both of the test signals produce signals that fall within the passband of the IF sideband filter. This is done so that the intermod rejection characteristic of all the IF/RF AGC-controlled stages combined can be determined.

Test Procedure

1. Set up the equipment as shown in Figure 7-8.
2. Tune generator *A* to the receiver frequency +/− 1,000 Hz (+1,000 Hz for the USB mode or −1,000 Hz for the LSB mode).
3. Tune generator *B* to the receiver frequency +/− 1,600 Hz (+1,600 Hz for the USB mode or −1,600 Hz for the LSB mode).
4. Set the output level of both generators to 158,000 μV.
5. Set the spectrum analyzer to cover the appropriate band of frequencies, depending upon the channel, sideband mode, and IF frequencies involved.
6. Note the level of the highest third order intermod signal. While observing the level of the highest intermod product relative to one of the desired tones, slowly reduce the level of both generators together until the output level of both generators is at the 10-dB $S+N/N$ level. At some point within this range, the difference in the level of the highest level intermod and the reference tone reaches a minimum. At this point of minimum intermod rejection, the difference (in dB) between the intermod signal and the reference tone is called "the AGC intermodulation distortion figure of merit," and according to the EIA's standard for 27-MHz CB transceivers this figure of merit shall be no less than 20 dB; that is, the intermod product at the point of minimum rejection shall be attenuated *at least 20 dB* below the reference tone.

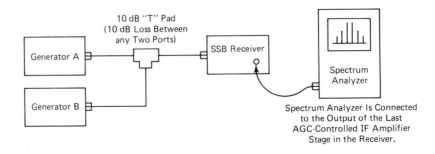

FIGURE 7-8. This test setup is used to determine the amount of intermod produced by two on-channel signals.

Although this standard was developed for CB transceivers, its usefulness is not limited to CB transceivers. The principle will apply to other SSB
receivers in other services and other frequency bands.

Off-Channel Intermodulation Rejection   In channelized communications
systems—that is, communications systems in which the specified operating
frequencies are separated by a fixed frequency spacing—intermod interference is common. This is because the odd-order difference frequencies fall
in-band.

For the sake of illustration, suppose that the frequency band from
14.200 MHz to 14.300 MHz is divided into eleven separate channels with a
uniform frequency spacing of 10 kHz (see Table 7-2). Suppose that two

Channel	Frequency
A - - - - - - - - - -	14.200 MHz
B - - - - - - - - - -	14.210 MHz
C - - - - - - - - - -	14.220 MHz
D - - - - - - - - - -	14.230 MHz
E - - - - - - - - - -	14.240 MHz
F - - - - - - - - - -	14.250 MHz
G - - - - - - - - - -	14.260 MHz
H - - - - - - - - - -	14.270 MHz
I - - - - - - - - - -	14.280 MHz
J - - - - - - - - - -	14.290 MHz
K - - - - - - - - - -	14.300 MHz

TABLE 7-2

transmitters, one on channel E and the other on channel F, are overloading
the front end of a nearby receiver. The overloaded receiver becomes a
nonlinear mixer resulting in the generation of many new sum and difference
frequencies. The even-order frequencies (such as $F - E$, $F + E$, $2F - 2E$,
$2F + 2E$, etc.) will fall far out-of-band. Also, the odd-order sum frequencies
(such as $2F + E$, $2E + F$, $3E + 2F$, $3F + 2E$, etc.) fall far out-of-band. The
out-of-band intermod signals are greatly attenuated by the receiver's selective circuits so that they usually pose no problem. However, the odd-order
difference frequencies (such as $2E - F$, $2F - E$, $3E - 2F$, $3F - 2E$, etc.)
fall in-band and will cause interference to in-band signals. For example, one
of the third-order intermods:

$$2E - F = 2(14.240 \text{ MHz}) - 14.250 \text{ MHz} = 14.230 \text{ MHz}$$

will produce interference on channel D (14.230 MHz). The other third-order
intermod:

$$2F - E = 2(14.250 \text{ MHz}) - 14.240 \text{ MHz} = 14.260 \text{ MHz}$$

will cause interference on channel G. Higher-order intermod products such as fifth, seventh, ninth, etc., can also be produced in a similar manner. For example, one of the fifth-order intermods produced by signals E and F is:

$$3E - 2F = 3(14.240 \text{ MHz}) - 2(14.250 \text{ MHz}) = 14.220 \text{ MHz}$$

which falls on channel C. The other fifth-order intermod is:

$$3F - 2E = 3(14.250 \text{ MHz}) - 2(14.240 \text{ MHz}) = 14.270 \text{ MHz}$$

which falls on channel H. Thus, it is clear that the nonlinear mixing of two (or more) signals in a given band results in odd-order difference intermod signals that fall back into the same band, thus causing interference to neighboring channels within that band.

Intermod can also result when two signals differing in frequency by an amount equal to the first IF frequency of the receiver are sufficiently strong to overload the receiver's front end. In this case, the intermod is an even-order (second order) intermod difference frequency. An example of this is found in the EIA's (Electronic Industries Association) standard publication RS-424.[2] Their basic test procedure is outlined here.

Test Procedure

1. Set up the equipment as shown in Figure 7-5.

2. Set the frequency of generator A offset $+/-$ 1 kHz from the receiver frequency ($+1$ kHz for the USB mode or -1 kHz for the LSB mode).

3. Adjust the output level of generator A to produce a 10-dB $S+N/N$ audio output signal. Note this output level of generator A.

4. Adjust the receiver volume control for an audio output power 6 dB below the rated receiver audio power (one-quarter rated power).

5. Retune generator A below the receiver frequency by an amount equal to one-half the first IF frequency and set the level of generator A to 1,000 μV at the receiver's antenna input. Since the 10-dB "T" pad is between the signal generator and the receiver's antenna terminals, the signal generator output must be 10 dB *above* 1,000 μV or 3,162 μV.

6. Turn on generator B and tune it above the receiver frequency by an amount equal to one-half the first IF frequency and fine-tune generator B to produce the standard 1-kHz beat note in the receiver output.

---

[2] EIA Standard RS-424, Electronic Industries Association, Washington, D.C., 1975, p. 7, para. 10.2 (see footnote 1 also).

7. Increase the output level of generator B until the audio output reaches the reference level (one-quarter rated audio power). Note the level of generator B.

8. The difference (in dB) between the level of generator B (step 7) and the level of generator A (step 3) is the EIA's definition of RF intermodulation rejection figure for 27-MHz SSB CB receivers.

If the receiver under test has a first IF of 10.7 MHz, generator A is tuned $10.7/2 = 5.35$ MHz *below* the receiver frequency. Generator B is tuned 5.35 MHz *above* the receiver frequency. Then the difference between the two generator frequencies is 10.7 MHz, the first IF frequency. Now, in order to produce the standard beat note (1,000 Hz) in the receiver audio output, generator B is tuned to an offset of $+/-1$ kHz, depending upon the sideband mode and the receiver's first LO frequency. If the receiver's first LO frequency is above the receiver's resonant frequency, generator B is offset -1 kHz for the USB mode or $+1$ kHz for the LSB mode. If the receiver's first LO frequency is below the receiver's resonant frequency, generator B is offset $+1$ kHz for the USB mode or -1 kHz for the LSB mode.

Another common test procedure that can be used to determine the receiver's immunity to intermodulation is to use two signal generators tuned to two different in-band channels and the receiver tuned to one of the third-order intermod signals. The basic procedure is to tune one of the signal generators to the channel next to the receiver's resonant frequency and the other generator is tuned to two channels away from the receiver's resonant frequency. It doesn't make any difference whether the generators are tuned above or below the receiver's resonant frequency as long as both generators are tuned to the *same side* of the receiver's resonant frequency. For example: If the receiver is designed to operate over the frequency band shown in Table 7-1, one of the signal generators could be tuned to channel F (14.250 MHz) and the other generator tuned to channel G (14.260 MHz). Then one of the third-order intermod signals would fall on channel E and the other third-order intermod would fall on channel H.

There are basically two ways in which the test can be performed. The first method uses the 10-dB $S+N/N$ sensitivity level as the reference. Two generators are used to produce the intermod signal, which is adjusted to produce the same output as the 10-dB $S+N/N$ reference signal. Then the generator levels are compared to the generator level required to produce the 10-dB $S+N/N$ signal. The procedure is summarized below.

Test Procedure

1. Set up the equipment as shown in Figure 7-5.

2. Set generator B to standby.

3. Set generator A to the 10-dB $S+N/N$ level of the receiver and adjust the receiver volume control to produce a reference output (one-half rated power or so). Note the level of generator A.

4. Tune generator A to one of the channels adjacent to the receiver's desired channel.

5. Tune generator B to a channel two channels removed from the desired channel and on the same side of the desired channel as generator A.

6. In order for the resulting intermod signal to produce the reference 1 kHz beat note in the receiver's audio output, the frequency of generator B must be offset $+/-$ 1 kHz. The direction of offset $(+/-)$ depends upon the sideband mode in which the receiver is operating. If the receiver is operating in the USB mode, generator B must be set to an offset in the negative direction for the intermod to fall in the USB. Conversely, if the receiver is in the LSB mode, generator B must be set to an offset in the positive direction for the intermod to fall in the LSB.

7. With both generators set to the proper frequency, the two are coherently increased in amplitude until the reference audio output is reached.

8. At this point, the difference (in dB) between the output level of generator A in step 3 and the final level setting of generator A (or generator B) in step 6 is the intermod rejection figure of the receiver for this test method.

An alternate procedure is to use *three* signal generators, one of which provides a signal on the desired channel. The desired signal is set to produce a 1-kHz, 10-dB Sinad signal as measured on a distortion analyzer or a special Sinad meter. The other two signal generators are then set to produce an intermod signal that falls on the desired channel and thus degrades the reception of the desired signal. The basic test procedure is summarized below.

Test Procedure

1. Set up the equipment as shown in Figure 7-7.

2. Tune generator A to the receiver's resonant frequency offset by $+$ 1,000 Hz for the USB mode or $-1,000$ Hz for the LSB mode. Fine-tune generator A to produce maximum null on the Sinad meter.

3. Set the output level of generator A to produce a 10-dB Sinad audio output signal.

4. Tune generator B to one of the channels adjacent to the receiver's resonant frequency.

5. Tune generator C to a frequency two channels removed from the receiver's resonant frequency and on the same side of the receiver frequency as generator B. Fine-tune generator C to an offset of

approximately 400 Hz. If the receiver is being tested on the LSB mode, generator C is offset in the positive direction. If the receiver is being tested on the USB mode, generator C is offset in the negative direction.

6. Increase the output of generators B and C coherently until the Sinad meter indicates a degradation of the desired signal. At this point, generator C is fine tuned to produce maximum degradation of the Sinad reading.

7. The output level of generators B and C are then coherently adjusted until the Sinad reading is degraded from 10 dB Sinad to 6 dB Sinad.

8. At this point, the difference (in dB) between the output level of generator A and generator B (or generator C) is the receiver's intermod rejection figure.

AGC TESTS

AGC Threshold

In most receivers, the AGC circuit doesn't become active until the RF input voltage reaches a certain level. The point at which the AGC circuit becomes active is called the *AGC threshold.* The AGC threshold can be determined in the following manner.

Test Procedure

1. Set up the equipment as shown in Figure 7-9.

2. Set the generator to the receiver's resonant frequency offset by 1 kHz ($+1$ kHz for the USB mode or -1 kHz for the LSB mode).

3. With the generator on standby, measure the static (no signal) AGC voltage.

4. Turn on the generator and increase the output level until the AGC voltage just begins to change.

5. At this point, the output level of the signal generator is the AGC threshold.

AGC Effectiveness Test

The purpose of this test is to determine how well the AGC circuit maintains a fairly uniform audio output under widely varying input signal conditions. If desired, a graph of input level versus output level change (in dB) can be made.

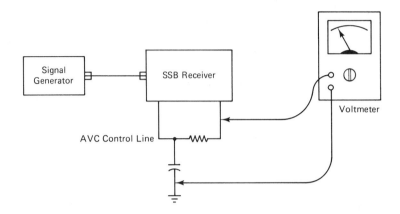

FIGURE 7-9. This test setup is used in performing the AGC tests.

Test Procedure

1. Set up the equipment as shown in Figure 7-2.
2. Set the generator output to approximately 1 μV and the frequency to the receiver's resonant frequency offset 1 kHz (+1 kHz for the USB mode or −1 kHz for the LSB mode).
3. Adjust the receiver's volume control to produce a reference reading of approximately one-quarter rated audio power. This is the 0-dB reference level. For convenience in working with the dB units, the audio output meter should have a dB scale.
4. Gradually increase the generator output level up to the desired maximum level, typically 50,000 μV to 100,000 μV or higher. As the generator output level is varied, the audio level changes are noted. As mentioned, a graph can be made for a record if desired. The amount of change of audio output level is a measure of the AGC effectiveness. A large change in the audio output indicates a poor AGC effectiveness, while a small change in the audio output indicates good AGC effectiveness.

DETERMINING THE RECEIVER'S RESONANT FREQUENCY

The resonant frequency of any receiver can be determined by using a "sniffer" coil in conjunction with a frequency counter or other frequency measuring instrument. The counter or other instrument is very loosely coupled to the first LO through the sniffer coil and a frequency measurement is taken (see Figure 7-10).

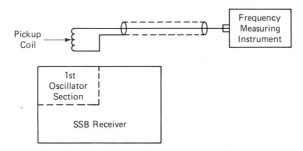

FIGURE 7-10. The first oscillator frequency can be measured by placing a "sniffer" coil near the first oscillator in the receiver. The receiver frequency can then be determined by *adding* or *subtracting* the first IF frequency from the oscillator frequency.

Once the LO frequency is known, the receiver's resonant frequency can be determined by:

$$Fr = Flo - \text{IF (for high-side injection)}$$

or

$$Fr = Flo + \text{IF (for low-side injection)}$$

In an SSB receiver, it is important that all of the oscillators in the receiver be on frequency. This is necessary to ensure that the proper beat note is produced in the receiver output. It is possible that a proper beat note can be produced in a receiver in which two (or more) oscillators are off frequency but one is off in a direction that compensates for the other. However, this is not desirable because, while the proper beat note may be obtained in the end (receiver output), the IF probably will not be located on the proper point of the response curve of the bandpass filter. This can cause both improper frequency response in the desired sideband mode *and* insufficient rejection of the unwanted sideband.

For best results, the BFO and LO frequencies should be set on frequency beginning with the BFO and working toward the front end. The oscillator frequency should be checked on both the USB and LSB modes if the oscillator frequency *changes* between the two modes.

If all the oscillators except the first LO are *known* to be on frequency the first LO frequency can be set by one of the following procedures.

Test Procedure 1

1. Set up the equipment as shown in Figure 7-11.
2. Tune the generator to the desired receiver frequency offset $+1$ kHz for the USB mode or -1 kHz for the LSB mode.

3. Set the generator output to a moderate level: 50 to 100 μV or so.

4. Set the oscilloscope to *external* sweep.

5. Adjust the receiver's volume control and/or the scope's vertical gain control for proper vertical deflection. Adjust the tone generator level control and/or the scope's horizontal gain control for proper horizontal deflection. (Tone generator frequency is 1 kHz.)

6. Tune the receiver's first LO for a stationary Lissajous pattern.

Although a 1-kHz tone was used as the reference in the previous test any audio frequency can be used as long as the amount of generator offset *exactly equals* the tone generator frequency.

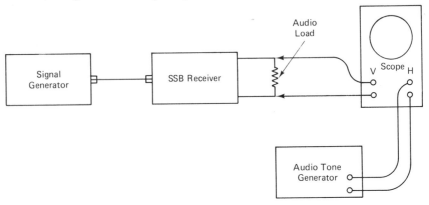

FIGURE 7-11. This method can be used to determine if the receiver is tuned to the proper frequency. If the signal generator is offset from the receiver's resonant frequency by an amount equal to the frequency of the audio tone generator, a stationary (or nearly so) Lissajous pattern will appear on the scope *if* the receiver is tuned to the proper frequency.

Test Procedure 2

1. Set up the equipment as shown in Figure 7-12.

2. Set the audio generator frequency to 1,500 Hz or so (not critical).

3. Set the RF signal generator to external modulation and adjust the generator's modulation control and/or the audio generator's level control for a high level modulation (90% or so).

4. Tune the RF generator to the desired receiver frequency and set to moderate output level (50 to 100 μV).

5. Adjust the receiver's volume control and/or the scope's vertical gain control for proper vertical deflection. Adjust the scope's horizontal gain control for proper horizontal deflection.

6. Tune the receiver's first LO for a stationary Lissajous pattern on the scope.

7. Without changing any other controls, switch the receiver to the other sideband mode. The Lissajous pattern should remain the same.

In test procedure 2, the audio-tone generator is used to amplitude-modulate the signal generator. This produces an upper and lower sideband separated from the carrier by an amount equal to the audio generator frequency. In this manner, both the USB and LSB modes can be tested without changing the frequency of the signal generator. In step 7, if the Lissajous pattern changes a great deal between the two sideband modes, this indicates that one of the other oscillators is operating off frequency.

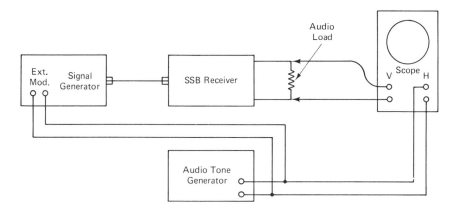

FIGURE 7-12. In this test setup, the signal generator is amplitude-modulated by the audio-tone generator. In this manner, the signal generator frequency must not be offset. If the receiver is tuned to the frequency of the signal generator, a stationary (or nearly so) Lissajous pattern will appear on the scope. Both sideband modes can be checked in this manner without retuning the signal generator.

Test Procedure 3

1. Set up the equipment as shown in Figure 7-13.

2. Tune the signal generator to the desired receiver frequency, offset $+1$ kHz for the USB or -1 kHz for the LSB.

3. Set the RF signal generator output to a moderate level (50 to 100 μV).

4. Tune the receiver's first LO for *exactly* 1 kHz on the frequency counter. The opposite sideband also should be checked by tuning the generator to a 1-kHz offset in the other direction. It should not be necessary to readjust the receiver's first LO to get the 1-kHz audio beat note.

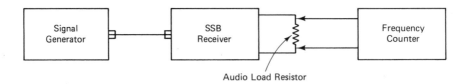

FIGURE 7-13. This test setup can be used to determine whether or not the receiver's oscillators are properly tuned.

NOISE BLANKER EFFECTIVENESS TEST

The purpose of this test is to determine the effectiveness of a receiver's noise blanker. The noise pulse generator used in this test is the standard EIA noise pulse generator as described in the EIA standard publication RS-424.[3]

The noise pulse generator produces 100 pps (pulses per second) with a pulse width of 1 μs and rise and fall times of less than 10 nanoseconds (ns) (10^{-9} seconds). This pulse signal will produce interference over a broad RF range.

Test Procedure

1. Set up the equipment as shown in Figure 7-14.
2. Set the signal generator to the receiver frequency, offset +1 kHz for the USB mode or −1 kHz for the LSB mode.
3. With the receiver's noise blanker off, set the signal generator output level to produce a 10-dB $S+N/N$ audio output signal from the receiver. Note the signal generator output level.

---

[3] EIA Standard #RS-424, Electronic Industries Association, Washington, D.C., 1975, p. 10, para. 14.2 (see footnote 1 also).

4. Set the receiver's volume control to produce a reference audio output of approximately one-quarter rated receiver audio power.

5. Remove the signal from the signal generator and turn on the noise pulse generator.

6. Increase the output of the noise pulse generator until the receiver's audio output level reaches the reference level.

7. Turn on the noise blanker and again set the signal generator output level to produce the 10-dB $S+N/N$ level at the receiver output. Note the signal generator output level.

8. The difference in the generator-level setting (in dB) between steps 3 and 7 is a measure of the effectiveness of the noise blanker of the receiver. A higher dB figure indicates less effectiveness of the noise blanker circuit.

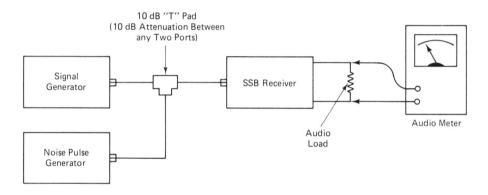

FIGURE 7-14. This test setup can be used to determine the effectiveness of the receiver's noise blanker circuit.

S-METER TEST AND CALIBRATION

The calibration of the S-meter varies from one manufacturer to another. On some receivers, the S-9 point corresponds to a signal input of 50 μV. Other manufacturers may use 100 μV as the S-9 calibration point. You may find still other values of input signal used to calibrate the S-9 reading. Another important factor is the way the S-meter tracks with various input signal levels. Although it isn't "carved in stone," it is generally understood that a change of 1 S-unit indicates a change of approximately 6 dB in the input signal level. The following test procedures can be used to determine the S-meter characteristic of a receiver.

Test Procedure 1—S-9 Calibration

1. Set up the equipment as shown in Figure 7-15.
2. If the receiver is equipped with an RF gain control, set it to maximum gain.
3. Tune the signal generator to the receiver frequency offset by approximately 1,000 Hz to 1,500 Hz (offset in the positive direction for the USB mode or negative direction for the LSB mode).
4. Set the signal generator output level to produce an S-9 reading on the S-meter.
5. At this point, the generator output level is the input signal level corresponding to an S-9 reading.
6. If the input level in step 5 differs from the manufacturer's specified S-9 level, the generator should be set to the specified S-9 input level and then the S-meter calibration control adjusted to produce an S-9 reading on the S-meter.

FIGURE 7-15. This simple test setup is used to perform tests on the receiver's S-meter and/or to calibrate the S-meter.

Test Procedure 2—Tracking Test

Note: The first three steps are the same as the previous test procedure.

4. Set the signal generator output level to produce an S-1 reading on the S-meter. Note the generator output level.
5. Increase the generator output level until an S-2 reading is obtained on the S-meter. Note the generator output level.
6. Continue in this manner until all S-units from S-1 to S-9 are covered and compare the generator output levels required to produce the various S-units.

If the above test showed that the S-meter tracking is not linear, it may be desirable to graph the response of the S-meter for reference purposes, especially if the receiver is to be used in any tests requiring a calibrated S-meter.

AUDIO FREQUENCY RESPONSE

The audio frequency response of an SSB receiver is dependent to a large degree upon the bandpass of the IF bandpass filter. For this reason, the receiver should be tested for overall frequency response.

Test Procedure

1. Set up the equipment as shown in Figure 7-2.
2. Tune the signal generator to the receiver frequency, offset by $+1$ kHz for the USB mode or -1 kHz for the LSB mode.
3. Set the signal generator to a moderate output level (500 μV or so).
4. Adjust the receiver's volume control to produce a reference reading on the audio meter (one-quarter rated audio power or so).
5. Vary the generator offset frequency from 0 to 5,000 Hz and note the audio output level at the various frequencies throughout the range.
6. Using the reference level in step 4 as the 0-dB point, plot a graph of frequency versus change-in-output (in dB).

BIBLIOGRAPHY

American Radio Relay League, Newington, Connecticut. *Single Sideband for the Radio Amateur.* ©1970 American Radio Relay League.

Collins Radio Company, *Amateur Single Sideband,* (1st ed., 2nd printing). Greenville, N.H.: The Ham Radio Publishing Group, 1977.

Electronic Industries Association, Standard #RS-424, *Minimum Standards, Citizens Radio Service Operating in the 27 MHz Band.* Washington, D.C., 1975.

Hoonton, Harry D., *Single Sideband Theory and Practice.* New Augusta, Ind.: Editors & Engineers, 1967.

Miller, Gary M., *Handbook of Electronic Communications.* Englewood Cliffs, N.J.: Prentice-Hall, Inc., 1979.

Pappenfus, E. W.; Bruene, Warren B.; and Schoenike, E. O. *Single Sideband Principles and Circuits.* New York: McGraw-Hill, 1964.

8

FM TRANSMITTER
TESTS AND
MEASUREMENTS

Although it is assumed that the reader already possesses a working knowledge of FM transmitter principles and circuitry, a brief review is presented here for reference purposes.

The object of an FM transmitter is to convert the audio input signal (from the microphone) to a frequency-modulated RF output signal at a specified frequency. Basically, there are two different methods used to produce the frequency-modulated signal. One process is called the *direct method.* In the direct method, the frequency of the oscillator is varied directly by the audio modulating signal, usually by applying the audio signal across a varactor that is part of the frequency determining circuit of the oscillator. A simplified block diagram of a direct FM transmitter is shown in Figure 8-1. This is typical of commercial FM transmitters in current use. The microphone audio signal is fed to a *pre-emphasis network,* which emphasizes or favors the higher audio frequencies. The purpose of pre-emphasis is to compensate for poorer signal-to-noise ratio at the higher audio modulating frequencies. More about pre-emphasis later. The pre-emphasized signal is amplified and then passed through a peak limiter or *clipper* which limits or clips the peaks of the audio signal. This clipping process creates audio harmonic distortion which is filtered by the "splatter" filter. The audio signal is then applied to the varactor of the oscillator at a level determined by the setting of the peak deviation control. This control limits the amount of frequency modulation caused by the audio signal. The oscillator operates at a frequency much lower than the transmitter output frequency. The frequency multiplier multiplies the oscillator frequency to reach the operating frequency of the transmitter. The amount of deviation or modulation is also multiplied. The output of the

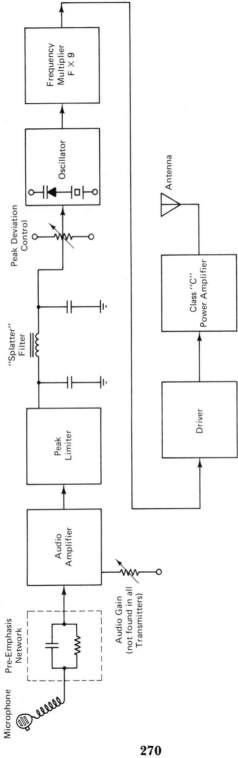

FIGURE 8-1. A simplified block diagram of an FM transmitter of the direct FM type.

multiplier is then fed to an RF driver amplifier and then to the final power amplifier. Since the signal is frequency-modulated rather than amplitude-modulated, a class C RF power amplifier can be used resulting in better efficiency.

The other method of producing an FM signal is the *indirect method.* The indirect method employs the phase-modulation process, The process of phase modulation produces frequency modulation. However, there are certain important differences between frequency and phase modulation which must be recognized. The most important are: (1) with true FM the deviation or modulation is proportional to the amplitude of the modulating signal and independent of the modulating frequency, (2) with true phase modulation the deviation is proportional to both the amplitude and frequency of the modulating signal.

An indirect or phase-modulation type of FM transmitter is shown in Figure 8-2. Notice that a pre-emphasis circuit is shown here also, even though in phase modulation high frequency emphasis occurs naturally. To offset this, a de-emphasis circuit is added following the peak limiter or clipper circuit. The de-emphasis circuit functions just the opposite of the pre-emphasis circuit; that is, the higher frequencies are attenuated more than the lower frequencies. A natural question is: Why use a pre-emphasis circuit and de-emphasis circuit in the same signal processing chain? The answer lies in the fact that the pre-emphasis in conjunction with the peak clipper circuit produces a better overall response to a wide range of audio input signals thus allowing higher average modulation without exceeding the limit. The de-emphasis circuit then basically is used to compensate for the natural emphasis in the phase modulator.

Notice also that in the phase-modulation type of transmitter the audio modulating signal is not applied directly to the oscillator but rather to a buffer stage. Since phase modulation produces less deviation than direct FM, more frequency multiplication is necessary. From the frequency multiplier on to the transmitter output, the phase-modulation and frequency-modulation type of FM transmitters are identical.

POWER MEASUREMENT

The RF power output of an FM transmitter can be measured with an average-reading type wattmeter. The amplitude of the RF output signal will not change with modulation, so the wattmeter reading should not be affected at all from full modulation to zero modulation. RF power measurements for FM transmitters follow quite closely the procedures outlined for measuring RF output of AM transmitters as described in Chapter 4. Just remember that it doesn't matter if the transmitter is modulated or unmodulated, the power measurement will be the same. For the same reason, peak-power measurement has no place in FM transmitter work.

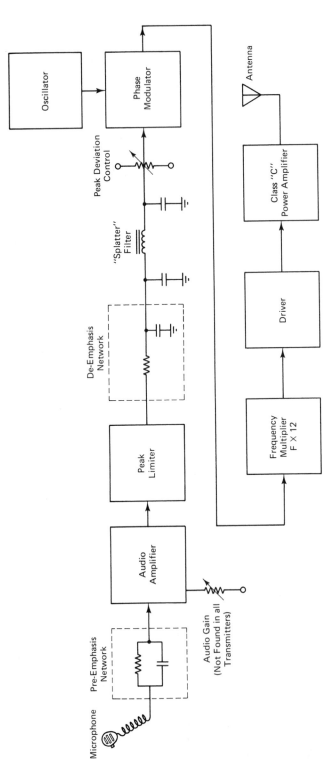

FIGURE 8-2. A simplified block diagram of an FM transmitter of the indirect FM (phase modulation) type.

Dc power input and efficiency measurement or calculation is also the same as described in Chapter 4 for AM transmitters. Almost without exception, FM communications equipment manufacturers provide test sockets in their transceivers to facilitate testing. They also sell test sets that can be plugged into these sockets to check the various stages of the transmitter. The collector (or plate) voltage and current of the "final" stage can be determined by switching the test set to the proper position. The dc power input can then be determined by multiplying the voltage and current values. Figure 8-3 shows a typical arrangement to provide the dc collector current indication on the test set meter. The voltage dropping resistor, R1, is very small, on the order of 0.1 Ω or less. Such a small resistor drops a very small voltage, and thus its presence in the circuit has a negligible effect on the operation of the "final" stage. The small voltage drop across R1 is used to give an indication on the microammeter of the test set. The multiplier resistor, R2, limits the current flow to the range of the microammeter. Usually the meter reading is multiplied by a certain factor to get the collector current in amperes.

The power output levelling test described in Chapter 4 also applies to multichannel FM transmitters, especially when the frequency separation between the upper and lower channel is large.

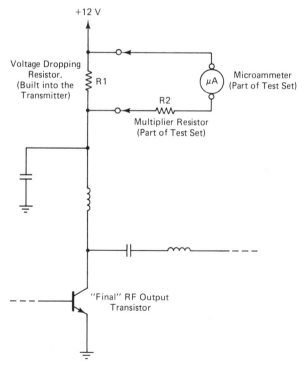

FIGURE 8-3. Simplified schematic showing the metering method used to determine the dc input power to the "final" stage of a transmitter.

MODULATION OR DEVIATION TESTS

In FM communications terminology, modulation is referred to as *deviation,* deviation meaning the peak change in frequency above or below the center frequency. In narrow band FM communications transmitters, the peak deviation is limited to $+/-5$ kHz. This means that the instantaneous frequency of the transmitter is not permitted to exceed 5 kHz above or below the carrier frequency. The following tests and measurements are all related to checking the performance of the sections of the transmitter which are related to modulation or deviation.

Peak Deviation Measurement As mentioned, the amount of peak deviation is the maximum instantaneous frequency change in either direction, above or below the center or carrier frequency. Ideally, the peak deviation above the carrier frequency (positive direction) should equal the peak deviation below the carrier frequency (negative direction). However, in the practical world, slight imperfections often will cause the positive and negative deviation to differ slightly. This is called modulation or deviation *dissymmetry.* When the transmitter deviation is checked, it should be checked on both the positive and negative sides. The higher of the two measurements is taken as the peak deviation of the transmitter.

The peak deviation should be measured with an audio input signal large enough to ensure full limiting. Dedicated deviation monitors such as those described in Chapter 3 can be used to measure the deviation. These deviation monitors usually provide a "scope" output so that the waveform can be seen. The scope can then be calibrated to read the deviation directly. The dynamic response of the scope is far superior to any meter. The shape of the waveform as seen on the scope can point out defects which might never be detected on a meter. Service monitors that are very popular around land mobile radio shops almost always have a built-in scope that doubles as the deviation indicator.

Test Procedure

1. Set up the equipment as shown in Figure 8-4.
2. Set the audio frequency generator to approximately 1 kHz.
3. Key the transmitter and increase the audio generator level until the meter reading or scope waveform levels off. Increase the audio generator still more to ensure proper clipping. The deviation meter reading should not change much and the waveform on the scope should look like Figure 8-5.
4. The peak deviation is now indicated by the deviation meter or scope (if the scope is calibrated).

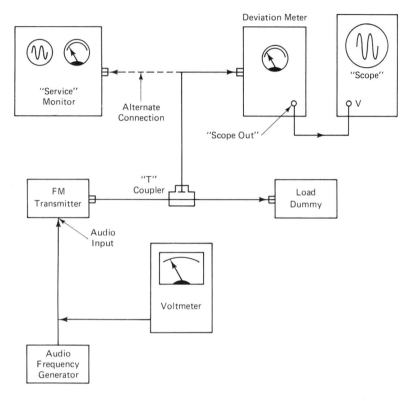

FIGURE 8-4. A typical setup used to perform deviation (modulation) tests on a FM transmitter.

The deviation should be checked on both sides of the carrier, the largest reading being taken as the peak deviation. In some deviation meters, the scope waveforms will show both positive and negative deviation. This is true of service monitors that use a discriminator as the detector.

Percent Deviation Dissymmetry Deviation dissymmetry in percent can be determined by measuring both the positive and negative peak deviation and then applying the following formula:

$$\% \text{ dissymmetry} = \frac{D1 - D2}{D1} \times 100$$

where D1 = larger deviation (kHz) and D2 = smaller deviation (kHz). For example, suppose the positive deviation is 4.7 kHz and the negative deviation is 4.9 kHz. The percent of dissymmetry is then determined from the formula as follows:

$$\% \text{ dissymmetry} = \frac{4.9 - 4.7}{4.9} \times 100 = 4.08\%$$

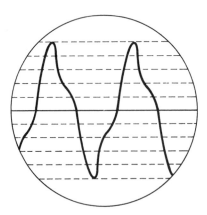

FIGURE 8-5. The waveform shown here is an audio signal as recovered by a deviation meter. Notice the departure from a true sinewave. This indicates that the audio modulating signal was driven well beyond the clipping threshold.

Modulation Sensitivity　The peak deviation of an FM communications transmitter must be set or measured using an audio input test signal of sufficient amplitude to exceed the clipping level of the clipper circuit. This ensures that higher level signals will not cause the modulation or deviation to exceed the allowable limits. In Figure 8-1 or 8-2, notice that the peak deviation control is located at the end of the speech signal-processing chain. Since it is this control that sets the peak modulation or deviation, it is imperative that this control be adjusted only with a fully clipped audio signal. The manufacturer's technical manual usually specifies the audio signal level necessary to produce a fully clipped audio signal with which the peak deviation is set. The audio frequency usually used is 1 kHz. Also, the manufacturer usually specifies the audio signal level required to produce 60% peak deviation.

Test Procedure

1. The equipment setup shown in Figure 8-4 is used for this test.
2. Set the audio generator to 1 kHz.
3. Key the transmitter and increase the audio generator level until the peak deviation point is reached. The peak deviation point is indicated by a levelling off of the meter reading and a scope waveform shown in Figure 8-5. If the peak deviation is incorrect, the peak deviation control should be adjusted to produce the proper deviation level. Remember to check both the positive and negative deviation and use the higher

measurement. The audio input level at this point can be taken as the full *clipping sensitivity* of the transmitter.

4. Reduce the audio generator level until the deviation meter indicates 60% peak deviation. For 5-kHz deviation systems, the deviation is reduced to 3.0 kHz (60% of 5 kHz).

5. The audio input level at this point is what most manufacturers refer to as the transmitter audio sensitivity, usually expressed in millivolts.

Voice Modulation Tests

Since the entire purpose of the transmitter is to transmit voice signals, some means of gauging the deviation with voice modulation is very helpful in analyzing the performance of the speech processing section of the transmitter. While the peak deviation may be set to a proper level by the peak deviation control, the average deviation may be quite low with a normal voice signal input. On the other hand, the average deviation may be quite high if the operator has a loud voice. Never try to compensate for the operator's voice by resetting the peak deviation control. Besides being illegal, it usually only makes matters worse. Any compensation for the operator's voice must be made at a point prior to the speech clipper or limiter to ensure that overmodulation can't occur. It is the average modulation that determines (to a great extent) the effective range of the transmitter.

Having established the importance of average modulation, let's now consider some of the methods that can be used to gauge the average deviation under voice signal modulation conditions. The degree of movement of the needle of a deviation meter for voice signal modulation will vary considerably from one instrument to another. However, it is entirely possible that once a technician has become familiar with the response of a particular meter he or she may be capable of determining whether or not a voice signal is producing "good" average modulation. Average modulation also can be gauged with some degree of accuracy by observing the scope waveforms. The average modulation can be gauged by the relative number of peaks occurring at or near the peak deviation line.

These methods of gauging the average modulation will vary from one technician to another as well as from one instrument to another.

Modulation Density   The term *modulation density* is not part of the standard terminology of FM radio communications. This term is associated with a particular instrument which has been specifically developed to aid the technician in determining and setting the average modulation of an FM transmitter. The instrument is the Autopeak™ modulation monitor, a product of Helper Instruments Company. The instrument is pictured in Figure 8-6. Notice that the Autopeak™ modulation monitor has two meters. The one on the right indicates the peak deviation, while the one on the left indicates the

FIGURE 8-6. The Autopeak™ modulation monitor from Helper Instruments Company helps to make proper adjustments to the transmitter audio system. The scanner is used to tune in the desired frequency to be monitored. Courtesy of Helper Instruments Company.

average degree of modulation or "modulation density" that occurs under normal voice modulating conditions.

Let's take a closer look now at some of the unique features and details of this modulation monitor. A cost-saving feature of this instrument is that it is used in conjunction with a receiver that is tuned to the frequency of the transmitter to be checked. In Figure 8-6 the instrument is shown along with a popular programmable-type scanner. The Autopeak™ modulation monitor is designed to operate from the low IF (455 kHz or so) of the receiver. The programmable scanner simply provides the means of tuning to a broad range of frequencies, its low IF being coupled to the input of the modulation monitor. Figure 8-7 shows a simplified partial block diagram of the Autopeak™ modulation monitor. The audio from the discriminator passes through a special circuit called an *absolute value circuit*, which leaves the positive alternations unchanged but converts the negative alternations to postive pulses so that the output looks like a full-wave rectified signal. The output of the absolute value circuit feeds both the peak measurement circuit and the average measurement circuit. The special meter "hang" circuits provide a fast-rise and slow-fall characteristic so that meter indications can be read with actual speech modulation. Also, two LED peak flashers (not shown) are used to indicated peak deviations. One flasher is set for 5 kHz and the other for 4.5 kHz. If the transmitter deviation is not symmetrical, the action of the absolute value circuit will produce a higher positive pulse for the higher deviation. Since the peak measurement circuit responds only to the peak of the waveform, the greater deviation value will be indicated on the meter.

The average measurement circuit is the one that indicates modulation density. The time constant of the average measurement circuit is such that the output in conjunction with the meter hang circuit will give an indication of the average value (not to be confused with true average) of the speech signal.

The Autopeak™ also has a built-in 1-kHz tone generator that can be used to modulate the transmitter. The general test procedure for using the Autopeak™ modulation monitor is described below.

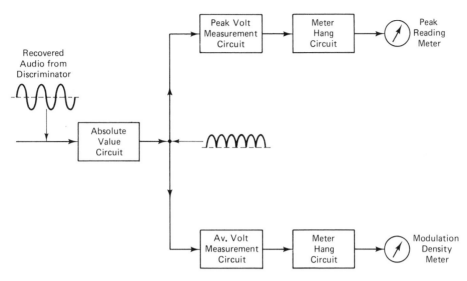

FIGURE 8-7. A partial block diagram of the Autopeak™ modulation monitor from Helper Instruments Company. Courtesy of Helper Instruments Company.

Test Procedure

1. Set up the equipment as shown in Figure 8-8.

2. Key the transmitter and increase the audio tone level to a level well past the clipping point of the transmitter speech circuit.

3. The Autopeak™ should indicate a 5-kHz peak deviation. If not, adjust the peak deviation control to produce 5-kHz deviation.

4. Remove the 1-kHz modulating tone and modulate the transmitter using a normal voice level and proper microphone technique. The modulation density will be indicated by the maximum reading that is repeatedly reached by the meter's pointer.

The modulation density meter is calibrated on a scale of 1 to 10. A reading of 4 or less would indicate insufficient gain preceding the clipper stage. A reading of 9 or more would indicate excessive gain ahead of the clipper. A reading of 6 to 8 units would indicate proper modulation density. Remember: The peak deviation is closely related to the setting of the peak deviation control that follows the clipper, while the modulation density is closely related to the gain (or lack of it) that precedes the clipper. Stated another way, the modulation density is proportional to the degree of clipping.

Many FM communications transmitters do not provide any means of changing the gain in the speech amplifier preceding the clipper. The peak deviation control is often the only control found in the speech section of the

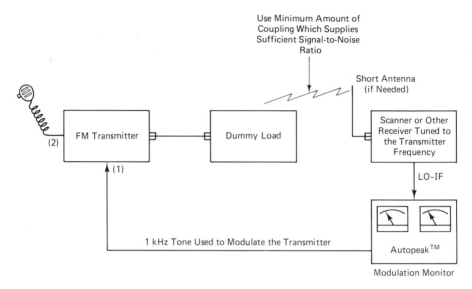

FIGURE 8-8. A typical test setup using the Autopeak™ modulation monitor to (1) set the peak deviation and (2) check the modulation density with voice modulation.

transmitter. Since the peak deviation control cannot (legally) be adjusted beyond the point that produces the maximum specified deviation (5 kHz for narrow band FM) with a properly clipped audio signal, there is no way to change the modulation density without changing something in the speech section preceding the clipper. If no input gain control is provided and the modulation density is too high or too low, some part must be defective and troubleshooting is indicated.

If a transmitter is operated via remote control telephone lines, the audio output of the remote control console must be adjusted properly to produce the proper modulation density. Usually, such remote control consoles have built-in compression amplifiers that increase the average audio level while maintaining the proper peak level. The adjustments within the console must be set for the best modulation density.

Another case is the repeater operation in which the audio output of a receiver directly feeds the audio input to another transmitter. The audio gain of the receiver must be adjusted properly to produce the proper modulation density.

Modulation Bandwith

The modulation bandwidth of an FM transmitter is defined as the bandwidth which contains 99% of the total radiated energy of the transmitter. A rough calculation of the bandwidth can be made from the following formula:

$$\text{Bandwidth (in kHz)} = 2D + 3(Fmax)$$

where D = deviation in kHz and *Fmax* = highest modulating frequency. As an example, suppose that the highest audio modulating frequency is 3 kHz. The bandwidth in this case is roughly equal to:

$$\text{Bandwidth} = 2(5 \text{ kHz}) + 3(3 \text{ kHz}) = 10 \text{ kHz} + 9 \text{ kHz} = 19 \text{ kHz}$$

Figure 8-9 shows an envelope outline for the transmitter sideband spectrum for an FM transmitter with a deviation of 5 kHz. All emissions from the transmitter must fall within the envelope. The 0-dB reference point is the unmodulated carrier level. Notice that at frequencies greater than $+/-$ 25 kHz from the carrier the attenuation is given as $-43 \log P$ or -80 dB, whichever is the smaller. For example, if the transmitter carrier power is 45 W, the attenuation at frequencies more than $+/-$ 25 kHz from the carrier must be at least:

$$\text{Attenuation (dB)} = \log(45) = 43(1.65) = 70.95 \text{ or } 71 \text{ dB}$$

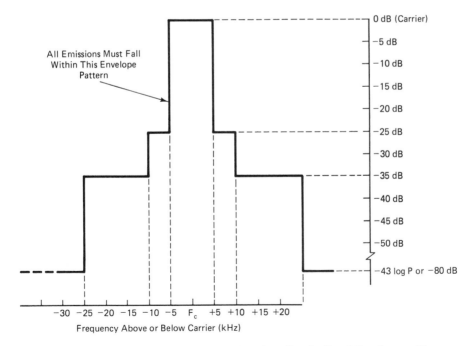

FIGURE 8-9. An envelope outline showing the limits of the transmitter sideband spectrum of a narrow-band FM transmitter in the land-mobile radio service. All transmitter emissions must fall within the envelope outline.

The sideband spectrum of a transmitter can be checked by modulating the transmitter with a relatively high frequency audio signal and observing the display on a spectrum analyzer. The step-by-step procedure is described below.

Test Procedure

1. Hook up the equipment as shown in Figure 8-10. It is important that the spectrum analyzer not be too tightly coupled to the transmitter in order to prevent overloading the front end of the spectrum analyzer.

2. Set the audio generator frequency to approximately 2,500 Hz or so. Key the transmitter and increase the audio generator output level until the deviation meter indicates full system deviation ($+/-$ 5 kHz for narrow-band transmitters).

3. Increase the audio generator level several dB and observe the modulation spectrum on the spectrum analyzer. All components should fall within the limits set by the envelope of Figure 8-9. Don't forget to check the harmonic frequencies.

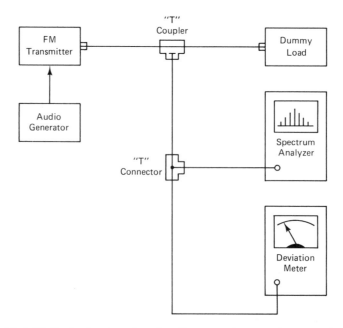

FIGURE 8-10. Typical setup showing the use of a spectrum analyzer to check the transmitter sideband spectrum.

AUDIO FREQUENCY RESPONSE

The audio frequency response of an FM transmitter is a measure of the transmitter deviation versus modulating frequency. The frequencies between 300 Hz and 3,000 Hz are pre-emphasized at a rate of 6 dB/octave. This is illustrated in Figure 8-11. For transmitters operating between 25 MHz and 450 MHz, the frequencies between 3,000 and 15,000 Hz shall be attenuated more than the attenuation at 1 kHz by at least $40 \log(f/3)$ dB, where $f =$ frequency in kHz. For example, the attenuation of a 10 kHz signal must be at least $40 \log(10/3) = 40 \log 3.33 = 21$ dB greater than the attenuation at 1 kHz. Above 15 kHz, the attenuation must be at least 28 dB greater than the attenuation at 1 kHz.

For transmitters operating in the frequency ranges 450 to 512 MHz, 806 to 821 MHz, and 851 to 866 MHz, the frequencies from 3,000 to 20,000 Hz must be attenuated by at least $60 \log(f/3)$ dB, where $f =$ frequency in kHz. At frequencies above 20,000 Hz, the attenuation shall be at least 50 dB greater than the attenuation at 1 kHz.

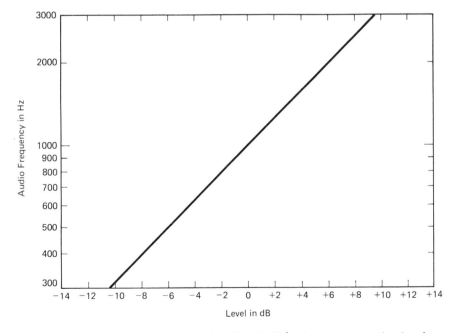

FIGURE 8-11. This graph illustrates the 6 dB/octave pre-emphasis of audio frequencies between 300 Hz and 3,000 Hz. The 0-dB reference point is 1,000 Hz.

In-Band (300 to 3,000 Hz) Audio Response Test

The purpose of this test is to check the transmitter response to normal audio frequencies from 300 to 3,000 Hz.

Test Procedure

1. Hook up the equipment as shown in Figure 8-12.

2. Set the audio generator to 1 kHz.

3. Key the transmitter and adjust the audio generator level to produce a deviation well below the clipping point. A deviation level of 20% to 25% of the peak system deviation is fine. For narrow-band FM transmitters, a deviation level of approximately 1 kHz is fine. Note the audio generator level. This generator level will serve as the 0-dB reference level for the rest of the test procedure.

4. Set the audio generator frequency to 300 Hz. Adjust the audio output level to produce the reference deviation (20 to 25% peak deviation). Note the audio output level required to produce this deviation. If the meter doesn't have a dB scale compute the dB change by using the formula dB = 20 log(E1/E2), where E1 is the 0-dB reference level established in step 3 and E2 is the audio voltage at the frequency being checked.

5. Repeat step 4 for several different frequencies ranging from 300 Hz to 3,000 Hz and plot a graph of frequency versus amplitude. The resulting graph should not deviate much from the graph shown in Figure 8-11.

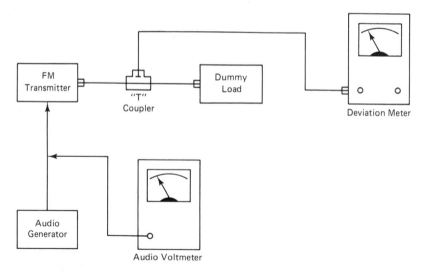

FIGURE 8-12. This setup can be used to check the normal in-band (300 Hz to 3,000 Hz) audio response of the transmitter.

Out-of-Band Response Test (Above 3,000 Hz)

Audio-modulating frequencies above 3,000 Hz must be attenuated in accordance with the specifications found on p. 283. The test procedure for this measurement is necessarily a little different from the procedure for *in-band* audio response testing. The FM deviation meter used in this test must have good response up to the maximum frequency used in this test. Usually, from a practical standpoint, it is unnecessary to go beyond 10 kHz or so. It is also important that the FM deviation meter have an audio output that is not de-emphasized.

Test Procedure[1]

1. Set up the equipment as shown in Figure 8-13.
2. Set the audio generator to 1 kHz.
3. Key the transmitter and adjust the audio generator output level to produce standard deviation (60% peak system deviation or $+/- 3$ kHz for narrow-band FM). Now, increase the audio generator output level by 16 dB (multiply the voltage level by 6.31 for a 16-dB increase). Note this audio input level. It is necessary to maintain this input level at each frequency where it is desired to make a measurement.
4. If the FM deviation meter has a de-emphasis circuit, make sure it is switched out for this test. Measure the audio output level of the FM deviation meter. This will serve as the 0-dB reference level for the remainder of this test.
5. Tune the audio generator to 3,000 Hz. If necessary, readjust the generator output level to maintain the reference input level established in step 3. Note and record the audio output level from the FM deviation meter. (The transmitter must be keyed for this.)
6. Repeat step 5 at several discrete frequencies up to and including the highest frequency desired. The data obtained from this test may be used to plot a graph of frequency versus amplitude. The attenuation of frequencies above 3,000 Hz should be at least as much as the attenuation shown on the graph in Figure 8-14: $40 \log(f/3)$ for frequencies from 25 to 450 MHz or $60 \log(f/3)$ for frequencies above 450 MHz, where f = audio-modulating frequency in kHz.

---

[1] This procedure is based on IEEE Standard 377-1980, "IEEE Recommended Practice for Measurement of Spurious Emission from Land-Mobile Communication Transmitters" p. 17, para. 5.1.2.1 "Audio Characteristic" published by The Institute of Electrical and Electronics Engineers, Inc. 345 E. 47th St. New York, NY 10017, November 14, 1980.

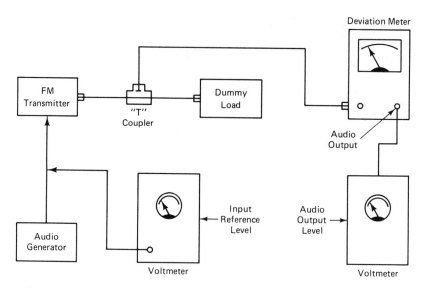

FIGURE 8-13. This setup can be used to check the out-of-band (above 3,000 Hz) audio response of the transmitter.

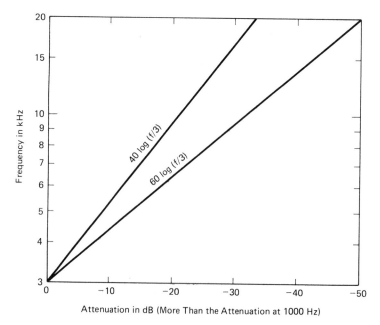

FIGURE 8-14. The transmitter audio response at frequencies above 3,000 Hz must comply with one of the two graphs shown here, depending upon the transmitter operating frequency. (See text.)

FM HUM AND NOISE LEVEL

The FM hum and noise level measurement is a measurement of the relative amount of hum and/or noise modulation that exists in the absence of an audio signal at the transmitter input.

Test Procedure

1. Set the test equipment up as shown in Figure 8-15. The scope and/or headset provide a means of determining the type of hum or noise being measured on the audio meter.
2. Set the audio generator to 1 kHz. Key the transmitter and adjust the audio generator level to produce a deviation of 60% peak system deviation ($+/-$ 3 kHz for narrow-band FM).
3. Note the audio output level from the deviation meter's audio output terminals. This will serve as the 0-dB reference level.
4. Remove the audio generator from the transmitter input and connect a resistor across the transmitter audio input. The resistance value should match the transmitter audio input impedance.

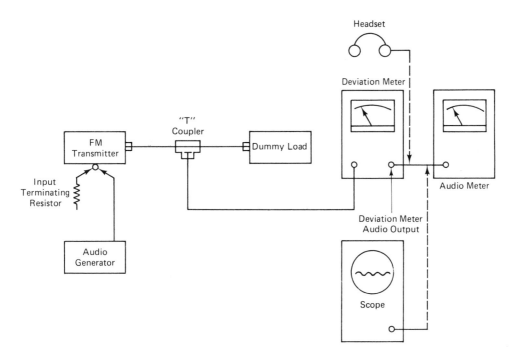

FIGURE 8-15. Typical test setup for checking hum and noise modulation.

5. Key the transmitter and note the hum and/or noise level. Use the scope or headset to determine the type of noise or hum present. If the meter has no dB scale, convert the voltage readings to dB by the formula dB = 20 log(E1/E2), where E1 = the reference voltage and E2 = the hum and/or noise voltage.

TRANSMITTER AUDIO DISTORTION

This test is basically a measure of the distortion produced by the audio stages of the transmitter. However, improper adjustment of a phase modulator can also produce distortion which will show up in this measurement.

Test Procedure

1. Set up the equipment as shown in Figure 8-16A or 8-16B.
2. Key the transmitter and increase the 1-kHz tone level until the deviation meter indicates 60% full system deviation. Measure the distortion on the distortion meter (or Sinad meter). On the Sinad meter, the distortion is indicated as so many dB below the composite (fundamental plus distortion) signal. Figure 8-17 correlates distortion in percent with the Sinad meter reading in decibels.

As mentioned above, if the transmitter uses a phase modulator the phase modulator stage can be tuned for minimum distortion with this same equipment setup. It may be necessary to compromise between the best distortion figure and adequate deviation level.

The Bessel-Zero Method

The Bessel-zero method can be used to set the deviation of an FM transmitter to a specific level. The procedure (when properly used) is so accurate that it can also be used to check the calibration of FM deviation meters. It can be proven through a complicated mathematical computation involving Bessel functions that for certain values of modulation index (deviation/modulating frequency) the carrier is completely nulled out. Table 8-1 lists the values of modulation index for the first five carrier nulls. The carrier null is more easily determined with a spectrum analyzer but a receiver with a BFO or some other means of producing a beat note can be used. The spectrum analyzer method is described here first.

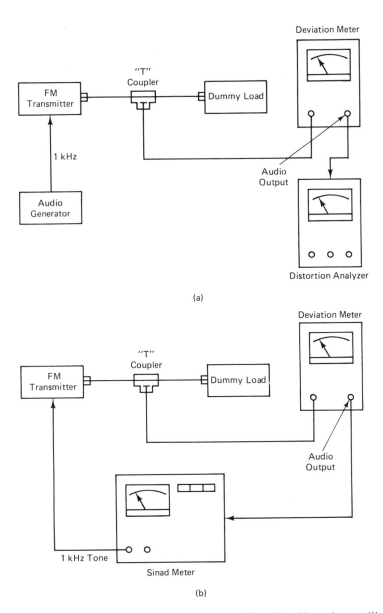

FIGURE 8-16. Typical equipment setups for checking transmitter audio/modulation distortion. At (A) a conventional distortion analyzer is connected to the audio output of the deviation meter. At (B) a Sinad meter is used to measure the distortion while the built-in 1,000 Hz tone generator supplies the modulating signal.

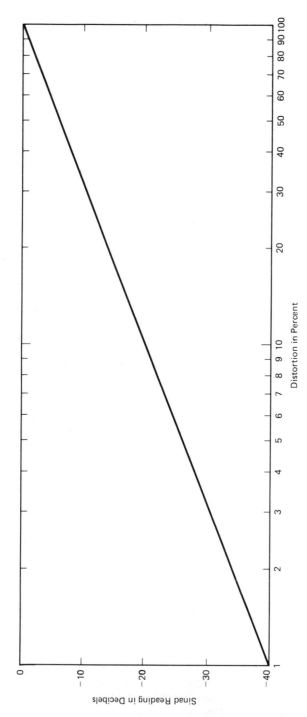

FIGURE 8-17. The Sinad meter reads distortion in *dB below the reference level*. This graph correlates this dB figure with "percent" distortion.

	Null	Modulation Index
1	. . . . .	2.405
2	. . . . .	5.520
3	. . . . .	8.653
4	. . . . .	11.792
5	. . . . .	14.931

TABLE 8-1

The Spectrum Analyzer Method

In Chapter 2, some mention was made concerning carrier nulls. Since the spacing of the sidebands of an FM signal is equal to the modulating frequency the spectrum analyzer should have a resolution sufficient to separate two signals (carrier and nearest sidebands) that differ in frequency by the lowest modulating frequency used. Good resolution is especially important to be able to determine the exact carrier null.

The basic procedure for using the spectrum analyzer in the Bessel-zero application is described below. This procedure concerns setting the deviation of the FM transmitter to a specified level. For example, in narrow-band FM the deviation should be set just below $+/- 5$ kHz deviation.

Test Procedure

1. Set up the equipment as shown in Figure 8-18.

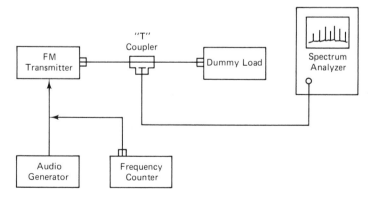

FIGURE 8-18. A setup for using a spectrum analyzer to locate the carrier null points when using the Bessel method of checking deviation.

2. Determine the frequency of the first carrier null by referring to Table 8-1 for the modulation index and then using this modulation index to determine the correct modulating frequency to be used. As an example: The first carrier null occurs at a modulation index of 2.405. If the deviation is to be set to $+/-$ 5 kHz, the modulating frequency will be 5 kHz/2.405 = 2,079 Hz.

3. Set the frequency of the audio generator to the exact frequency calculated in step 2. If necessary, monitor the output of the audio generator with a frequency counter to ensure an accurate frequency setting.

4. Key the transmitter and starting with the audio generator output at zero, gradually increase the output from the audio generator while closely monitoring the spectrum analyzer display. At some point, the carrier will start to decrease in amplitude. Adjust the generator output level for minimum carrier level. Any residual at the carrier location is probably an IM product and not due to incomplete carrier null.

5. At the point where the first carrier null occurs, the deviation is 2.405 $\times$ Mf, where Mf is the audio modulating frequency.

The sequence of events is illustrated in Figures 8-19A through 8-19E. In Figure 8-19A, the unmodulated carrier is set to the reference line. In Figure 8-19B, the audio modulating signal is modulating the transmitter at a low level. In Figure 8-19C, the modulating signal is increased causing a substantial increase in the sidebands. In Figure 8-19D, a further increase in the modulating signal causes the carrier to approach the null. In Figure 8-19E, the carrier has completely nulled. The residual component at the carrier location is an IM product.

The accuracy of the Bessel-zero method is as good (or as bad) as the care used to perform the test. Inaccuracy can also result from using a very distorted audio signal to modulate the transmitter.

The Beat-Frequency Method   The beat-frequency method requires some means of producing a beat note resulting from the carrier. The beat note must be recognizable in the presence of other tones when the transmitter is being modulated. There are several ways that this beat note can be obtained: (1) a receiver with a BFO; (2) a heterodyne-frequency meter tuned slightly off the carrier frequency; and (3) a signal generator can be used to produce a beat note in the receiver if the receiver has no built-in BFO.

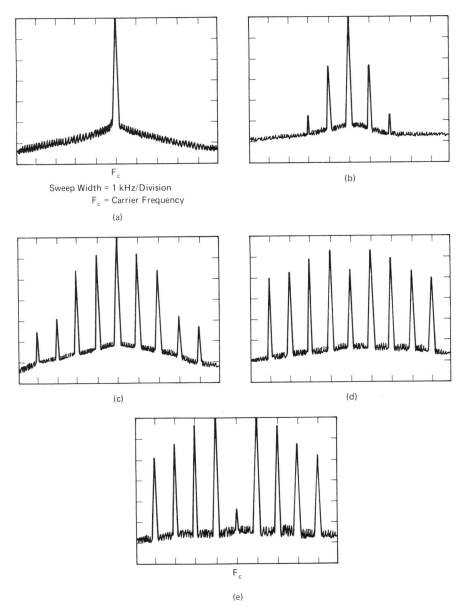

Sweep Width = 1 kHz/Division
F_c = Carrier Frequency

FIGURE 8-19. (A) Unmodulated carrier is indicated by absence of sidebands. (B) Low-level modulation, notice sidebands start to appear. (C) Increased modulation, notice higher level sideband components. (D) Increasing the modulation further causes the carrier to start toward the null. (E) At this modulation level the carrier has nulled; further increases in modulation will bring the carrier back up. The cycle will be repeated with further increases in the modulation.

Test Procedure

1. Set up the equipment as shown in Figure 8-20.
2. Set the frequency of the audio generator to: desired deviation/2.405. This is for the first null.
3. Key the transmitter (don't modulate it yet). If a receiver is used to monitor the signal, adjust the BFO (A) or the signal generator (B) to produce a beat note that you can easily recognize in the presence of other beat tones. The exact beat frequency isn't important as long as you can recognize it. If the heterodyne frequency meter is used, it is simply tuned away from the carrier until the desired beat note is produced. It may take some practice at first to be able to recognize the beat note in the presence of other tones that will occur with modulation applied.

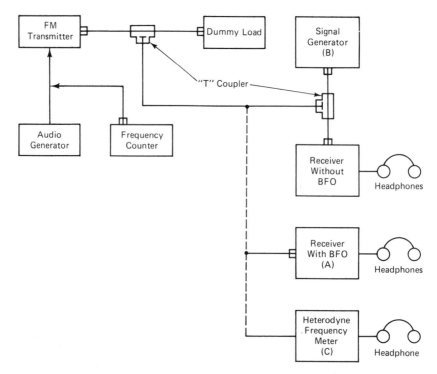

FIGURE 8-20. This setup shows several approaches which may be used to perform the Bessel method for setting deviation by using a beat frequency indicator to locate the carrier null points.

4. With the transmitter keyed, increase the audio generator level until the reference beat note disappears. Disregard the other beat notes.

5. At the point at which the reference beat note disappears, the deviation is equal to $2.405 \times Mf$, where Mf is the modulating frequency.

FREQUENCY MEASUREMENTS

Frequency measurement procedures for FM transmitters are the same as those outlined for AM transmitters in Chapter 4 and for that reason are not repeated here.

BIBLIOGRAPHY

Detwiler, William, "Journey Through a Speech Amplifier," *Communications Magazine,* June 1977.

———, "SINAD—A Valuable Measurement System," *Communications Magazine,* March 1979.

———, "What the Modulation Monitor Has Missed," *Communications Magazine,* March 1980.

FCC Rules: Part 90, *Private Land-Mobile Radio,* Federal Communications Commission, Washington, D.C.

Philco Corporation, *Radio Communication System Measurements* © 1952, Philco Corporation.

The Institute of Electrical and Electronics Engineers, Inc. IEEE Std. 377-1980. New York: The Institute of Electrical and Electronics Engineers, Inc., 1980.

Titchmarsh, R. S., *Modulation Measurements with a Spectrum Analyzer.* From: MI MEASURETEST, an application note from MARCONI INSTRUMENTS, Northvale, New Jersey.

Zeines, Ben, *Electronic Communication Systems.* Englewood Cliffs, N.J.: Prentice-Hall, Inc., 1970.

9

FM RECEIVER
TESTS AND
MEASUREMENTS

This chapter describes in great detail many test and measurement procedures that have been developed for testing the performance of FM communications receivers. A simplified block diagram of a typical FM communications receiver is shown in Figure 9-1. This is a dual-conversion superheterodyne receiver. Much of the circuitry is very similar to an AM or SSB receiver. However, a different type of detector must be used to detect the FM signal. One of the most popular types of FM detectors is called a *discriminator*. The discriminator is preceded by two or more limiter stages which serve to remove any amplitude variations from the signal.

As a servicing convenience, most FM communications receivers provide a metering socket which provides direct access to key test points within the receiver. Many manufacturers market a special test set which can be plugged directly into the metering socket of their radios. Some of these test sets have one meter which is switched to various test points, others provide a separate meter for each test point. If such a test set is not available, a VOM can be used instead, though the convenience factor is lost.

SENSITIVITY TESTS AND MEASUREMENTS

The purpose of the sensitivity test is to determine just how much input signal is needed to produce a certain signal-to-noise ratio in the audio output. This is usually one of the first tests performed on a receiver during routine checks or performance verification tests. Most manufacturers' specifications list two sensitivity figures in the "specs" chart. These are: (1)

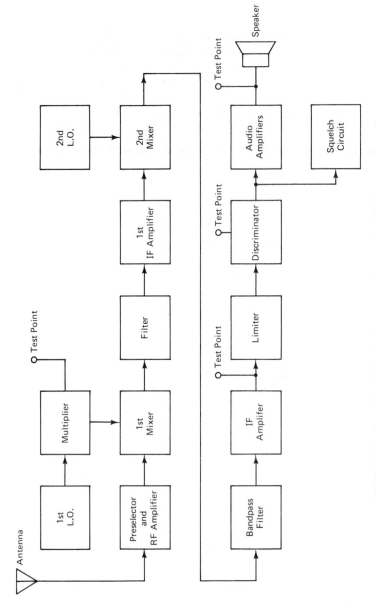

FIGURE 9-1. Simplified block diagram of a typical FM communications receiver.

the 20-dB quieting sensitivity test and (2) the 12-dB Sinad sensitivity. These are the two basic methods presented here. The effective sensitivity test, which is performed with an antenna connected, can be done by either the 20-dB quieting method or the 12-dB Sinad method.

The 20-dB Quieting Method

In the absence of an input signal and with the audio stage unsquelched, there will be a great deal of noise at the receiver's audio output. When an unmodulated (CW) on-channel signal is applied to the receiver input, the receiver will quiet down. The extent of quieting is dependent upon the level of the input signal. The 20-dB quieting test is based on this principle.

This and all other sensitivity tests should be performed in as noise-free an environment as possible—that is, one that is free from electrical disturbances and static. An often troublesome source of such static noise is the fluorescent light, which is usually hanging just above the workbench! These noise sources, if severe enough, can completely invalidate your test results and, if not recognized, can mislead you into thinking the radio receiver is at fault when it may not be.

Test Procedure

1. Hook up the equipment as shown in Figure 9-2.
2. Adjust the squelch control so that the receiver is completely unsquelched.
3. Set the volume control for a noise output reference level on the voltmeter.
4. Adjust the signal generator CW level until the noise output voltage drops to one-tenth the reference level. This represents a 20 dB drop.
5. At this point the output level from the signal generator (in microvolts or dBm) is the 20 dBm quieting sensitivity of the receiver under test.

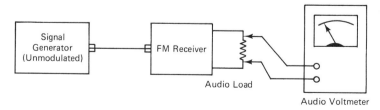

FIGURE 9-2. Test setup for the 20-dB quieting test.

Typical figures for 20-dB quieting sensitivity range from 0.3 μV to 1 μV. If the receiver requires more than the specified level, corrective action such as realignment or troubleshooting may be required. Since the signal generator is unmodulated, the major significance of this test is to prove (or disprove) that the receiver has sufficient gain in the RF and IF sections.

The 12-dB Sinad Method

The 12-dB Sinad test has been widely adopted by the communications industry as the best way to measure receiver sensitivity properly. To be effective, the receiver must be capable of recovering the modulation from weak signals and faithfully reproducing it in the audio output. In processing the signal, the receiver must keep the generation of noise and distortion products to a minimum, since an excess of these will seriously impair the reception of weaker signals. The Sinad measurement is a method that takes into account the signal, noise, and distortion. In fact, the word Sinad is an acronym for *S*ignal + *n*oise *a*nd *d*istortion. Although the Sinad test really isn't new, it has been only in the last few years that special equipment has been developed that has facilitated the use of the Sinad measurement in the service shop. These special Sinad instruments have succeeded in taking the "pain" out of performing the Sinad test, and as a result communications shops throughout the country are now using the Sinad test as a standard procedure in receiver testing. In addition to testing or specification verification, these instruments are extremely helpful in receiver alignment work also.

Certain standards to be used in performing the Sinad test proper have been prescribed by the Electronic Industries Association (EIA).[1] Some of those requirements are:

1. The *standard* modulating frequency is 1 kHz. This is the signal that is used to modulate the signal generator.

2. The *standard* level of modulation is 60% of the peak deviation used. For example, in narrow-band FM where the peak deviation is $+/-5$ kHz the modulation level will be 60% of 5 kHz or $+/-3$ kHz.

3. The audio output level at which the Sinad sensitivity measurement is taken must be at least 50% of rated receiver audio output.

4. The audio output of the receiver is terminated in a resistive (dummy) load that matches the output impedance of the receiver.

---

[1] EIA Standard RS-204-C, "Minimum Standards for Land Mobile Communication FM or PM Receivers, 25–947 MHz," Electronic Industries Association, Washington, D.C. 1982, p. 10, para. 7.0. This document is available in its entirety from: Electronic Industries Association, 2001 Eye St. N.W., Washington, D.C. 20006, Phone (202) 457-4900.

When a modulated signal is fed to the input of a receiver, the *composite* audio output signal will consist of: (1) the original modulating tone or *signal component*; (2) *noise components*; and (3) *distortion components*. At high input signal levels, the noise components are negligible, leaving only the signal and distortion components. However, at lower input signal levels (at which sensitivity measurements are made) noise makes a significant contribution to the composite signal. The distortion is due in part to distortion produced in the audio stages of the receiver and in part to distortion produced in the RF and IF stages (largely the IF stages) of the receiver. Improper alignment of the narrow bandpass filters or improper placement of the IF signal within the bandpass will contribute to distortion.

The ratio of the composite audio signal to the noise plus distortion components is called the Sinad ratio. The receiver sensitivity measurement is taken at the 12-dB Sinad point—that is, the point at which the noise plus distortion is 12 dB below the composite (signal + noise + distortion) audio signal. Figure 9-3 will serve to illustrate the meaning of the 12-dB Sinad ratio. The signal generator is frequency-modulated by a 1-kHz tone from the audio generator. This FM signal is then fed to the receiver input. A composite signal then appears at the audio output. The amplitude of the composite

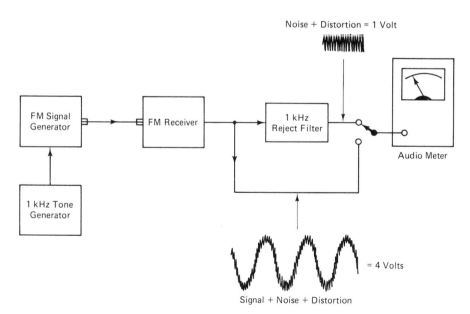

FIGURE 9-3. The meaning of 12-dB Sinad is illustrated here. The *composite* signal (signal + noise + distortion) is 4 V. The filter removes the 1-kHz signal component leaving only the noise + distortion components at the output of the filter. The amplitude of the noise + distortion components is 1 V. Thus, the $S+ N+ D/N+ D$ ratio is 4/1. A 4/1 voltage ratio equals 12 dB.

signal, as measured by a voltmeter, is 4 V. The composite signal is also fed to a filter that effectively removes the 1-kHz *signal* component, leaving only the noise plus distortion components. This noise plus distortion is then measured by the voltmeter, which indicates 1 V. Thus, the Sinad ratio is 4 to 1, or simply 4. This is shown by the formula:

$$\text{Sinad ratio} = \frac{\text{signal} + \text{noise} + \text{distortion}}{\text{noise} + \text{distortion}} = \frac{4}{1} = 4$$

This means that the noise plus distortion components represent 25% of the composite signal. If this composite signal were analyzed by a distortion analyzer, the distortion analyzer would indicate 25% distortion.

The Sinad ratio is almost always stated in terms of decibels. This ratio can be converted to decibels by the formula:

$$\text{Sinad (dB)} = 20 \log (E1/E2)$$

Substituting into the formula we have:

$$\text{Sinad (dB)} = 20 \log (4/1) = 20 \ (0.6) = 12 \ \text{dB}$$

Thus, it is shown that a Sinad ratio of 4 equals 12 dB Sinad and that 12 dB Sinad corresponds to 25% distortion (distortion + noise). The basic principle of using a filter to extract the noise + distortion as shown in Figure 9-3 is used in distortion analyzers and *dedicated* Sinad meters.

Two methods of performing the 12-dB Sinad sensitivity measurement are presented here. One method shows how to use a common distortion analyzer to perform the test while the other method shows how special dedicated Sinad meters simplify the measurement.

The Distortion Analyzer Method[2]

1. Set up the equipment as shown in Figure 9-4.

2. Adjust the output level of the 1-kHz audio generator to produce 60% of the peak deviation. For +/− 5 kHz (narrow-band FM), this is +/− 3 kHz. Thus, the signal generator is modulated +/− 3 kHz for narrow-band FM equipment.

3. Tune the signal generator to the receiver frequency and adjust the signal generator level to 1,000 μV.

4. Set the distortion analyzer function switch to "Voltmeter." Adjust the receiver's volume control to produce the full rated audio output power. Use the graph in Appendix D to determine the voltage necessary to produce the rated audio output power.

---

[2] This procedure conforms to the general requirements of EIA Standard RS-204-C, p. 10, para. 7.0 (see also footnote 1).

5. Set the distortion analyzer function switch to "Set Level." Adjust the distortion analyzer level control for a full-scale reference indication.

6. Switch the distortion analyzer function switch to "Distortion." Tune the distortion analyzer for maximum null (minimum meter indication). The indication at the maximum null point should be well below the 25% distortion (12 dB) point. If not, the receiver audio stage(s) must be contributing excessive distortion. This must be corrected before a valid 12-dB Sinad sensitivity test can be conducted.

7. If the distortion measured in step 6 is well below the 25% mark, decrease the signal generator output level until the distortion meter indicates 25% distortion. (This is really more noise than distortion, but since the noise is undesired it is treated as distortion.)

8. When the distortion meter indicates 25% (12 dB Sinad), the signal generator output level (in microvolts or dBm) is the 12-dB Sinad sensitivity of the receiver. Make sure that the audio output is at least 50% of the rated power. If not, increase the signal generator output level until the audio output power is at least 50% rated power and use this signal generator level as the 12-dB Sinad sensitivity of the receiver.

Almost all distortion analyzers provide a decibel scale. This scale can be used to set the level and measure the Sinad directly, rather than using the percent distortion scale as described earlier. The reference setting and Sinad measurement are made on the decibel scale. The difference between the reference setting and the final reading must be 12 dB. For example, if the reference is set to the +2-dB mark on the scale, the 12-dB Sinad point will be at the −10-dB mark on the scale—a difference of 12 dB, corresponding to 12 dB Sinad.

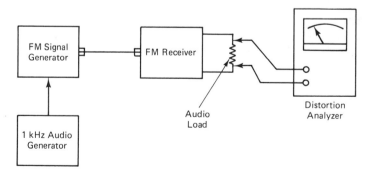

FIGURE 9-4. Test setup for performing the 12-dB Sinad test with a standard distortion analyzer.

The Sinad Meter Method

As mentioned previously, special Sinad meters have been designed to overcome some of the difficulties of performing the 12-dB Sinad test. Among the advantages of the special Sinad meter are: (1) automatic level-setting circuitry; (2) fixed-tuned 1-kHz reject filter eliminates need for null tuning; and (3) many models provide a 1-kHz accurate tone to be used for modulating the signal generator.

A typical Sinad meter is shown in Figure 9-5. This is the Sinadder 3™ which is manufactured by Helper Instruments Company. The block diagram of this instrument is shown in Figure 9-6. Briefly, the operation of the instrument is as follows. With the function switch set to the Sinad position, the input signal is fed to the input of an AGC amplifier. The gain of the AGC amplifier is controlled by a feedback circuit consisting of an absolute value circuit and the AGC control amplifier. This feedback arrangement acts to keep the composite signal at the output of the AGC amplifier at a constant level regardless of changes in the input level (from 10 mV to 10 V). The composite signal level at the output can consist of any combination of signal, noise, and distortion. This composite signal appears at the input to the 1-kHz reject filter (point *A* on the block diagram). Any signal component (1 kHz) that may be present at point *A* does not show up at point *B*. It is not passed by the reject filter. The only components that appear at point *B* are the noise plus distortion components. These noise plus distortion compo-

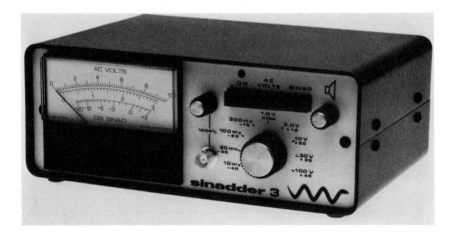

FIGURE 9-5. This Sinad meter provides a 1-kHz tone which can be used to externally modulate a signal generator to perform the 12-dB Sinad test. The instrument also features an audio voltmeter and a built-in speaker so that the signal being measured can be heard in the speaker. Level-setting is automatic. Courtesy of Helper Instruments Company.

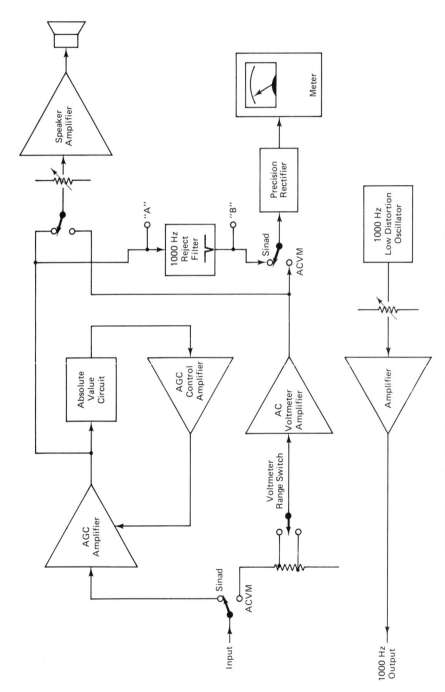

FIGURE 9-6. Simplified block diagram of the Sinadder 3™. Courtesy of Helper Instruments Company.

nents are then rectified and fed to the meter, which has a scale calibrated in dB.

Figure 9-7 shows a test setup using the Sinadder 3™ for the 12-dB Sinad test. Notice that the 1-kHz tone used to modulate the signal generator is supplied by a tone generator within the Sinadder 3™. This 1-kHz tone is highly accurate (+/− 1 Hz) and low in distortion. The 1-kHz tone must be accurate, otherwise it will not be properly filtered by the reject filter in the Sinadder 3™ and will be treated as noise plus distortion. This would cause erroneous results. Now let's see how the actual procedure is performed with the Sinad meter. The procedure is essentially the same as the one just described using the common distortion analyzer, except that the Sinad meter is used instead.

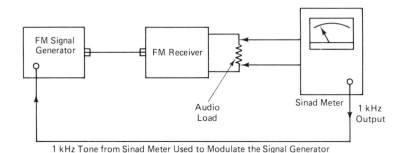

1 kHz Tone from Sinad Meter Used to Modulate the Signal Generator

FIGURE 9-7. A typical test setup using the Sinad meter to perform the 12-dB Sinad test.

Test Procedure

1. Set up the equipment as shown in Figure 9-7.

2. Adjust the 1 kHz tone output from the Sinad meter to produce 60% of the peak deviation used for the particular service. For narrow-band FM the peak deviation is +/− 5 kHz so the generator is modulated +/− 3 kHz.

3. Unsquelch the receiver and feed a 1,000-μV signal into the input.

4. Adjust the volume control of the receiver to produce an audio output power equal to the rated audio output of the receiver. Many Sinad meters (including the Sinadder 3™) have a built-in ac voltmeter which can be used to set the audio output to the rated power level. Use the graph in Appendix D to correlate the voltage level required to produce the correct amount of audio power.

5. At this point, the Sinad reading should be well below the 12-dB mark

on the dB scale. If not, this indicates that some stages are contributing excessive distortion which will make the 12-dB Sinad sensitivity measurement impossible.

6. If the Sinad reading is well below the 12-dB mark on the scale, decrease the signal generator output level until the Sinad meter indicates 12 dB Sinad. If the audio output at this point is at least 50% rated receiver audio power, the signal generator output level at this point is the 12-dB Sinad sensitivity of the receiver. If the audio output is less than the 50% level, increase the output of the signal generator until the audio power is 50% rated power. Use this signal generator output level as the 12-dB Sinad sensitivity of the receiver.

For a better understanding of what is going on, let's analyze what happens within the Sinad meter during the above procedure. In step 3, when 1,000 μV modulated signal is fed into the receiver input the audio output (composite signal) from the receiver should consist of very little noise and distortion (assuming proper receiver operation). The signal is amplified by the AGC amplifier in the Sinad meter. The absolute value circuit along with the AGC control amplifier controls the gain of the AGC amplifier so that the output level is kept at a constant value or reference level. The noise and distortion components are separated from the composite signal by the 1-kHz reject filter. The noise and distortion components appear at point B and are rectified and measured by the meter circuit. The meter indicates the level of the noise and distortion components as so many dB below the reference level (established by the composite signal). In step 3, the noise and distortion components should be very low with a 1,000-μV signal fed to the receiver input. As the signal generator level is decreased the percentage of noise in the composite signal increases. This shows up as a higher reading on the Sinad meter scale. The point at which the noise (and distortion) reach the 12-dB mark on the scale is the point at which the 12-dB Sinad sensitivity is measured.

In summary, both the distortion analyzer method and the Sinad meter method use the same principle. Essentially what is done is to set the reference level using the composite signal + noise + distortion, then extract the noise + distortion components and compare the level of the noise + distortion components to the composite signal (signal + noise + distortion). The primary advantage of the Sinad meter is to keep the reference level constant with changes in the input signal level.

Most technicians don't set the audio level quite as high as the EIA proper procedure specifies. This really just shows how much distortion the audio output stages are producing at the higher audio levels. It really isn't necessary to run the audio so high if you are mainly concerned with the receiver sensitivity or if you are using the test as a means of alignment. If you are interested in audio distortion, by all means run up the audio level.

"Effective" Sensitivity Test

The effective sensitivity test can be performed by either the 20-dB quieting method or the 12-dB Sinad method. The basic principle of the effective sensitivity test is described in Chapter 4 under "Effective Sensitivity Test." It is only necessary to substitute either the 20-dB quieting test or the 12-dB Sinad test. Otherwise, the procedure is the same.

SENSITIVITY VARIATION BETWEEN CHANNELS

When a receiver is operated at more than one channel (frequency), a sensitivity test should be conducted on all channels. This is especially true if the channels are separated by a wide frequency margin. Where a great number of channels are used the receiver sensitivity can be "spot-checked" at the low end, middle, and high end of the frequency range over which the receiver is operated. Preferably, the 12-dB Sinad method should be used.

Sensitivity Variation with Signal Frequency

The purpose of this test is to determine just how much effect receiver or transmitter frequency drift will have on reception. In this case, the signal generator is tuned to one side and then the other to simulate receiver or transmitter oscillator frequency drift.

Test Procedure [3]

1. Set up the equipment as shown in Figure 9-7.
2. Perform the 12-dB Sinad test and leave the signal generator at the 12-dB Sinad sensitivity level.
3. Increase the generator output level by 6 dB (this amounts to doubling the microvolt indication of step 2). This will cause the Sinad reading to drop below 12-dB.
4. Tune the signal generator above the receiver frequency very slowly while observing the Sinad reading. Adjust the signal generator frequency until the Sinad reading returns to the 12-dB mark. Note the signal generator frequency.
5. Tune the signal generator below the receiver frequency to produce the 12-dB Sinad reading again. Note the signal generator frequency.
6. Calculate the change in frequency in steps 4 and 5. The smaller figure is defined as the minimum useable bandwidth.

---

[3] EIA Standard #RS-204-C, Electronic Industries Association, Washington, D.C., 1982, p. 11, para. 11.0 (see also footnote 1).

SQUELCH TESTS

The purpose of the squelch circuit is to eliminate the constant noise output from the receiver in the absence of a received signal. Without the proper functioning of the squelch circuit, it would be quite fatiguing to listen to the noise for even a short period of time. The squelch circuit senses the signal and turns on the receiver audio stage(s) so that the signal can be heard in the speaker. The squelch control, which is usually made accessible to the operator, can be adjusted to set the sensitivity of the squelch circuit— that is, the squelch can be set to *open* on various signal strengths. If desired, weaker (and noisy) signals can be squelched out, thus allowing only the stronger (and clearer) signals to be heard in the speaker.

Squelch circuits can cause a variety of problems and symptoms when they are malfunctioning. Constant noise (receiver open all the time), no reception (receiver won't open at all), reception at reduced volume, receiver shuts off during voice modulation peaks. All of these can be caused by a malfunctioning squelch circuit. The following test procedure provides a thorough check of the squelch circuit sensitivity.

Threshold (Critical) Squelch Sensitivity Test[4]

1. Set up the equipment as shown in Figure 9-8.

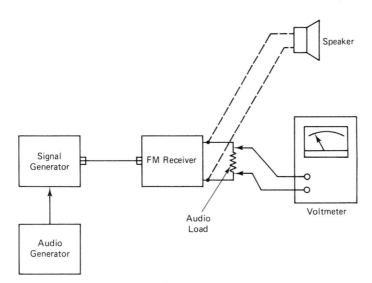

FIGURE 9-8. A test setup showing how squelch tests are performed. The speaker is used for the squelch blocking test.

---

[4] EIA Standard RS-204-C, Electronic Industries Association, Washington, D.C., 1982, p. 20, para. 17.2 (see also footnote 1).

2. Set the signal generator to the receiver frequency, set the output level to 1,000 μV, and modulate the generator with a 1-kHz tone at a deviation level of 60% peak system deviation. This is $+/-$ 3 kHz for narrow-band FM.

3. Set the receiver volume control to produce the rated output from the receiver.

4. Remove the input signal (by adjusting the generator for minimum output).

5. Adjust the squelch control to the point at which the audio output just shuts off. This is the threshold or critical point. The audio output should be reduced at least 40 dB (1/100 times the original output voltage).

6. *Gradually* increase the signal generator output level while observing the output meter (and listening). Set the generator output level to the minimum level which produces a steady audio output level which is near the rated power output level (not more than 10 dB below the rated power output level). Note this signal generator output level.

7. Remove the input signal. Note the reduction in the audio output level. If the audio output level is not reduced by at least 40 dB, readjust the squelch control for more squelch and repeat steps 6 and 7. The threshold squelch sensitivity is the value recorded in step 6.

Tight (Maximum) Squelch Sensitivity

The same basic procedure described for the threshold squelch sensitivity is also used for the tight squelch sensitivity test. The only difference is that the squelch control is adjusted to the maximum squelch position.

Squelch Blocking (Clipping) Test

The squelch circuit is designed to operate from a noise signal from the detector. The normal audio frequencies below 3 kHz are prevented from entering the squelch circuit by a special high-pass filter, which readily passes the noise generated by the receiver.

Under normal conditions, voice modulation will not have any adverse affect on the squelch circuit, unless the signal being received is excessively overmodulated. However, certain problems can arise within the receiver which can cause the squelch circuit to block the receiver output during voice modulation. This type of problem is most often caused by insufficient IF bandwidth in the receiver or by improper location of the IF signal in the IF passband. This can cause either the positive or negative peak excursions to fall outside the bandpass resulting in the clipping of that portion of the signal which falls out of the passband. This clipping results in the creation of noise and distortion components which the squelch circuit recognizes as noise, thus shutting off the receiver audio stage(s). The following test procedure will expose any squelch-blocking tendency of the receiver.

Test Procedure

1. Set up the equipment as shown in Figure 9-8.
2. Set the squelch control to the tight or maximum squelch position.
3. Set the signal generator output level approximately *10 to 15 dB higher than the tight squelch sensitivity* of the receiver. (The tight squelch sensitivity test was described previously.)
4. Set the audio generator to 1 kHz and adjust the output level to modulate the signal generator to a low level (approximately 20% of peak system deviation or $+/-$ 1 kHz for narrow-band FM).
5. Adjust the receiver's volume control for a comfortable listening level.
6. While monitoring the audio output of the receiver, increase the signal generator modulation level (by turning up the audio generator output level) until peak system deviation is reached ($+/-$ 5 kHz for narrow-band FM). If the receiver audio output cuts off as the deviation level is increased to maximum, squelch blocking is occurring.
7. Step 6 should be repeated at several frequencies throughout the 300 Hz to 3,000 Hz frequency range.

Squelch blocking can usually be determined by simply monitoring an on-the-air signal that is voice modulated (but not overmodulated), preferably one in which the voice peaks are regularly hitting the peak deviation point. If the signal is fairly strong and the receiver audio seems to turn on and off with modulation, a squelch blocking problem exists.

MODULATION ACCEPTANCE BANDWIDTH

Modulation acceptance bandwidth is a measure of the maximum amount of modulation (deviation) on the input signal which the receiver can accept before serious distortion results in the audio output signal. This is dependent upon the IF bandwidth of a receiver. Usually, the modulation acceptance bandwidth of a receiver is at least $+/-1$ kHz more than the peak system deviation. For narrow-band FM (where the peak system deviation is $+/-5$ kHz), the modulation acceptance bandwidth is usually at least $+/-6$ kHz. In any case, the modulation acceptance bandwidth must be at least equal to the peak system deviation in order to accomodate fully modulated signals.

Test Procedure

1. Set up the equipment as shown in Figure 9-7.
2. Perform the 12-dB Sinad sensitivity test (as previously described in this chapter).
3. Increase the signal generator level by 6 dB above the 12-dB Sinad sensitivity level. (This amounts to two times the microvolts required to produce the 12-dB Sinad level.) This will result in a better than 12-dB Sinad reading.
4. Increase the signal generator deviation by increasing the audio generator output level until the Sinad meter again indicates exactly 12 dB Sinad.
5. The generator deviation at this point is the modulation acceptance bandwidth of the receiver.

If desired, the IF bandpass can be aligned by modulating the signal generator to the full system deviation level (or even slightly more) and then tuning the bandpass circuits for the best Sinad reading.

AUDIO SENSITIVITY

The audio sensitivity of a receiver has to do with both the efficiency of the FM detector and the gain of the audio amplifier. It is simply a measurement of the deviation required to produce the full-rated audio power from the receiver.

Test Procedure

1. Set up the equipment as shown in Figure 9-8.
2. Set the receiver volume control to maximum and the squelch control to minimum (receiver fully open).
3. Set the signal generator output to a high level to ensure a high S/N ratio (full quieting).
4. Set the audio generator to 1 kHz.
5. Beginning with the deviation at minimum, gradually increase the deviation until the audio meter indicates full audio output power. Note the deviation level of the signal generator. This is the audio sensitivity of the receiver.

HARMONIC DISTORTION FIGURE

1. Set up the equipment as shown in Figure 9-4.
2. Set the signal generator output level high enough to produce full quieting. Set the audio generator frequency to 1 kHz and set the audio generator output level to produce standard test modulation from the signal generator[5] ($+/-3$ kHz for narrow-band FM).
3. Set the receiver volume control to produce full-rated audio output from the receiver.
4. Use the distortion analyzer to measure the harmonic distortion. The harmonic distortion should also be measured at a lower output power level such as 20% rated power.

RESIDUAL HUM AND NOISE MEASUREMENT

When a very strong unmodulated signal is applied to a receiver, the audio output from the receiver should be minimal (ideally zero) but a certain amount of residual hum and/or noise level is always present. This test compares the amplitude of these residual hum and noise components to the amplitude of a reference signal.

Test Procedure

1. Set up the equipment as shown in Figure 9-8.
2. Set the signal generator output level to produce full receiver quieting (maximum quieting).
3. Set the audio generator frequency to 1 kHz and the audio generator output level to produce standard test modulation (60% peak system deviation from the signal generator). This is $+/-3$ kHz for narrow-band FM.
4. Set the receiver squelch control for minimum squelch.
5. Set the receiver volume control to produce full-rated audio output power.
6. Remove the audio-modulating signal and note the reduction (in dB) of the audio output level. This is the residual hum and noise level. If desired, a scope can be used to identify the type of hum or noise present.

---

[5] Standard test modulation per EIA and IEEE standards is 60% of the peak system deviation.

AUDIO FREQUENCY RESPONSE

The audio output of FM communications receivers is de-emphasized on a 6-dB/octave scale. This compensates for the pre-emphasis of the audio modulating signal at the transmitter end. Figure 9-9 shows the standard 6-dB/octave de-emphasis curve. The following procedure is used to plot a curve of the receiver's audio response, which can then be compared to the receiver's specification.

Test Procedure[6]

1. Set up the equipment as shown in Figure 9-8.

2. Set the signal generator output to 1,000 μV, the audio generator frequency to 1 kHz, and the output level of the audio generator to produce a deviation of 60% peak system deviation ($+/-3$ kHz for narrow-band FM).

3. Set the receiver volume control to produce 50% rated audio output power.

4. Reduce the generator deviation to 20% peak system deviation ($+/-3$ kHz for narrow-band FM). Note the audio output level. This is the 0-dB reference level.

5. Slowly tune the audio generator from 300 Hz to 3,000 Hz while observing the audio output level. Note the audio output level at several different frequency points and use the data to plot a graph of the audio response and compare it with the graph of Figure 9-9. Be sure to maintain the deviation at a constant level as the audio-modulating frequency is tuned through the range.

SELECTIVITY TESTS

The selectivity of a receiver is just as important (if not more so) as the sensitivity of a receiver. Due to the overcrowding of the various communications frequency bands, the trend has been toward reducing the channel spacing within those bands to create more available channel frequencies. This requires a very high performance receiver, one that is sensitive enough to receive weak signals on its frequency and selective enough to reject the many undesired signals that may be near the desired frequency.

Several test procedures have been devised to test various aspects of receiver selectivity. These include adjacent channel selectivity, intermodulation rejection ability, and spurious response attenuation. Several test procedures and tips are presented here in regard to receiver selectivity.

---

[6] EIA Standard RS-204-C, Electronic Industries Association, Washington, D.C., 1982, p. 12, para. 9.2 (see also footnote 1).

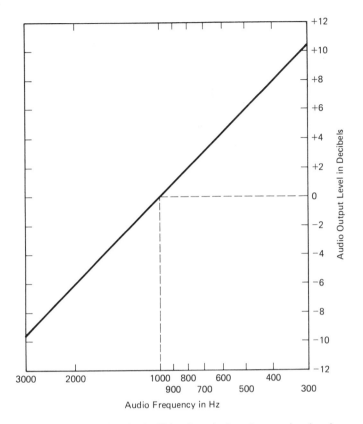

FIGURE 9-9. The detected FM signal is de-emphasized on a 6-dB/octave scale as shown in the graph.

Adjacent Channel Selectivity and Desensitization

Three different methods are presented here. The 20-dB quieting method is probably the oldest method used to specify a selectivity figure. It also yields the highest selectivity figure—deceptively high. It is presented here because many of the receivers that have a selectivity specification figure based on this method are still around.

Test Procedure 1—The 20-dB Quieting Method

1. Set up the equipment as shown in Figure 9-2.
2. Set the receiver squelch control for minimum squelch.
3. Adjust the volume control to produce a noise output equal to 25% of the rated audio output.
4. Set the signal generator to the receiver frequency. Do not modulate it.

5. Gradually increase the signal generator output level until the noise level drops 20 dB (one-tenth the reference level voltage). Record the signal generator level.

6. Tune the generator frequency to the offset specified. For example, if the receiver specification for selectivity states "Selectivity $= -100$ dB @ $+/-$ 15 kHz," the signal generator is tuned to an offset of 15 kHz above the receiver frequency. The signal generator level is then increased until the 20-dB quieting point is again reached. The signal generator level is noted. The difference between this generator level and the 20-dB sensitivity level should be at least as great as the specification calls for. In the example above, this difference should be at least 100 dB.

7. Repeat step 6 by tuning the generator to the specified frequency offset below the receiver frequency.

The second test procedure presented here appeared in a publication of the IEEE.[7] A procedure identical to this one in all essential respects was adopted by the EIA as a standard test procedure at one time.[8] The EIA has since changed to a modified procedure, but many radio receivers whose specification was based on this second procedure are still around.

Test Procedure 2—The Sinad Method

1. Set up the equipment as shown in Figure 9-10.

2. With generator B turned off (or standby mode), set up generator A as follows: Tune generator A to the receiver frequency, set the modulation (deviation) to 60% maximum system deviation ($+/-$ 3 kHz for narrow-band FM), set the output level of generator A to produce 12-dB Sinad on the Sinad meter.

3. Tune generator B to one of the adjacent channels (above or below) and set the deviation to 60% maximum system deviation ($+/-$ 3 kHz for narrow-band FM) using a 400-Hz tone from the audio generator.

4. Increase the output level of generator B until the Sinad reading is degraded by 6 dB (to an indication of 6 dB Sinad).

5. The difference in the output levels of the two signal generators (in dB) is the adjacent channel selectivity of the receiver.

6. Perform this test again with generator B tuned to the other adjacent channel and use the minimum figure as the adjacent channel selectivity.

---

[7] IEEE Test Procedure for Frequency-Modulated Mobile Communications Receivers, Institute of Electrical and Electronics Engineers, New York, 1969, p. 10, para. 4.2.1. Complete document is available from IEEE, 345 East 47 St., New York, NY 10017

[8] EIA Standard RS-204-B, p. 11, para. 11, Electronic Industries Association, Washington, D.C. This standard was superseded by RS-204-C.

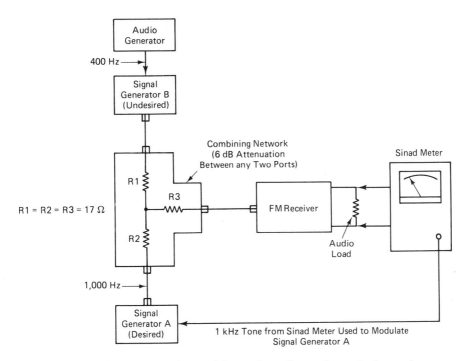

FIGURE 9-10. This setup is used to perform the adjacent channel selectivity tests.

This third procedure is a slightly modified version of the second procedure and is the one currently adopted and recommended by the Electronic Industries Association.[9]

Test Procedure 3—Updated Sinad Method

1. Set up the equipment as shown in Figure 9-10.

2. With generator *B* off, set up generator *A* as follows: Tune generator *A* to the receiver frequency, set the modulation (deviation) to 60% maximum system deviation (+/− 3 kHz for narrow-band FM), set the output level of generator *A* to produce a 12-dB Sinad reading.

3. Increase the output level of generator *A* by 3 dB. This will improve the Sinad reading to better than 12 dB Sinad.

4. Tune generator *B* to one of the adjacent channels (above or below) and set the deviation to 60% maximum system deviation (+/− 3 kHz for narrow-band FM), using a 400-Hz tone from the audio generator.

[9] EIA Standard RS-204-C, p. 16, para. 14.2, Electronic Industries Association, Washington, D.C., 1982 (see also footnote 1).

5. Increase the output level of generator *B* until the Sinad reading returns to 12 dB Sinad.

6. The difference in the output levels of the two signal generators (in dB) is the adjacent channel selectivity of the receiver.

7. Repeat the test with generator *B* tuned to the other adjacent channel and use the minimum figure as the adjacent channel selectivity.

Intermodulation Rejection

The overcrowded condition of FM communications channels has created another real interference problem—intermodulation interference. Intermod, as it is called, is not limited to FM communications. It can occur in any type of radio service. See also Chapters 5 and 7 for additional information on intermod. A complete, in-depth discussion of intermod is beyond the scope of this book. Some excellent articles concerning intermod interference appear from time to time in magazines such as *Communications Magazine* and *Mobile Radio Technology*. Two articles that I recommend are: "What to Do When You're Up to Your Ears in Intermod" by William Detwiler (February 1977, *Communications Magazine*) and "The Other Side of Intermod" by Dennis G. Hohman (September 1983, *Communications Magazine*).

The possibility of encountering intermodulation interference is very good, especially in the larger metropolitan areas. Mountaintops are popular locations for transmitter sites, especially when those mountains are located near and are convenient to metropolitan areas. (see Figure 9-11).

There are many possible mixing combinations by which two or more signals can produce an intermod interfering signal. In fact, so many intermod-producing combinations are possible at multiple-transmitter sites that computers are often employed to run an intermodulation study to find any possible combinations that might cause interference to one of the receivers at or near the site. Table 9-1 shows an example of such an intermod study. The various frequencies are fed into the computer, which then prints out the data showing potential intermod problems. The chart in Table 9-1 shows a total of eight transmitters and ten receivers. The computer printout lists five combinations that can easily produce intermod interference to one of the receivers.

The mixing together of the various signals must occur in some nonlinear device, such as an overloaded receiver front-end. Receiver front-end design is therefore very important in determining just how much intermod rejection ability a receiver has. We are concerned here strictly with receiver-produced intermod. If the intermod is formed at some point external to the receiver, the receiver can't be held responsible! The following test procedures can be used to check the intermod-rejection ability of a receiver.

FIGURE 9-11. Mountaintops such as Paris Mountain (near Greenville, S.C.) are very popular locations for transmitter sites. Notice the number of towers. Such a concentration of transmitters will usually lead to intermodulation interference unless the site is properly controlled. Severe intermod interference was experienced by the "lookout" tower (far right). The interference was "cleaned up" by installing a triple-section cavity filter set up as a series-bandpass filter between the antenna and the transceiver. Photo by Don Wood.

Test Procedure 1[10]

1. Set up the equipment as shown in Figure 9-12. Signal generator A represents the desired signal. Generators B and C provide the signals that form the intermod signal within the receiver.

2. With generators B and C turned off, perform the 12-dB Sinad sensitivity test with generator A. Increase the output of generator A by 3 dB. This will result in a better than 12-dB Sinad indication.

3. Tune generator B to one of the adjacent channels (above or below the desired channel). (Generator B frequency = receiver frequency +/− channel spacing.)

---

[10] This is the latest procedure adopted by the Electronic Industries Association. EIA Standard RS-204-C, p. 16, para. 14.2, Electronic Industries Association, Washington, D.C., 1982. See also footnote 1.

4. Tune generator C to the second channel above or below the desired channel (but on the same side of the desired channel as generator B). (Generator C frequency = receiver frequency $+/-$ *twice* the channel spacing.)

5. Coherently increase the output level of generators B and C until the Sinad indication is degraded to 12-dB Sinad. (Generators B and C are unmodulated.)

6. Carefully fine-tune the frequency of generator B (or C) to *maximize* the degradation (minimum Sinad) and, if necessary, coherently readjust the output levels of generators B and C to return the Sinad reading to 12-dB Sinad.

7. At this point, the difference in the output levels (in dB) of generators A and B (or A and C) is the intermodulation rejection figure of the receiver.

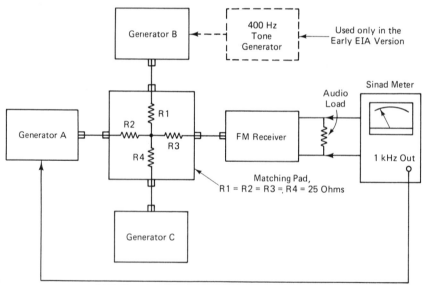

FIGURE 9-12. This setup can be used to test a receiver for intermodulation rejection figure. The 400-Hz audio generator is used only for intermod test procedure 2.

```
                    * INTERMODULATION STUDY PROGRAM (IMS) VERSION 1.31 *
TODAYS DATE IS: APRIL 20, 1983
SITE NAME OR REFERENCE NUMBER: GREAT BIG HILL

TX 1 =     1.5200          RX  1 = 157.5200
TX 2 = 156.0000          RX  2 = 154.4800
TX 3 = 154.0000          RX  3 = 159.0400
TX 4 = 157.0000          RX  4 = 151.0000
TX 5 = 462.0000          RX  5 = 467.0000
TX 6 =   31.0000          RX  6 = 153.0000
TX 7 =   31.5000          RX  7 = 152.0000
TX 8 =   39.0000          RX  8 = 156.0000
                         RX  9 = 160.5600
                         RX10 = 154.0000

*** THE EQUATIONS SOLVED WILL BE INDICATED BY THE FOLLOWING
   ***—*** HITS HAVE BEEN FOUND PRINT HEADING

*** NO ENTRIES UNDER THE PRINT HEADING INDICATES THAT NO HITS
   WERE FOUND FOR THAT EQUATION
* * * * * * * * * * * * * * * * * * * * * * * * * * * * * * * * * * * * * * *

              THE FOLLOWING *** A-B *** HITS HAVE BEEN FOUND- - -
  FREQ A       FREQ B       FREQ IM       FREQ RX       TEST WINDOW
   (MHZ)        (MHZ)        (MHZ)         (MHZ)          (+/−MHZ)
 156.0000      1.5200      154.4800      154.4800         0.0170
* * * * * * * * * * * * * * * * * * * * * * * * * * * * * * * * * * * * * * *

              THE FOLLOWING *** 2A-B*** HITS HAVE BEEN FOUND- - -
  FREQ A       FREQ B       FREQ IM       FREQ RX       TEST WINDOW
   (MHZ)        (MHZ)        (MHZ)         (MHZ)          (+/−MHZ)
 154.0000    156.0000      152.0000      152.0000         0.0210
 154.0000    157.0000      151.0000      151.0000         0.0210
* * * * * * * * * * * * * * * * * * * * * * * * * * * * * * * * * * * * * * *

              THE FOLLOWING *** A+B+C *** HITS HAVE BEEN FOUND- - -
 FREQ A      FREQ B      FREQ C      FREQ IM      FREQ RX      TEST WINDOW
  (MHZ)       (MHZ)       (MHZ)       (MHZ)        (MHZ)         (+/−MHZ)
 154.0000   156.0000   157.0000    467.0000     467.0000        0.0210
* * * * * * * * * * * * * * * * * * * * * * * * * * * * * * * * * * * * * * *

              THE FOLLOWING *** A+B−C *** HITS HAVE BEEN FOUND- - -
 FREQ A      FREQ B      FREQ C      FREQ IM      FREQ RX      TEST WINDOW
  (MHZ)       (MHZ)       (MHZ)       (MHZ)        (MHZ)         (+/−MHZ)
 154.0000   156.0000   157.0000    153.0000     153.0000        0.0210
* * * * * * * * * * * * * * * * * * * * * * * * * * * * * * * * * * * * * * *

              THE FOLLOWING 3A − 2B HITS HAVE BEEN FOUND- - -
  FREQ A       FREQ B       FREQ IM       FREQ RX       TEST WINDOW
   (MHZ)        (MHZ)        (MHZ)         (MHZ)          (+/−MHZ)
 156.0000    157.0000      154.0000      154.0000         0.0320
```

TABLE 9-1 This is a typical computer printout showing the possible combinations which will produce on intermod signal which can interfere with one of the on-site receivers. Courtesy *Communications Magazine.*

Test Procedure 2[11]

1. Set up the equipment as shown in Figure 9-12. Signal generator *A* represents the desired signal. Generators *B* and *C* provide the signals that form the intermod signal within the receiver.

2. With generators *B* and *C* turned off, perform the 12-dB Sinad sensitivity test with generator *A*. Leave the generator output set to produce the 12-dB Sinad reading.

3. Tune generator *B* to one of the adjacent channels (above or below the desired channel). (Generator *B* frequency = receiver frequency +/− channel spacing.) Modulate generator *B* with a 400-Hz tone at 60% maximum system deviation (+/− 3 kHz for narrow-band FM).

4. Tune generator *C* to the second channel above or below the desired channel (but on the same side of the desired channel as generator *B*). (Generator *C* frequency = receiver frequency +/− twice the channel spacing.) Do not modulate generator *C*.

5. Coherently increase the output level of generators *B* and *C* until the Sinad indication is degraded to 6 dB Sinad.

6. Carefully fine-tune generator *B* to maximize the degradation (minimum Sinad) and, if necessary, coherently readjust the output level of generators *B* and *C* to return the Sinad reading to 6 dB Sinad.

7. At this point the difference in the output levels (in dB) of generators *B* and *A* (or *C* and *A*) is the intermodulation rejection figure of the receiver.

Test Procedure 3—Spectrum Analyzer Method

1. Set up the equipment as shown in Figure 9-13.

2. Tune the spectrum analyzer to the receiver's first IF frequency and connect the spectrum analyzer to the output of the first mixer. Use a probe that won't cause a severe loading effect.

3. With generator *B* off, tune generator *A* to the receiver frequency and set the output level of generator *A* to the minimum level that will produce a good reference indication on the spectrum analyzer.

4. Note the generator level and the level of the displayed signal.

5. Tune generator *A* to the adjacent channel above or below the receiver frequency. (Generator *A* frequency = receiver frequency +/− channel spacing.)

---

[11] EIA Standard RS-204-B, p. 13, para. 13, Electronic Industries Association, Washington, D.C. This is an older version of the intermod test and is superseded by RS-204-C, p. 16, para. 14.2. It is presented here only because many receivers whose specification figure was derived from this test are still around and probably will be for some time to come.

6. Tune generator *B* to the second channel above or below the receiver (but on the same side of the desired channel as generator *A*). (Generator *B* frequency = receiver frequency $+/-$ twice the channel spacing.)

7. Coherently increase the output level of generators *A* and *B* until an intermod signal appears on the display (at the same point previously occupied by the reference signal). Increase the output level of generators *B* and *A* until the intermod signal reaches the reference level. Note the output level of either generator.

8. The difference in this output level and the reference level of step 4 is the intermod rejection figure.

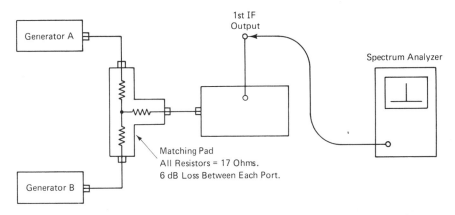

FIGURE 9-13. This test setup is helpful in conducting intermodulation tests with a spectrum analyzer.

Identifying Receiver-Produced Intermod

The following test procedure can be used to identify receiver-produced intermod. The principle upon which this test is based is described in detail in Chapter 5.

Test Procedure

1. Set up the equipment as shown in Figure 9-14. Monitor the IF or limiter test point. Use a VOM if a proper test set is not available.

2. Start with the antenna connected and the attenuator out. When the intermod signal is heard, note the meter reading. This is reference reading *A*.

3. Insert the attenuator and note the new meter reading when the intermod signal is present. This is reference reading *B*.

4. Disconnect the antenna and attenuator. Connect a signal generator to the receiver input. Set the generator to the receiver frequency and adjust the output level to produce reference reading *A* on the meter (see step 2).

5. Insert the attenuator between the signal generator and the receiver. Note the meter reading. If the meter reading is *significantly* higher than reference B (step 3), the intermod signal is being produced within the receiver. If there is no significant difference in the meter readings, the intermod must be produced at some point external to the receiver.

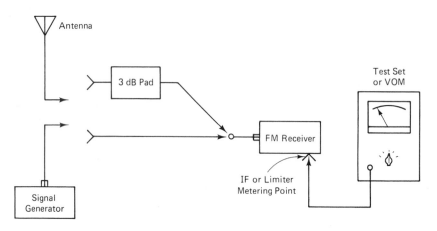

FIGURE 9-14. This setup is used to identify *receiver-produced* intermod interference.

SPURIOUS RESPONSE TESTS

The purpose of this test is to locate those frequencies at which the receiver exhibits an unexpected and undesired response. A typical front-end of an FM receiver is shown in Figure 9-15. Several possible spurious responses are listed below for this particular mixing scheme.

1. LO +/− IF = 27.2111 MHz and 5.8111 MHz. The 27.2111 MHz frequency falls in the middle of the 27-MHz citizens band service. Imagine the potential for interference from this!

2. 3(LO) +/− IF = 38.8333 MHz and 60.2333 MHz.

3. 2(LO) +/− IF = 43.7222 MHz and 22.3222 MHz.

4. 18(LO) +/− IF = 286.4998 MHz and 307.8998 MHz.

These are just a few of the possibilities. In order to test a receiver for spurious, it is obvious that the test must be run over a very wide range of frequencies. Generally, the test includes those frequencies from the lowest

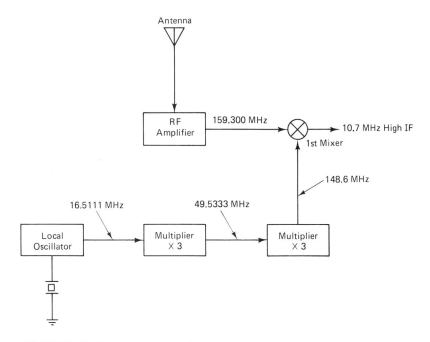

FIGURE 9-15. Block diagram of a typical front end of an FM–VHF communications receiver. The local oscillator operates at a frequency which is one-ninth the injection frequency. The oscillator and multiplier stages must be well shielded to minimize spurious responses.

radio frequency amplified in the receiver to a frequency several times the desired receiver frequency. Make sure that any spurious response is not caused by signal generator spurious, such as harmonics of the generator oscillator. A spectrum analyzer should be used to check the spectral purity of the signal generator to avoid possible confusion. The frequencies between the two adjacent channels can be omitted since the adjacent channel selectivity test covered this area.

The basic setup for this test is shown in Figure 9-16. The spectrum analyzer is simply used as an aid in locating spurious responses. If a spectrum analyzer is not available, a scope connected to the output of the IF chain may be helpful in locating spurious responses.

Test Procedure[12]

1. Set up the equipment as shown in Figure 9-16.
2. With generator B off, tune generator A to the receiver frequency and perform the 12-dB Sinad sensitivity test. Note and record the 12-dB Sinad sensitivity level (in dBm).

---

[12] This procedure is based on EIA Standard RS-204-C, p. 15, para. 13.2 (Electronic Industries Association, Washington, D.C., 1982). See also footnote 1.

3. Turn off generator *A* so that it won't interfere with the spurious search.

4. Set the output level of generator *B* to a high level (30,000 μV or so). Modulate generator *B* 60% of maximum system deviation using a 400-Hz audio-modulating tone. (This deviation level is +/− 3 kHz for narrow-band FM.)

5. Tune the signal generator over the range of interest while watching for any indication of a spurious response.

6. When a spurious response frequency is located, perform the following steps:

   A. Reduce the output of generator *B* to minimum.
   B. Set up generator *A* as in step 2, except set the output level *3 dB higher* than the level recorded in step 2. This will produce a better than 12-dB Sinad reading.
   C. Increase the output of generator *B* until the Sinad reading is degraded to 12 dB Sinad. Fine-tune generator *B* to maximize the degradation. If necessary, readjust the generator *B* output level to produce exactly 12 dB Sinad.
   D. At this point, the difference in the output level of generators *A* and *B* (in dB) is the spurious response attenuation of the receiver at this particular frequency.
   E. Repeat steps 3 through 6 until all spurious responses within the range of interest are found and measured.

Hint: If you are only interested in the spurious responses that are above a specified level, the procedure can be shortened by setting the generator to that particular level and using the spectrum analyzer as an indicator. For example, if you are only interested in those spurious responses that are attenuated less than 80 dB, the generator level is set 80 dB higher than the 12-dB Sinad sensitivity level of the receiver. A signal at the desired frequency and at the 12-dB Sinad level is fed into the receiver. The spectrum analyzer (at the first mixer output) is then adjusted to produce a reference display. Then the signal generator is set to a level 80 dB above the 12-dB Sinad sensitivity level and tuned through the frequency range of interest, while the spectrum analyzer display is carefully watched. Any spurious response that produces a display near or above the level of the reference display should be studied more carefully by the step-by-step procedure outlined in the test procedure.

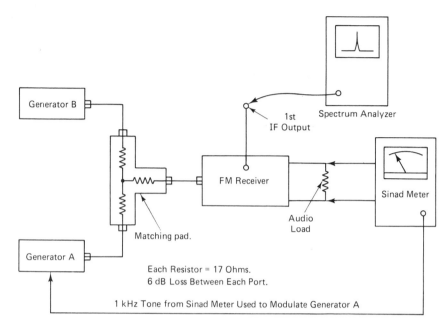

FIGURE 9-16. Test setup for conducting spurious response tests. The spectrum analyzer is used to locate spurious responses, then the step-by-step procedure is followed to determine the spurious response rejection figure at that particular frequency.

DISCRIMINATOR RESPONSE TESTS

The frequency response and proper alignment of the discriminator is very important in order for proper demodulation to be obtained. Improper frequency response in the discriminator can cause severe distortion, resulting in a poor Sinad ratio at the output.

The response pattern of a typical FM discriminator detector is shown in Figure 9-17. Notice that at one point the curve crosses the *zero-axis*. The point at which the *S-curve* crosses the zero-axis should correspond to the low IF center frequency, typically 455 kHz. The response on the linear portion of the S-curve should have a bandwidth greater than the maximum system deviation in order to accommodate fully modulated signals. The response of a discriminator can be determined in basically two ways: (1) by using a swept frequency in conjunction with a scope to display the response curve, and (2) by manually tuning the signal generator to various frequencies above and below the center frequency and using a voltmeter to measure the voltage at the discriminator output.

Before performing the procedures outlined below the discriminator should be aligned to the low IF frequency by feeding a precise low IF signal into the low IF chain and then zeroing the discriminator to this IF frequency. The discriminator is zeroed by tuning the secondary of the discriminator transformer.

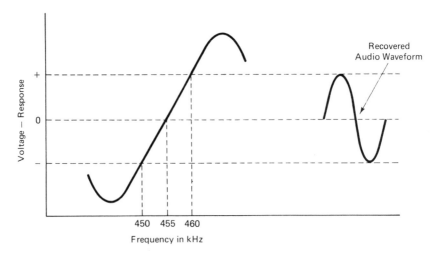

FIGURE 9-17. A typical response curve of a discriminator.

Test Procedure 1—Swept Frequency Method

1. Set up the equipment as shown in Figure 9-18.
2. Tune the signal generator to the receiver frequency.
3. Set the output level of the generator to a fairly high level—100 μV or more.
4. Set the scope horizontal sweep to *internal.*
5. Set the scope's coupling to "DC."
6. With no signal applied to the scope input set the trace to the zero centerline.
7. Connect the vertical input of the scope to the output of the discriminator.
8. If the trace moves up or down from the zero point, tune the receiver frequency-adjust to return the trace to the zero point.
9. Set the scope horizontal sweep to external.
10. Modulate the signal generator at maximum system deviation (+/− 5 kHz for narrow-band FM) using a 20-Hz modulating tone. The same modulating tone is also used to sweep the scope horizontal.

11. The scope display should look like Figure 9-19A.

12. Increase the generator deviation until the display looks like Figure 9-19B.

13. Decrease the generator deviation until the nonlinear portion of the curve and the noise are excluded from the trace. The noise is caused by the generator deviation exceeding the IF bandpass limits.

14. At this point, the generator deviation is the maximum linear range of the discriminator.

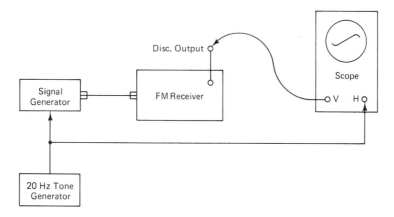

FIGURE 9-18. This setup can be used to run a discriminator response test with an FM signal generator. The 20-Hz tone is used to modulate the signal generator and to sweep the scope's horizontal.

Test Procedure 2—Nonswept Method

1. Set up the equipment as shown in Figure 9-20.

2. Set the signal generator to the receiver frequency and set the level to 100 μV or more. Do not modulate the signal generator.

3. The meter should indicate zero. If it doesn't, tune the receiver frequency-adjust until it does.

4. Tune the generator in 1-kHz increments above and below the center frequency, noting the meter reading at the various frequency points. Tune the generator above or below the center frequency until the meter reading starts back toward the zero level.

5. Use the data obtained to plot a discriminator response curve.

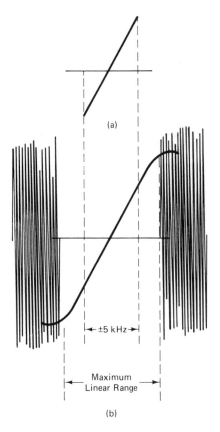

(a)

±5 kHz

Maximum
Linear Range

(b)

FIGURE 9-19. The display at A shows the response of a discriminator to a signal swept +/−5 kHz around the center frequency. The display at B shows how the response is affected by increasing the sweep width around the center frequency. The maximum linear range of the discriminator is determined by decreasing the sweep width of the generator (modulation level) until only the linear portion of the curve is displayed on the scope. The deviation at that point is the maximum linear range of the discriminator.

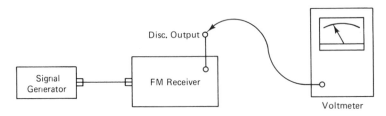

Disc. Output

Signal
Generator

FM Receiver

Voltmeter

FIGURE 9-20. This test setup can be used to determine the response of the discriminator by manually tuning the signal generator above and below the center frequency and using the data to plot a graph showing the response.

BIBLIOGRAPHY

Cushman Electronics, Inc. *Mobile Radio Testing—Using the Cushman CE-50A Series Communications Monitors—Part 2—Receivers.* San Jose, Calif., 1982.

Detwiler, William. "Sinad—A Valuable Measurement System" *Communications Magazine,* March 1979, Denver, Colo.

Electronic Industries Association. Standard #RS-204-C: "Minimum Standards for Land Mobile Communication FM or PM Receivers, 25-947 MHz." Washington, D.C., January 1982.

Institute of Electrical and Electronics Engineers, Inc. *"IEEE Test Procedure for Frequency-Modulated Mobile Communications Receivers."* New York, 1969.

General Electric Mobile Communications Division. "Mobile Radio Datafile Bulletin #1000-6, Subject: Receivers, General" Lynchburg, Va., April 1978.

Philco Corporation. *"Radio Communication System Measurements"* ©1952 Philco Corporation.

Zeines, Ben. *Electronic Communication Systems.* Englewood Cliffs, N.J.: Prentice-Hall, Inc., 1970.

10
ANTENNA AND TRANSMISSION LINE TESTS AND MEASUREMENTS

This chapter does not go into the theory of antennas and transmission lines as such. The concern of this chapter is with those tests and measurements that are useful in checking the performance of the antenna *system*, which includes the transmission line. Where practical, several different methods are presented for performing the various tests. Emphasis is placed upon the tests that are practical, avoiding those that are extremely difficult if not impossible to perform in the field.

ANTENNA RESONANCE

There are several ways in which the resonant frequency of an antenna can be determined. Some of the more practical methods are presented here. This assumes that the *approximate* resonant frequency of the antenna is known, by either the dimensions of the antenna or frequency range specified by the manufacturer.

The Noise Bridge Method

The operation of a basic noise bridge was described in Chapter 3. This instrument is very useful in antenna work.

Test Procedure

1. Hook up the equipment as shown in Figure 10-1A. If at all possible, the antenna should be connected directly to the noise bridge. If a long transmission line must be used, it should be one-half wavelength long (or integral multiples thereof) to produce accurate results.

2. Set the reactance dial to zero and the resistance dial to the *anticipated* radiation resistance of the antenna (typically, 50 Ω).

3. A hiss will be produced in the receiver by the broadband noise generator. Tune the receiver frequency around the expected resonant frequency until a null is found. The frequency at which a null is found is the resonant frequency of the antenna.

An alternative to using a communications receiver to find the null is to use a spectrum analyzer as shown in Figure 10-1B. The null point in the noise is the resonant frequency of the antenna.

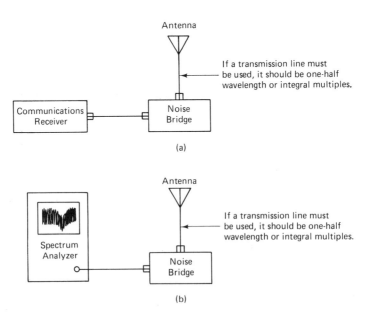

(a)

(b)

FIGURE 10-1. Two methods of using a noise bridge to find an antenna's resonant frequency are shown here. At A the null is found by listening to the audio output of the receiver as the receiver is tuned around the expected resonant frequency of the antenna. At B the null in the noise is visible on the analyzer when the analyzer is set to sweep across the antenna's resonant frequency.

The SWR Bridge Method

A *VSWR bridge* such as the one shown in Figure 10-2 provides a useful means of testing antennas and transmission lines. The VSWR bridge separates the forward wave from the reflected wave. The frequency at which the reflected wave drops to a minimum is taken as the resonant frequency of the antenna. There are several methods by which a VSWR bridge can be used to find the antenna's resonant frequency. Several methods are presented here.

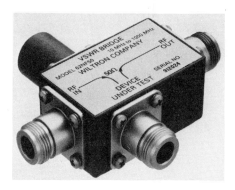

FIGURE 10-2. A VSWR bridge such as this one can be used to perform a variety of antenna and transmission line tests and measurements. Photo courtesy Wiltron Company.

The VSWR Bridge/Sweep Generator Method:

1. Set up the equipment as shown in Figure 10-3.
2. Set up the sweep generator to sweep through a range of frequencies around the expected resonant frequency of the antenna.
3. The resonant frequency of the antenna is indicated by the point of maximum dip in the response curve. A signal generator can be used to place a marker signal on the response curve to identify the resonant frequency (see Figure 10-4).

The VSWR Bridge/Signal Generator/RF Voltmeter Method:

1. Set up the equipment as shown in Figure 10-5. The signal generator output and the RF voltmeter response should be reasonably flat over the range of frequencies required for the test.
2. Tune the signal generator through a range of frequencies around the expected resonant frequency of the antenna.
3. The RF voltmeter will show a minimum indication when the signal generator frequency is equal to the resonant frequency of the antenna.

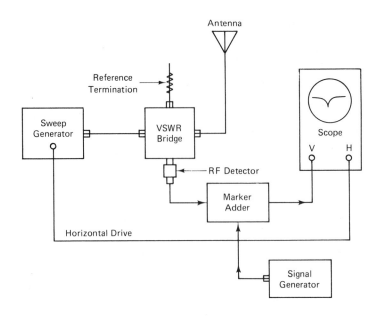

FIGURE 10-3. This setup will produce a graph of reflected signal amplitude versus frequency on the scope. The point of minimum response is the resonant frequency of the antenna.

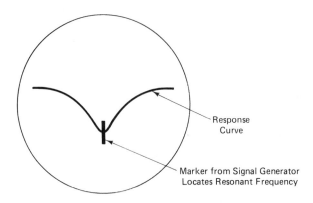

FIGURE 10-4. A marker signal is placed on the response curve to locate the antenna resonant frequency.

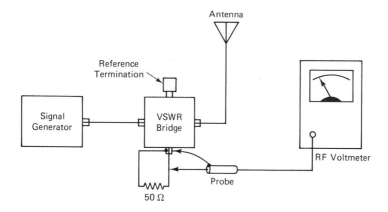

FIGURE 10-5. This setup shows how an RF voltmeter and signal generator can be used in conjunction with an SWR bridge to perform antenna resonance tests.

The VSWR Bridge/Signal Generator/Receiver Method:

1. Set up the equipment as shown in Figure 10-6.
2. Tune the signal generator and receiver around the expected resonant frequency of the antenna.
3. The receiver S-meter can be used to locate the null point. If the receiver has no S-meter, the signal generator can be amplitude modulated by a tone generator and then the audio output level of the receiver can serve as the null indicator.
4. The frequency of the signal generator and receiver at the null point is the resonant frequency of the antenna.

The VSWR Bridge/Signal Generator/Spectrum Analyzer Method:

1. Set up the equipment as shown in Figure 10-6.
2. Tune the signal generator around the expected resonant frequency of the antenna, with the spectrum analyzer covering the same frequency range.
3. The point at which the signal generator is tuned to the resonant frequency of the antenna the height of the display on the spectrum analyzer will decrease to a minimum (see Figure 10-7).

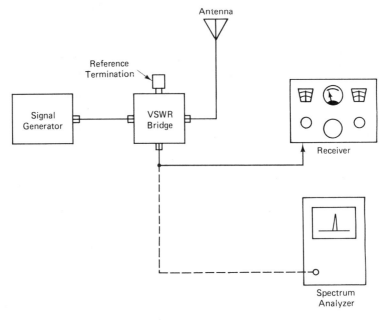

FIGURE 10-6. Typical setup for using a communications receiver (or spectrum analyzer) and a signal generator in conjunction with the VSWR bridge to perform antenna tests.

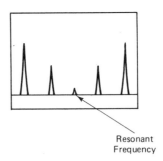

FIGURE 10-7. When the signal generator is tuned to the resonant frequency of the antenna the display signal will drop to minimum height.

The VSWR Bridge/Tracking Generator Method:

1. Set up the equipment as shown in Figure 10-8.
2. Set the tracking generator to cover the frequency range around the expected resonant frequency of the antenna.

3. The point at which the response curve reaches a minimum level corresponds to the resonant frequency of the antenna.

Another method of using a sweep generator to determine an antenna's resonant frequency is described later in this chapter (see "Locating Transmission Line Faults").

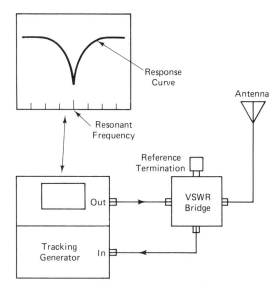

FIGURE 10-8. Typical setup showing how a tracking generator is used in conjunction with the VSWR bridge to perform antenna tests.

CHECKING THE ANTENNA SYSTEM MATCH

In order for maximum power transfer to occur, the antenna system (load) must be matched properly to the transmitter (source). If a mismatch occurs, power will be reflected back from the antenna toward the transmitter. By measuring the amount of power reflected from the antenna and comparing this with the amount of forward power travelling to the antenna, the degree of mismatch can be determined.

Several different terms are used to define the degree of mismatch. These are: (1) VSWR, (2) forward-to-reflected power ratio, (3) return loss, and (4) reflection coefficient. All of these terms can be determined from the forward and reflected power measurements. Now, let's look at each one of these terms in more detail.

Forward/Reflected Power Measurement

This is the method most commonly used by professional communications technicians in checking antenna mismatch. Directional wattmeters such as the one in Figure 10-9 are used to measure the forward power and reflected power separately. The higher the forward/reflected power ratio, the better the antenna match. Forward-to-reflected power ratios should be at least 10/1 or simply 10 in typical installations, and figures far better than this are easily attainable in properly adjusted antenna systems. Any problem in the antenna or connecting transmission line will show up as an increase in the *reflected* power.

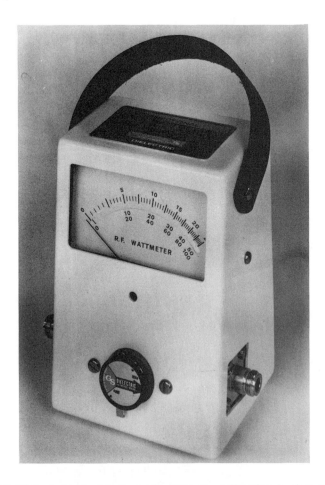

FIGURE 10-9. A typical in-line directional wattmeter is shown here. Photo courtesy Coaxial Dynamics, Inc.

Test Procedure with Directional Wattmeter:

1. Hook up the equipment as shown in Figure 10-10.

2. Measure the forward power by turning the coupling element so that the arrow points in the direction of forward power flow, from the transmitter toward the antenna. Note the reading.

3. Measure the reflected power by turning the coupling element so that the arrow points in the direction of the reflected power flow, from the antenna toward the transmitter. Note the reading.

4. Calculate the forward-to-reflected power ratio by:

$$\frac{\text{forward power}}{\text{reflected power}}$$

this ratio should be at least 10, preferrably more for a properly operating antenna system.

Hint: When the reflected power is very low in comparison to the forward power, a more accurate measurement of reflected power can be made by using a more sensitive element to measure the reflected power. Be careful not to measure the forward power with the more sensitive element in the meter.

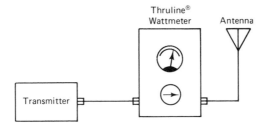

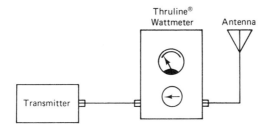

FIGURE 10-10. The typical hookup for using the directional wattmeter to measure forward and reflected power is shown here.

Some wattmeters, such as the Telewave™ 44A described in Chapter 3, provide a switch to measure forward and reflected power. Other wattmeters, such as the Heathkit HM-2140™ described in Chapter 3, provide two separate meters. One meter indicates forward power while the other meter indicates reflected power. Both readings can be observed simultaneously.

Long Transmission Lines versus Test Results   In the test just described, the forward power versus reflected power at the antenna feedpoint will not be substantially different from the measurements taken from the transmitter end, provided that the power loss in the transmission line is not significant. If the power loss in the transmission line is significant (as it will be in longer runs), the forward-to-reflected power ratio at the antenna feedpoint will be significantly worse than the levels measured at the transmitter end of the line. A couple of examples will serve to clarify this point.

Example: A 100-foot length of RG-58 is used to connect the transmitter to the antenna (see Figure 10-11). The transmitter operates at 150 MHz. The power loss of a 100-foot section of RG-58 at 150 MHz will be 6.5 dB (see Figure 10-22.) Suppose that forward and reflected power measurements at the transmitter end show the following: forward power = 50 W and reflected power = 1 W. Thus, the forward/reflected power ratio is 50/1 or just 50 at the transmitter end of the line. Normally, this figure is considered excellent. But, more importantly, what is the forward/reflected power ratio that actually exists at the antenna feedpoint?

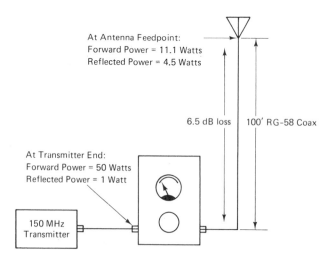

At Antenna Feedpoint:
Forward Power = 11.1 Watts
Reflected Power = 4.5 Watts

6.5 dB loss   100′ RG-58 Coax

At Transmitter End:
Forward Power = 50 Watts
Reflected Power = 1 Watt

150 MHz
Transmitter

FIGURE 10-11. This illustration shows how a lossy transmission line can hide true forward-to-reflected power ratios which exists at the antenna feedpoint. When the measurement is made at the transmitter end of the line the situation looks much better than it really is.

Solution: The forward power at the antenna feedpoint will be 6.5 dB below the forward power at the transmitter end, while the reflected power at the antenna feedpoint will be 6.5 dB above the reflected power measured at the transmitter end. Referring to Figure 1-2 in Chapter 1, a decibel change of 6.5 dB is equal to a power ratio of approximately 4.5. Thus, the forward power at the antenna end of the transmission line is 50/4.5 = 11.1 W. The reflected power at the antenna of the transmission line will be 4.5 × 1 = 4.5 W. Thus, the forward-to-reflected power ratio at the antenna feedpoint is 11.1/4.5 = 2.47/1. This is a very poor ratio. Compare this to the 50/1 ratio that was obtained at the transmitter end of the line. Normally, such a lossy line would not be used in such long runs. Let's see what would happen to our readings if a larger transmission line such as RG-8 is used.

Example: A 100-foot length of RG-8 has a loss of approximately 2.7 dB at 150 MHz. If the same conditions exist at the antenna feedpoint as existed in the previous example, calculate the forward/reflected ratio that exists at the transmitter end with the RG-8 cable instead of the RG-58 cable.

Solution: The forward/reflected power ratio at the antenna end will remain the same: 2.47/1. Again, the forward power measured at the transmitter end is 50 W. The forward power reaching the antenna end of the line is 50/1.9 = 26.3 W (1.9 is the power ratio which corresponds to 2.7 dB). The reflected power at the antenna end of the line will be 26.3/2.47 = 10.6 W. The reflected power reaching the transmitter end will be 10.6/1.9 = 5.6 W. Thus, the forward/reflected power ratio at the transmitter end will be 50/5.6 = 8.9/1, still significantly higher than the 2.47/1 ratio at the antenna feedpoint. This clearly demonstrates that forward and reflected power measurements taken at the transmitter end of a lossy transmission line can be very misleading if the line loss is not taken into consideration. In the first example, the forward/reflected power ratio at the transmitter end was 50/1 but only 2.47/1 at the antenna end. In the second example, with RG-8 cable, the forward/reflected power ratio at the transmitter end was 8.9/1, compared to 2.47/1 at the antenna end. The 8.9/1 ratio is much closer to the actual ratio that exists at the antenna feedpoint but still significantly different.

There is a short-cut method that can be used to determine the actual ratio that exists at the antenna feedpoint by determining the ratio at the transmitter end and then applying a factor determined by the transmission line loss. Perform the following steps:

1. Calculate the forward/reflected power ratio at the transmitter end.
2. Find the transmission line loss in dB from Figure 10-12. Then *double* this figure. (The figure is doubled because the signal must travel up and down the line.)
3. Find the dB figure on the power scale on the graph in Figure 1-2.

4. Find the ratio that corresponds to this dB figure.

5. Divide the figure in step 1 by this figure to get the actual forward-to-reflected power ratio that exists at the antenna feedpoint.

Example: Going back to the last example, the forward/reflected power ratio at the transmitter end was 8.9/1. The RG-8 cable has a 2.7 dB loss (one-way). This is multiplied by 2 to get a 5.4-dB loss. Looking on the power scale in Figure 1-2, the 5.4 dB equals a power ratio of 3.6. The actual forward/reflected power ratio at the antenna then is 8.9/3.6 = 2.47.

The graph in Figure 10-12 correlates the forward/reflected power ratio at the two ends of the transmission line for different cable losses.

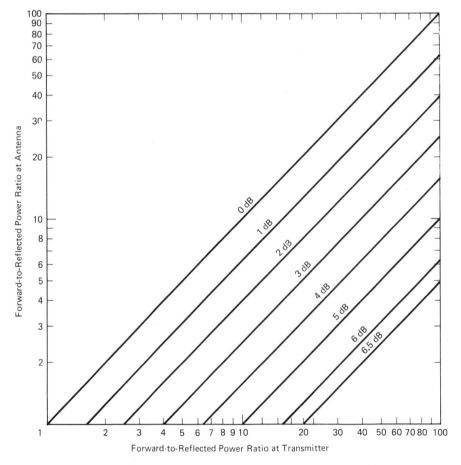

FIGURE 10-12. This graph can be used to convert forward/reflected power ratios at the transmitter to the actual forward/reflected power ratio at the antenna feedpoint if the line loss is known.

Example: A length of transmission line has a loss of 2 dB at a given frequency. If the forward/reflected power ratio is 20 at the transmitter end, what will be the forward/reflected power ratio at the antenna feedpoint?

Solution: Locate 20 on the horizontal scale. Move up this line vertically to intersect the 2-dB line. The corresponding forward/reflected power ratio at the antenna is then read from the vertical scale. It is 8.

VSWR MEASUREMENT

The term *VSWR* or simply *SWR* is often used to indicate the degree of antenna system match or mismatch. VSWR stands for *voltage-standing-wave-ratio.* This refers to points of maximum and minimum voltage that are set up on the transmission line due to the presence of a forward wave and reflected wave. The greater the mismatch, the greater the ratio of the maximum to minimum voltage or VSWR. In a perfectly matched antenna system, the VSWR will be 1/1. A VSWR of 1.5/1 is considered very good. Usually, there is no cause for concern unless the VSWR exceeds a 2/1 ratio. Even higher ratios are tolerable in some cases.

If the source and load impedance is known, the VSWR can be calculated by the formula:

$$VSWR = RL/RS \text{ or } RS/RL$$

where RS = source resistance or impedance and RL = load resistance or impedance. The larger value is used as the numerator so that the ratio is greater than 1.

Example: If a transmitter has a 50-Ω output impedance and is loaded into a 100-Ω load, the VSWR will be $100/50 = 2/1$.

There are several ways in which the VSWR can be determined. One method is to simply use a meter which is designed to indicate VSWR directly. Several types of instruments for measuring VSWR are available. The Heathkit HM-2140™ shown in Figure 3-7 measures SWR by using a sensitivity control to set the meter to full scale on the forward signal. The meter is then switched to read the reflected signal. The greater the reflected signal, the greater the VSWR reading.

An automatic computing SWR and power meter of the analog type is shown in Figure 10-13. No external calibrating control is needed. An automatic computing circuit simplifies the measurement. This is a great advantage when tuning an antenna for minimum SWR. Figure 10-14 shows a multifunction instrument that automatically computes SWR and displays it on a digital readout. This instrument also measures foward and reflected power, return loss (discussed later in this chapter), and amplitude modulation.

FIGURE 10-13. This instrument automatically computes the VSWR. The VSWR and reflection coefficient is read on one meter while the forward power is read on the other meter. Photo courtesy Signalcrafters, Inc.

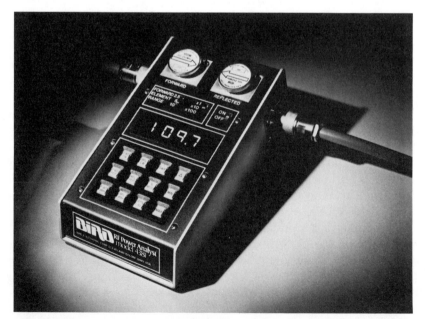

FIGURE 10-14. This sophisticated and versatile instrument can read forward power, reflected power, VSWR, return loss, and AM percent among other features. Photo courtesy Bird Electronic Corporation.

Computing VSWR from Forward and Reflected Power Measurements

If the forward and reflected power is known, the VSWR can be computed from the following formula:

$$VSWR = \frac{1 + \sqrt{1/R}}{1 - \sqrt{1/R}}$$

where R = forward/reflected power ratio.

Example: Suppose the forward power is 75 W and the reflected power is 5 W. What is the VSWR?

Solution: The ratio $(R) = 75/5 = 15$. Substituting into the formula we have:

$$VSWR = \frac{1 + \sqrt{1/15}}{1 - \sqrt{1/15}} = \frac{1 + 0.258}{1 - 0.258} = \frac{1.258}{0.742} = 1.695 \text{ or } 1.7/1$$

If a transmission line has a significant loss, the VSWR measured at the transmitter end will be significantly better than the VSWR that really exists at the antenna feedpoint. This follows the same principle discussed earlier for the forward/reflected power ratios which exist at either end of a lossy transmission line. An example will show how line loss affects VSWR measurement from the transmitter end.

Example: Suppose a transmitter is connected to an antenna by a length of transmission line that has a loss of 2 dB at the transmitter frequency. A directional wattmeter is used to take forward and reflected power measurements at the transmitter end. The forward power measures 45 W and the reflected power measures 2 W. The VSWR at the transmitter end is:

$$VSWR = \frac{1 + \sqrt{1/22.5}}{1 - \sqrt{1/22.5}} = \frac{1.21}{0.79} = 1.53/1$$

The VSWR at the transmitter end is 1.53/1, a good figure. But, what really counts is the VSWR at the antenna feedpoint. We can use the chart in Figure 10-12 to find the forward/reflected power ratio at the antenna. First, we must know the forward/reflected ratio at the transmitter. It is $45/2 = 22.5/1$. This corresponds to a forward/reflected ratio of 9/1 at the antenna feedpoint. Substituting into the formula, we have:

$$VSWR = \frac{1 + \sqrt{1/R}}{1 - \sqrt{1/R}} = \frac{1 + 0.333}{1 - 0.333} = \frac{1.333}{0.666} = 2.0/1$$

The VSWR at the antenna feedpoint is 2/1 and only 1.53/1 at the transmitter end. Thus, it is shown that VSWR measurements taken at the transmitter end of a long and/or lossy transmission line will show a better VSWR figure than that which actually exists at the antenna.

The graph in Figure 10-15 is very useful in converting VSWR at the transmitter end to VSWR at the antenna feedpoint for various degrees of cable loss.

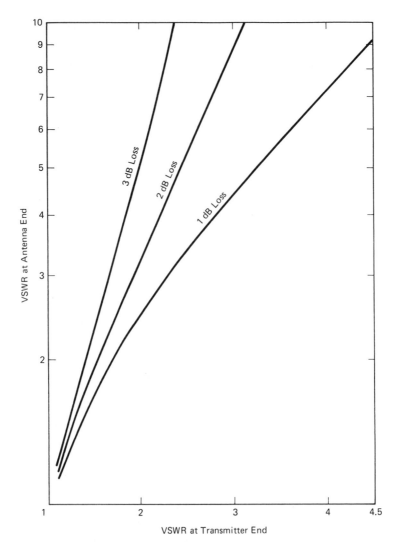

FIGURE 10-15. This graph correlates the measured VSWR at one end of the line with the actual VSWR which exists at the other end for the given values of line loss.

Example: Suppose a VSWR meter indicates a VSWR of 2/1 at the transmitter end of a transmission line. If the transmission line has a loss of 2 dB from one end to the other, what is the actual VSWR at the antenna feedpoint?

Solution: First, locate 2 on the horizontal scale of the graph in Figure 10-15. Move up this line vertically to intersect the 2-dB line. This corresponds to approximately 3.2 VSWR at the antenna feedpoint. Notice how much steeper the graph becomes as the transmission line loss factor increases. It approaches a vertical line at the higher loss factors.

The graph in Figure 10-16 converts forward/reflected power ratios to VSWR (and vice versa). The following formula can also be used to convert VSWR to the forward/reflected power ratio:

$$R = \frac{1}{\left(\dfrac{VSWR - 1}{VSWR + 1}\right)^2}$$

where R = forward/reflected power ratio.

Example: Use the formula to convert an SWR of 2.7/1 to the equivalent forward/reflected power ratio (R).

Solution: Substituting into the formula:

$$R = \frac{1}{\left(\dfrac{2.7 - 1}{2.7 + 1}\right)^2} = \frac{1}{\left(\dfrac{1.7}{3.7}\right)^2} = \frac{1}{(0.459)^2} = \frac{1}{0.211} = 4.74 = R$$

(This can be checked by referring to the graph in Figure 10-16.)

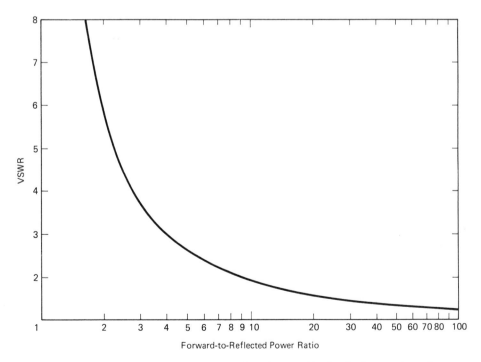

FIGURE 10-16. This graph correlates forward/reflected power ratio with the equivalent VSWR for ratios of 1 to 100.

The basic VSWR formula can be modified to find the VSWR that actually exists at the opposite end of a transmission line of *known* loss factor. The formula is:

$$VSWR \text{ (at antenna feedpoint)} = \frac{1 + \sqrt{K/R}}{1 - \sqrt{K/R}}$$

where K = power ratio (greater than 1), which is the equivalent of *twice* the dB loss of the cable. Twice the loss is used because the reflected signal is attenuated twice, once on the way up and once on the way down. The power ratio can be determined from Figure 1-2 or Table 1-1. R = forward/reflected power ratio as measured at the transmitter end of the line. $VSWR$ = actual $VSWR$, which exists at the antenna feedpoint.

Example: Forward and reflected power measurements are taken at the transmitter end of a transmission line which has a (one-way) loss of 3 dB. The forward power = 35 W, the reflected power = 2 W. What is the VSWR at the antenna feedpoint?

Solution: The 3 dB is multiplied by 2 = 6 dB. From Figure 1-2, 6 dB corresponds to a power ratio (K) of 4, and R = forward/reflected = 35/2 = 17.5. Substituting these values into the formula, we have:

$$VSWR \text{ (at antenna feedpoint)} = \frac{1 + \sqrt{4/17.5}}{1 - \sqrt{4/17.5}} = \frac{1 + \sqrt{0.228}}{1 - \sqrt{0.228}} = \frac{1.477}{0.523}$$

$$= 2.82$$

VSWR versus Transmitter Tuning

If a transmitter must be tuned up to an antenna which is showing a high VSWR, the wattmeter or VSWR meter should be connected into the transmission line by a one-half wavelength section of transmission line. This one-half wavelength section of line should include the *through-line* section of the wattmeter or VSWR meter (see Figure 10-17). If the antenna has a low VSWR, this precaution is not necessary. However, if the antenna shows a significant VSWR and the transmitter tuning is optimized using an in-line wattmeter connected with a cable which is not a half wavelength, the tuning of the transmitter will change when the wattmeter and connecting line is removed. Figure 10-18 on page 352 shows what the length of connecting cable should be when used with a Bird 43™ wattmeter. This also takes into account the length of the through-line section of line in the wattmeter. The length of cable required for any other meter can be calculated from the following formula:

$$\text{Cable length (half wave)} = \frac{5904 \, V}{F} - \text{L}$$

where cable length = inches, V = velocity factor, F = frequency in MHz, and L = length of in-line section of meter (inches).

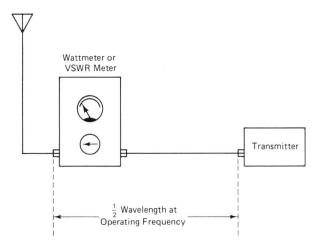

FIGURE 10-17. This illustrates how a one-half wavelength section of line (including the wattmeter line length) should be used when tuning a transmitter into a line showing a higher than normal VSWR.

The velocity factor of RG-8/U and RG-58/U coaxial cable is 0.66. The velocity factor of RG-8/U foam coaxial cable is 0.80 and 0.79 for RG-58/U foam cable.

Example: A certain power meter has an in-line section length of 3 1/2 inches. What length of RG-8/U foam coaxial cable is required to form a one-half wavelength section of line at a frequency of 160 MHz?

Solution: Substituting into the formula:

$$\text{Cable length} = \frac{5904 \times 0.8}{160} - 3.5 = 26 \text{ in.}$$

REFLECTION COEFFICIENT

The term *reflection coefficient* is often used to indicate the degree of mismatch. The reflection coefficient factor ranges from 0 (for a perfect match) to 1 (for the worse possible mismatch). The reflection coefficient can be determined from either the forward and reflected power measurements or from the VSWR measurement. The formulas are:

$RC = \sqrt{Pr/Pf}$, where RC = reflection coefficient, Pr = reflected power, and Pf = forward power

$RC = \sqrt{1/r}$, where RC = reflection coefficient and r = forward/reflected power ratio

$RC = \dfrac{VSWR - 1}{VSWR + 1}$

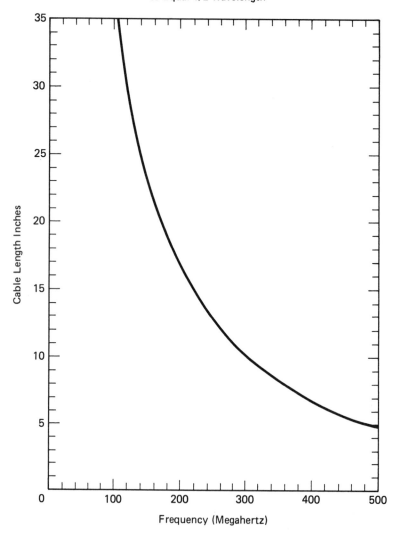

Required Length of RG-8/U Cable with Connectors
when Added to Bird Model 43 Thruline
to Equal 1/2 Wavelength

Note:

1. When using UHF Plug 259 from cable length is measured from
   tip to tip to center pin of plugs.
2. When using other connectors the cable length is measured
   from end to end of outer conductor of connectors.

FIGURE 10-18. This chart indicates the length of RG-8/U cable necessary to form a 1/2 wavelength section of line when used with the Bird
43 wattmeter. Courtesy Bird Electronic Corporation.

The graph in Figure 10-19 correlates forward/reflected power ratios from 1 to 100 with the equivalent reflection coefficient. The graph in Figure 10-20 correlates VSWR with the equivalent reflection coefficient. These graphs will enable you to convert forward/reflected power measurements or VSWR to the equivalent reflection coefficient at a glance. If a very high degree of accuracy is necessary or desirable, it is best to use the formula rather than the graph, since the graph is only a close approximation.

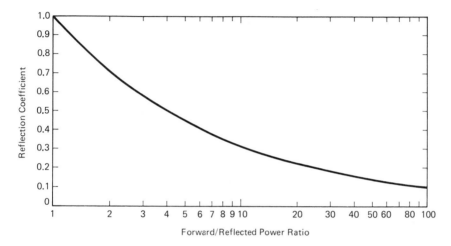

FIGURE 10-19. This graph converts forward/reflected power ratios from 1 to 100 to the equivalent reflection coefficient factor.

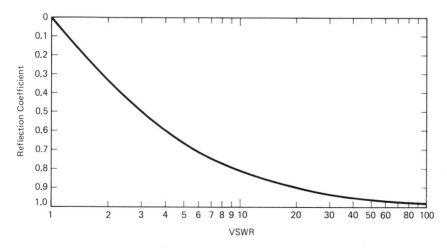

FIGURE 10-20. This graph converts VSWR to the equivalent reflection coefficient factor and vice versa.

RETURN LOSS

You may occasionally run into the term *return loss*. Return loss is a comparative measurement of the reflected signal or returned signal expressed as so many decibels below the level of the forward signal. Thus, the return-loss factor can be used to indicate the degree of mismatch. Several methods of determining return loss are described here.

Using the Directional Wattmeter to Determine Return Loss

Return loss can be determined by using a directional wattmeter to measure the forward and reflected power and then applying the following formula:

$$\text{Return loss (in dB)} = 10 \log \frac{Pf}{Pr}$$

where Pf = forward power and Pr = reflected power. In cases where the forward to reflected power ratio is very high, a more accurate measurement of the reflected power can be made by using a lower power (more sensitive) plug-in wattmeter element. In the case of the switchable type meter, the meter is switched to a lower scale when the reflected power is measured. CAUTION: *Make sure the arrow on the element is pointing toward the transmitter or switch is in the "reverse power" position!*

Example: Suppose the forward power measurement is 30 W and the reflected power measurement is 0.5 W. What is the return loss?

Solution: Substituting into the formula we have:

$$\text{Return loss (dB)} = 10 \log \frac{30}{0.5} = 10 \log 60 = 10(1.778) = 17.78 \text{ or } 17.8 \text{ dB}$$

One advantage of measuring the return loss in dB is that the loss of the transmission line (in dB) is simply subtracted from the return loss figure measured at the transmitter end of the line to give the actual return loss at the antenna feedpoint. However, the loss of the transmission line in dB must be doubled before being subtracted from the return loss figure.

Example: Suppose that in the preceding example the return loss of 17.8 dB is measured at the transmitter end of the transmission line and the transmission line has a (one-way) loss of 2 dB. What is the return loss at the antenna feedpoint?

Solution: The return loss at the antenna feedpoint is then:

$$17.8 \text{ dB} - 2(2 \text{ dB}) = 17.8 \text{ dB} - 4 \text{ dB} = 13.8 \text{ dB}$$

If it is desired to convert return loss in dB to VSWR, the following formula can be used:

$$VSWR = \frac{1 + \sqrt{\dfrac{1}{\text{antilog } (L/10)}}}{1 - \sqrt{\dfrac{1}{\text{antilog } (L/10)}}}$$

where L = return loss in dB.

Example: To convert a return loss of 22 dB to VSWR:

$$VSWR = \frac{1 + \sqrt{\dfrac{1}{\text{antilog } (22/10)}}}{1 - \sqrt{\dfrac{1}{\text{antilog } (22/10)}}} = \frac{1 + \sqrt{\dfrac{1}{\text{antilog } 2.2}}}{1 - \sqrt{\dfrac{1}{\text{antilog } 2.2}}} = \frac{1 + \sqrt{\dfrac{1}{158.5}}}{1 - \sqrt{\dfrac{1}{158.5}}}$$

$$= \frac{1 + 0.079}{1 - 0.079} = \frac{1.079}{0.921} = 1.17$$

The graph in Figure 10-21 gives a close approximation of return loss in dB versus VSWR.

Using the VSWR Bridge to Measure Return Loss

The principle of using the VSWR bridge in conjunction with a signal generator to find the resonant frequency of an antenna was described earlier in this chapter. The actual amount of return loss can be measured in the same manner with a VSWR bridge such as the one in Figure 10-2. The VSWR bridge can be used in conjunction with other instruments to determine the return loss.

Spectrum Analyzer/Signal Generator Method:

1. Set up the equipment as shown in Figure 10-6 but first disconnect the antenna from the test port.
2. Tune the signal generator (CW signal) to the frequency to be tested and tune the spectrum analyzer to the same frequency.
3. Increase the level of the signal generator until the spectrum analyzer display reaches the 0-dB reference mark at the top of the scale.
4. Connect the antenna to the test port and note the level of the signal displayed on the spectrum analyzer.
5. The difference in this signal level and the reference level (in dB) is the return loss at that point of measurement. If the transmission line has significant loss, the actual return loss at the antenna can be determined by subtracting twice the amount of cable loss from the return loss figure.

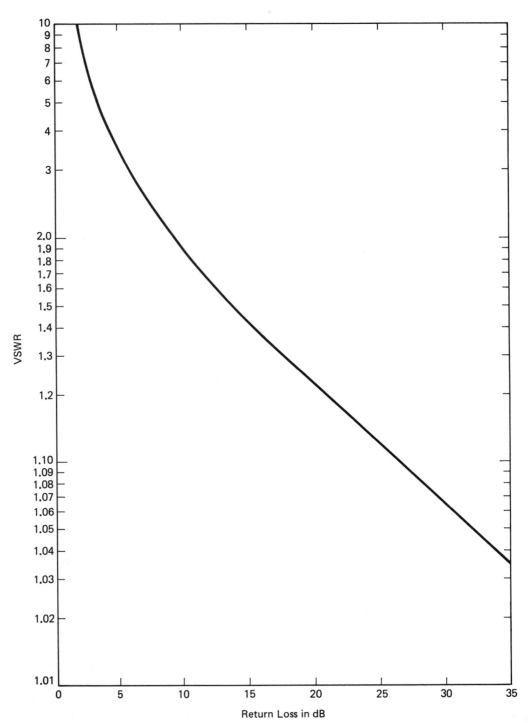

FIGURE 10-21. A conversion chart of VSWR versus return loss.

Communications Receiver/Signal Generator Method:

1. Set up the equipment as shown in Figure 10-6.

2. With the test port open (antenna disconnected), tune the signal generator to the frequency to be tested and increase the generator output level to establish a reference level on the receiver's S-meter. Note the generator output level and S-meter level.

3. Connect the antenna and increase the signal generator output level to produce the reference level on the S-meter. Note the signal generator output level.

4. The difference between the signal generator level in steps 2 and 3 (in dB) is equal to the return loss.

VSWR Bridge with Detector

Manufacturers of VSWR bridges also make detectors that can be used with the bridge. The detector plugs into the RF output port of the bridge. The output of the detector is then a dc voltage that can be measured with a dc voltmeter. The dc voltage is proportional to the *reflected* RF signal.

Test Procedure

1. Hook up the equipment as shown in Figure 10-22.

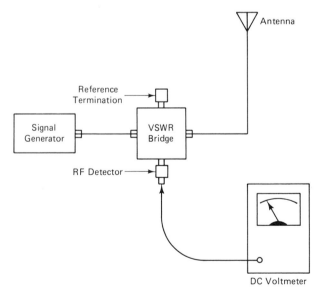

FIGURE 10-22. Typical setup showing how a signal generator and dc voltmeter can be used with a VSWR bridge and RF detector to determine return loss.

2. Connect the antenna to the test port and tune the signal generator to the desired test frequency.

3. Increase the output level of the signal generator to produce a reference level indication on the dc voltmeter. Note the generator level and the dc voltmeter reference level.

4. Disconnect the antenna, leaving the test port open. It may be necessary to switch the dc voltmeter to a higher range temporarily.

5. Decrease the signal generator output level until the dc voltmeter indicates the reference level. Note the signal generator output level.

6. The difference in the generator levels in steps 3 and 5 is equal to the return loss of the antenna.

MEASURING TRANSMISSION LINE LOSS

The chart in Figure 10-23 correlates typical values of attenuation for various types of transmission line at various frequencies. Cable loss also can

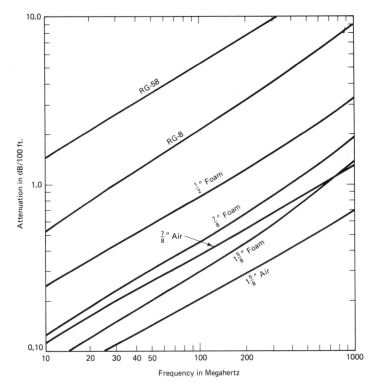

FIGURE 10-23. This graph shows typical losses of various types of transmission lines at frequencies between 10 MHz and 1,000 MHz. Courtesy Decibel Products, Inc.

increase as the cable deteriorates due to exposure to the elements over a long period of time. If water is getting into the cable, the loss will increase. Transmission line losses can be measured in several ways. The following methods are typical ways of measuring this loss.

The Directional Wattmeter Method 1

1. Set up the equipment as shown in Figure 10-24.
2. Measure the forward power entering the transmission line from the transmitter end, as shown in Figure 10-24A. Note the wattmeter reading.
3. Measure the forward power leaving the transmission line at the load end, as shown in Figure 10-24B. Note the wattmeter reading.
4. Compute the transmission line loss (in dB) from the following formula:

$$\text{Loss (dB)} = 10 \log(P1/P2)$$

where P1 = power entering transmission line and P2 = power leaving transmission line.

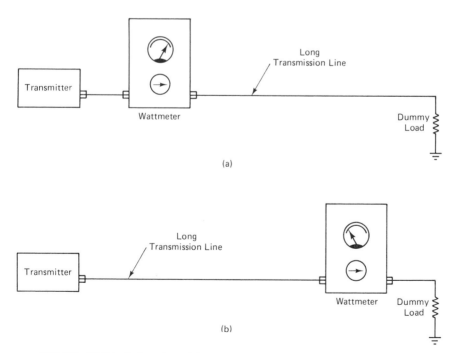

FIGURE 10-24. Setup in which an in-line wattmeter is used to: (a) measure the power entering the transmission line and (b) measure the power leaving the transmission line. The loss of the line can then be computed by dB loss = 10 log(P1/P2).

Directional Wattmeter Method 2

1. Set up the equipment as shown in Figure 10-25. The far end of the line is left open (not terminated) for this test.

2. Key the transmitter and measure the forward power. Note the wattmeter reading.

3. Key the transmitter and measure the reflected power. Note the wattmeter reading.

4. Compute the transmission line loss from the following formula:

$$\text{Line loss (in dB)} = \frac{10 \log (Pf/Pr)}{2}$$

where: Pf = forward power and Pr = reflected power. The divisor (2) must be used because the signal is attenuated as it travels in *both* directions, down the line and back. The attenuation is then the one-way loss from one end of the line to the other.

Attenuation per Unit Length: The attenuation of a transmission line in terms of dB per hundred feet can be computed from the following formula:

$$\text{dB/100 ft} = \frac{\text{dB loss} \times 100}{\text{line length in ft}}$$

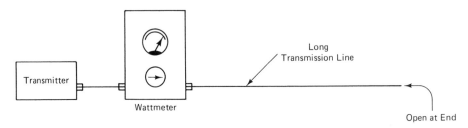

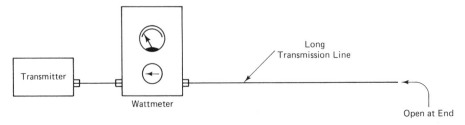

FIGURE 10-25. This setup shows how a transmission line loss can be determined from forward and reflected power measurements at one end of the line, the other end of the line being left open. The line loss in db is found by: dB loss = 10 log(P1/P2)/2.

Example: Suppose the attenuation of a 180-foot length of transmission line has been found to be 4.5 dB. Calculate the loss per 100 feet.

Solution: Substituting into the formula above yields:

$$dB/100 \text{ ft} = \frac{4.5 \times 100}{180} = \frac{450}{180} = 2.5 \text{ dB}/100 \text{ ft}$$

The VSWR Bridge/Spectrum Analyzer Method

1. Set up the equipment as shown in Figure 10-26.
2. Tune the signal generator and spectrum analyzer to the frequency to be tested.
3. With the test port of the SWR bridge open (the transmission line disconnected), increase the signal generator output level to produce a reference display on the spectrum analyzer. Note the reference level.
4. Connect the transmission line to the test port of the SWR bridge. Leave the far end of the line open.
5. Note the level of the display on the spectrum analyzer. The difference in this level and the reference level in step 3 is the two-way attenuation of the transmission line. The one-way attenuation of the transmission line is one-half this amount.

Other methods, such as those described for measuring return loss, can be applied to this situation as well. Just remember to take a reference reading with the VSWR bridge test port open (or shorted) and then connect the device under test (antenna or transmission line) to get the comparative reading.

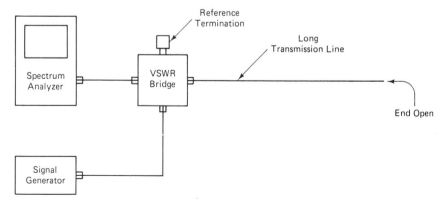

FIGURE 10-26. Typical setup showing how a signal generator and spectrum analyzer are used in conjunction with a VSWR bridge to determine transmission line loss.

LOCATING TRANSMISSION LINE FAULTS

If an open or a short occurs on a transmission line somewhere between the transmitter output and the antenna feedpoint, a sweep generator can be used to locate the approximate location of the fault (open or short). A typical setup is shown in Figure 10-27. Although the detector, scope preamp, and marker adder are shown as separate blocks, all of these are usually built into the sweep generator itself. The principle of this technique can best be analyzed and explained through the use of vectors. In the vector in Figure 10-28A, the forward wave is represented by Ef and the reflected wave is rep-

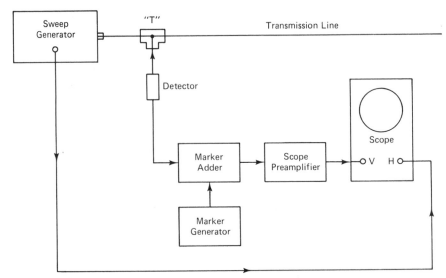

FIGURE 10-27. A typical setup showing how a sweep generator and auxillary equipment are used to determine distance from the line fault to the point of measurement.

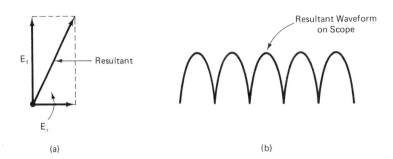

(a)   (b)

FIGURE 10-28. The vector diagram at *A* shows how the forward signal component (*Ef*) and the reflected signal component (*Er*) form the resultant signal which appears on the scope as shown at *B* of the figure.

resented by *Er*. The phase angle between vectors *Ef* and *Er* depends upon the frequency of the signal applied to the line and the distance to the fault. If we let vector *Ef* serve as the reference (fixed), vector *Er* will rotate as the frequency changes. The resultant signal, which is the one that appears on the scope, will vary in amplitude as *Er* rotates around *Ef*. The resultant signal as it appears on the scope is shown in Figure 10-28B. The height of this ripple waveform depends upon the amplitude of *Er*. A high amplitude ripple means a high VSWR.

The distance to the fault from the T-connector can be determined from the following procedure:

Test Procedure:

1. Set up the equipment as shown in Figure 10-27.
2. Set the sweep generator center frequency and sweep width to produce a minimum of two ripples on the scope.
3. Use the signal generator as a marker to locate the frequency of two adjacent peaks.
4. The distance from the T-connector to the fault is then determined from the following formula:

$$D = \frac{492\ V}{\Delta F}$$

where D = distance to fault in feet, V = velocity factor of the transmission line, and ΔF = frequency difference between ripple peaks.

Example: Suppose a length of RG-8 *foam* coaxial cable (velocity factor = 0.8) is being tested. The signal generator is first tuned to produce a marker on one of the ripple peaks. This generator frequency is about 150 MHz. The generator is then retuned to move the marker to the top of the next ripple peak. The new generator frequency is 153 MHz. Thus, $\Delta F = 153 - 150 = 3$ MHz. Substituting into the formula, we have:

$$D = \frac{492 \times 0.8}{3} = 131.2\ \text{ft}$$

A variation of the same principle involves the use of a *reference length* of transmission line. This eliminates the need for the formula. The procedure is as follows:

Test Procedure:

1. Set up the same equipment as shown in Figure 10-27.
2. First connect a reference length of transmission line to the T-connector. The reference line should be of the same type as the line to be

tested and the length should be at least 5% of the length of the line to be tested.

3. Connect the reference line section to the T-connector. Adjust the sweep generator center frequency and sweep width to produce one complete ripple cycle on the scope (see Figure 10-29).

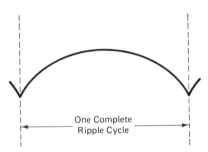

FIGURE 10-29. The sweep generator frequency and sweep width is adjusted to produce one complete ripple cycle on the screen of the CRT.

4. Remove the reference line section and connect the transmission line to be tested.

5. Count the number of ripple cycles on the scope. Multiply the number of ripple cycles by the length of the reference line section to get the distance to the fault.

Example: A transmission line 125-foot long is to be tested. A 10-foot section of line of the same type is used as the reference. The reference line is connected and the sweep generator is adjusted to produce one complete cycle of ripple as shown in Figure 10-29. The transmission line is then connected and the number of ripple cycles are 6 1/2. The distance to the fault is then 6 1/2 × 10 ft = 65 ft.

In a manner similar to the above procedure a variable frequency signal generator and an RF voltmeter may be used to perform the same test. The procedure is as follows:

Test Procedure:

1. Set up the equipment as shown in Figure 10-30.

2. Set the generator output to a fairly high level, 100,000 μV or more.

3. Vary the signal generator frequency until a null is found. The null is sharper than the peak and therefore easier to locate with a meter as an indicator. Note the frequency at which the first null is found.

4. Tune the signal generator frequency until the next (adjacent) null is found. Note the frequency of the second null.

5. ΔF = Difference between the two null frequencies (in MHz).

6. Use the formula $D = \dfrac{492\ V}{\Delta F}$ to determine the distance to the fault.

For a given ΔF, the spacing of the ripples will become closer as the distance to the line fault increases. Thus, when the line fault is very near the point of measurement, the ripples will be widely spaced, requiring a greater frequency change (ΔF) to locate the nulls. For example, if a transmission line with a velocity factor of 0.8 is being tested and a fault is located 10 feet from the measurement point, the $\Delta F = \dfrac{492 \times 0.8}{10} = 39.36$ MHz.

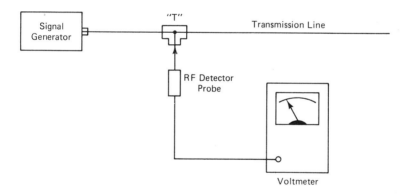

FIGURE 10-30. This setup shows how an ordinary signal generator and RF probe/voltmeter can be used to perform transmission line fault tests. For better accuracy a frequency counter can be used to measure the signal generator frequency.

Antenna Resonance

It may have already occurred to you that the approximate resonant frequency of an antenna can be determined from any one of the preceding cable fault location tests. Over the frequency range near the resonant frequency of the antenna, the ripple height decreases to a minimum and then increases on either side of the frequency range. Figure 10-31A shows a typical ripple pattern that occurs when the ΔF range includes the resonant frequency of the antenna. Figure 10-31B shows how the pattern is affected by a broadband termination such as a good dummy load.

(a)   (b)

FIGURE 10-31. The pattern at *A* is produced when the sweep generator is sweeping across the resonant frequency of the antenna. Note the reduction in ripple amplitude around the frequency at which the antenna is resonant. This is due to less reflected signal at this point. The pattern at *B* is typical of a transmission line terminated by a broadband dummy load.

FINE-TUNING THE ANTENNA

If an antenna is to be used at a single frequency or over a relatively narrow band of frequencies, it is possible to fine-tune the antenna for optimum performance at the desired frequency. Usually, base station antennas are pretuned by the manufacturer to the frequency specified on the customer order. On the other hand, mobile antennas usually must be tuned to the specific frequency by trimming the whip (rod) to the proper length. In this discussion, we will focus on mobile antennas. It is no problem to trim a whip to minimize the reflected power or SWR reading when we know that the whip is too long. But suppose a customer brings in a vehicle with the antenna already installed and the antenna rod trimmed by someone else. An SWR check shows that the antenna needs a little fine-tuning. But is the antenna too long or too short? To cut or not to cut? That is the question. It would be simple to trim the antenna and observe the effect on the meter reading, but if the antenna was already too short this would only make matters worse. Most antennas don't have enough adjustment to significantly raise the antenna with the set-screw adjustment.

Multifrequency Transmitters Can Provide Clue

If a transmitter has two channels that have a relatively wide frequency separation, perform the following test:

1. Check the SWR on the highest frequency available. Note the SWR reading (or reflected power).
2. Check the SWR on the lowest frequency available. Note the SWR or reflected power reading.
3. If the SWR was higher on the highest frequency, the antenna is too long and needs to be trimmed.
4. If the SWR was higher on the lowest frequency, the antenna is too short. Do not trim it more.

When an antenna is to be operated at several different frequencies over a wide spacing, such as 40 channel CB, the antenna should be tuned for minimum SWR at the middle of the band with a slight rise at each end of the band (see Figure 10-32A). The VSWR curve in Figure 10-32B indicates that the antenna is too long, and the curve in Figure 10-32C indicates that the antenna is too short.

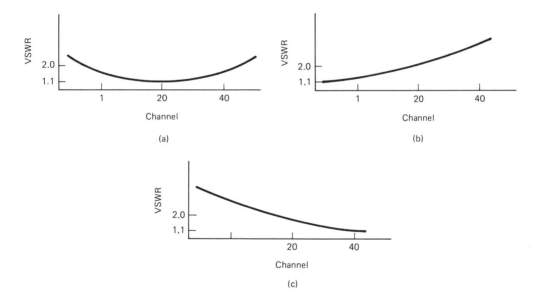

FIGURE 10-32. Typical VSWR curves. At *A* the antenna is properly tuned, since the lowest VSWR is at the center of the band. At *B* the antenna is too long, causing the VSWR to rise at higher frequencies. At *C* the antenna is too short, causing the VSWR to rise at the lower frequencies.

The Tuning Wand Test

If a transmitter is a single frequency type or operates over a relatively narrow range of frequencies, the test just described will not work. There is, however, another method that can be used to determine whether the antenna is too long or too short. This involves the use of a very simple homemade *tuning wand.* The tuning wand is constructed by simply wrapping a 3- or 4-inch width of aluminum foil around the end of a wood dowel (or broom handle, or any other insulated material) and then taping over the aluminum foil to completely insulate it (see Figure 10-33). The shop broom can be used for the tuning wand without interfering with the normal use of the broom!

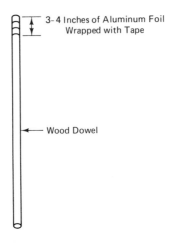

FIGURE 10-33. This simple homemade tuning wand aids in fine-tuning the antenna.

Test Procedure:

1. Refer to Figure 10-34.
2. Key the transmitter and, while observing the *reflected power reading,* place the end of the tuning wand at a point below the tip of the antenna and slowly move the end of the wand upward toward the tip of the antenna. If the reflected power increases as the tuning wand is moved near the tip of the antenna, the antenna is too long. If the reflected power decreases as the tuning wand is moved near the antenna tip, the antenna is too short.

When testing very long antennas, such as low frequency (25 to 50 MHz) quarter-wave whips, it may be necessary to add more surface area to the aluminum foil. This can be done by making a sphere of the foil. It is not necessary to insulate the foil with tape if direct contact with the antenna is

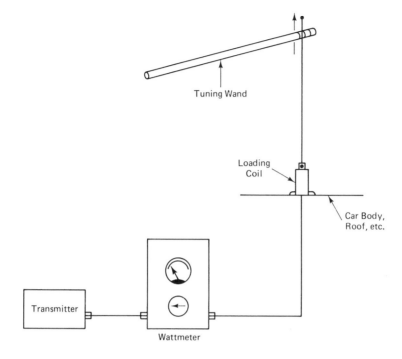

FIGURE 10-34. The antenna tuning wand is used in this manner to determine whether the antenna is too long or too short.

avoided. However, direct contact with the antenna can cause the SWR to increase drastically, possibly causing damage to the transmitter.

When testing very short antennas such as those used at UHF frequencies, it may sometimes work better to hold the wand slightly away from the antenna. Also, smaller areas of aluminum foil may be necessary.

The Noise Bridge Test

The noise bridge can be used to check the length of the antenna by utilizing the receiver itself. The transmission line between the noise bridge and the antenna feedpoint should be one-half wavelength or integral multiples thereof at the operating frequency.

Test Procedure:

1. Set up the equipment as shown in Figure 10-35.

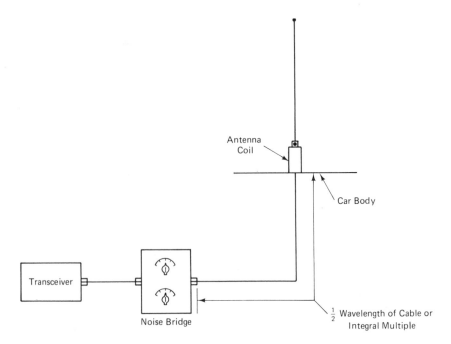

FIGURE 10-35. This setup shows how the noise bridge is used to test the antenna. For proper results the connecting length of transmission line should be one-half wavelength at the operating frequency.

2. Set the noise bridge resistance dial to 50 Ω, the expected radiation resistance of the antenna.

3. Turn on the receiver and noise bridge. Noise should be heard in the receiver.

4. Tune the reactance dial of the noise bridge to produce a null in the receiver noise.

5. If the reactance dial is on the XL side, the antenna is too long. If the reactance dial is on the XC side, the antenna is too short.

ANTENNA RADIATION PATTERN

The type of antenna and the location of the antenna on the vehicle will both have an affect on the radiation pattern of the antenna. A general idea of the radiation pattern of a vehicular antenna can be determined by using a relatively inexpensive portable test unit such as the one shown in Figure 10-36, which contains a relative field strength meter. As the name implies, the readings are strictly relative and the meter scale is usually calibrated from 1 to 10.

The meter should be used at a maximum distance from the vehicle which will still give a useable indication. The vehicle should be in an area

FIGURE 10-36. This portable tester contains a relative field strength meter. Courtesy of Simpson Electric Company, Elgin, IL 60120.

free of obstructions within a couple hundred feet of the vehicle. First, the direction of maximum signal is located. The distance is then adjusted so that the meter reads full scale. Note the distance from the vehicle and slowly walk around the vehicle at this distance while observing the relative field strength meter. The meter should be held in the same position relative to the body and the antenna throughout the test, keeping the antenna vertical at all times.

You will likely see quite a variation in the meter reading as you walk around the vehicle. It is not uncommon for the meter indication to drop to near zero at certain null points but such null points usually occupy only a relatively small portion of the total circle. Similar tests for base station antennas involve much greater distance and much better measuring equipment. A special (and expensive) field strength meter, not normally found in the technician's arsenal of equipment is the proper tool for the job. However, a spectrum analyzer can be used for this test by following the manufacturer's instructions very carefully and then plotting the results on polar graph paper. A procedure for using the spectrum analyzer for field strength measurement and radiation pattern tests is described below.

Getting the Field Strength Measurement in dBμV/m

Field strength is normally expressed in dBμV/m. However, the spectrum analyzer is calibrated in dBm (dB reference to 1 mW). So the spectrum analyzer reading in dBm must be converted to dBμV/m. This can be done in two steps. First, the dBm figure is converted to dBμV by simply adding 107 dB to the dBm figure. Thus,

$$-40 \text{ dBm} = -40 + 107 = 67 \text{ dB}\mu\text{V}$$

This is valid for spectrum analyzers with 50-Ω inputs. The next step is to convert the dBμV figure to dBμV/m. This step requires that the antenna correction factor (K) be known. The antenna factor (K) for 50-Ω systems can be determined from the formula:

$$K = 20 \log F - G - 29.8$$

where K = antenna correction factor, F = frequency in MHz, and G = antenna gain in dB. To convert dBμV to dBμV/m, simply add the antenna correction factor (K) to the dBμV figure. The graph in Figure 10-37 correlates antenna factor with antenna gain and frequency for 50-Ω systems. Antennas such as the one shown in Figure 10-38 can be constructed for use with the spectrum analyzer in making these tests. It will be necessary to use a 75-Ω matching transformer to match the 75-Ω antenna to the 50-Ω input impedance of the spectrum analyzer.

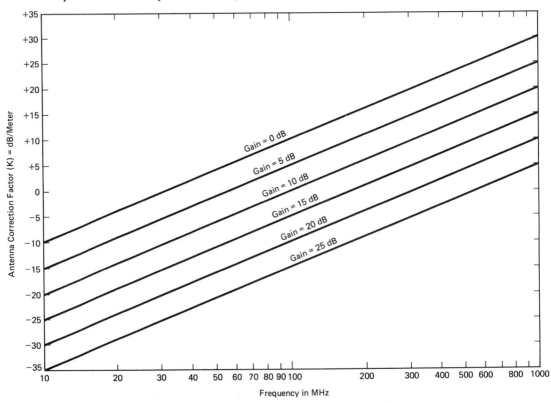

FIGURE 10-37. A graph of antenna correction factor (K) versus antenna gain and frequency.

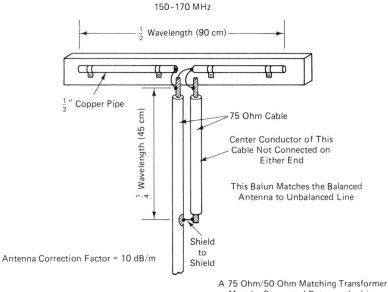

FIGURE 10-38. An antenna such as this can be constructed for use with the spectrum analyzer in making field strength measurements.

The step-by-step procedure is outlined below:

1. Set up the equipment as shown in Figure 10-39.
2. Rotate the antenna to get maximum signal amplitude on the analyzer display.
3. Read the signal amplitude in dBm from the analyzer.
4. Convert this dBm figure to dBμV/m by adding 107 dB *plus* the antenna factor K (for the antenna being used with the spectrum analyzer).
5. If desired, the dBμV/m figure can be converted to μV/m by using the following formula:

$$\mu V/m = \text{antilog } \frac{\text{dB}\mu V/m}{20}$$

Example: An antenna with a gain of 3 dB is used with a spectrum analyzer to check the field strength of a base station operating at 159 MHz. The spectrum analyzer display shows an amplitude of −45 dBm. What is the field strength in dBμV/m? In μV/m?

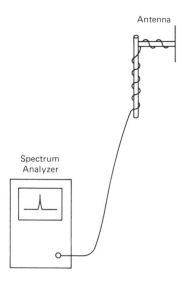

FIGURE 10-39. The spectrum analyzer along with a simple handheld antenna are used to make field strength measurements.

Solution: First, the antenna correction factor (K) is calculated by applying the formula:

$$K = 20 \log(159) - 3 - 29.8 = 20(2.2) - 3 - 29.8 =$$
$$44 - 32.8 = 11.2 \text{ dB/m}$$

The field strength in dBμV/m now can be found by adding $107 + K$ to the dBm figure.

$$\text{dB}\mu\text{V/m} = -45 + 107 + 11.2 = -45 + 118.2 = 73.2 \text{ dB}\mu\text{V/m}$$

The field strength in μV/meter can now be found from the formula:

$$\mu\text{V/m} = \text{antilog} \frac{\text{dB}\mu\text{V/m}}{20} = \text{antilog} \frac{73.2}{20} = \text{antilog } 3.66 = 4{,}571\mu\text{V/m}$$

Checking the Pattern

The spectrum analyzer can be used to check the radiation pattern of an antenna system. For checking base station antenna patterns, the measurement should be taken at a radius of approximately one mile from the antenna. The one-mile test point locations can be found by using a compass to draw a 1-mile radius on a map (using the scale of miles to determine radius length) and then use the intersection of the circle and various roads and other landmarks to locate the test points (see Figure 10-40). The measurement at those various points is performed in the same manner as just described

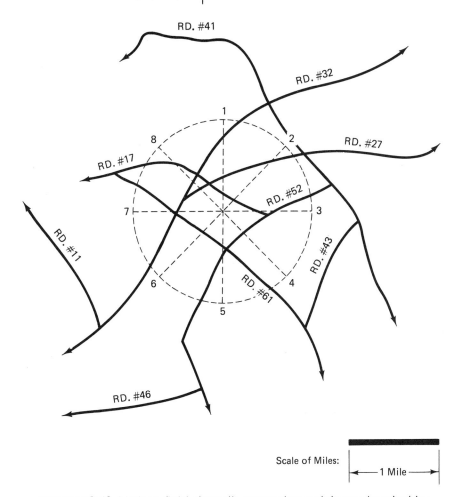

Measuring Point	Signal Strength
#1 (0°)	−40 dBm
#2 (45°)	−45 dBm
#3 (90°)	−41 dBm
#4 (135°)	−49 dBm
#5 (180°)	−54 dBm
#6 (225°)	−52 dBm
#7 (270°)	−48 dBm
#8 (315°)	−43 dBm

N

RD. #41

RD. #32

RD. #27

RD. #17

RD. #52

RD. #11

RD. #43

RD. #61

RD. #46

Scale of Miles:

◄— 1 Mile —►

FIGURE 10-40. Various field strength measuring points are located by drawing a circle around the antenna location and then using landmarks, road numbers, etc. to locate these points. The chart shows the various measurements taken.

for field strength measurements. The field strength is measured at several different points. The bearing of the measurement point also must be noted. It is usually better to take measurements at two or three different locations near each point and average the figures obtained.

Suppose that a total of eight readings are taken at an angular spacing of 45° as shown by the chart in Figure 10-40. We can graph this on a piece of polar coordinate graph paper by the following method. Determine the *maximum difference* between readings—for example, maximum reading *minus* minimum reading. In Figure 10-40, this will be −40 dB − (−54 dB), which yields a difference of 14 dB. Represent the maximum level as 0 dB and the other levels as so many dB below 0 dB. In Figure 10-39, 0 dB is −40 dB at the first measurement point. The second measurement is 5 dB lower, and so on. The completed graph is shown in Figure 10-41.

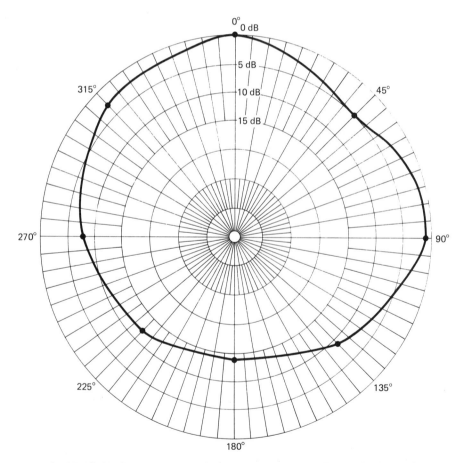

FIGURE 10-41. This is the resulting polar graph plotted with the information obtained from the measurements taken at the eight points shown in Figure 10-40.

BIBLIOGRAPHY

Bird Electronics Corporation, "Bird Model 43 Wattmeter Instruction Book" © Bird Electronics Corporation, Cleveland, Oh.

Carr, Joseph, *Elements of Electronic Instrumentation and Measurement* © Reston Publishing Company 1979, Reston, Va.

Cushman Electronics, Inc. "Mobile Radio Testing, Part 3—Spectrum Monitor and Tracking Generator." © 1982 Cushman Electronics, Inc., San Jose, Calif.

————. "Total System Testing." © 1982 Cushman Electronics, Inc., San Jose, Calif.

————. "Using the Spectrum Monitor." © 1978 Cushman Electronics, Inc., San Jose, Calif.

Decibel Products, Inc. "About R-F Transmission Lines" © 1973 Decibel Products, Inc., Dallas, Tex.

Tektronix, Inc. "Troubleshooting Two Way Radios with the Spectrum Analyzer" © 1978 Tektronix, Inc., Beaverton, Oreg.

11
MISCELLANEOUS RELATED TESTS AND MEASUREMENTS

This final chapter is used to present several test and measurement techniques and procedures that are of the more general nature. They are not necessarily related to each other but rather are grouped together here because they can be applied to all of the various communication modes (AM, FM, and SSB) discussed in the preceding chapters.

REMOTE CONTROL LINE TESTS

This discussion is limited to the remote control line. This is the line (usually leased from the telephone company) that is used to control a base station from a remote location. If the control line is leased from the telephone company, the maintenance of the line is the sole responsibility of the telephone company. However, this does not mean that you will never need to perform any tests or measurements on that line. Whenever there is a problem in the operation of the base station from the remote location, the control line is often the first suspect. However, before turning the problem over to the telephone company (Telco), you should first perform the necessary tests to ascertain that the line is indeed at fault. If the line is not the problem, Telco will charge you for the time you caused them to waste in checking it. Another reason for checking the line yourself is to load yourself with information so that you may explain to Telco the exact problem of the line. This will usually result in getting the problem repaired faster than just calling them and simply reporting that the line is out of order.

There are basically two kinds of remote control lines: (1) *metallic lines,* which provide dc continuity from end-to-end (for dc remotes), and (2) those used for *tone remotes.* Both types of lines are *voice-grade lines,* which should pass voice frequencies from 500 to 2,500 Hz with minimum attenuation.

Tests of these remote control lines can be reduced to the following: (1) resistance and continuity checks, (2) voltage checks across line, (3) frequency response tests, and (4) signal-to-noise ratio measurement.

Resistance and Continuity Tests

Refer to Figure 11-1. The following resistance and continuity checks are appropriate for metallic lines only, which provide service for the dc remote control system.

1. Disconnect equipment from each end of the line to isolate the control line.

2. Measure the resistance from each leg of the control line to ground (leg 1 to ground and leg 2 to ground). Theoretically, this resistance should be infinity but a lesser value must be tolerated in practice. A lower than normal resistance here may indicate a defective lightning protection device. These are usually installed on each end of the control line.

3. Measure the resistance between the two legs (leg 1 and leg 2) of the control line. This measurement should be made on the highest ohm-meter scale ($\times$ 100 K or so). When the meter is first connected across the line, the needle may move down scale temporarily but should move swiftly back upscale, all of this happening in a flicker. This is normal and is due to the capacitance of the line. A low resistance reading means trouble!

4. Have someone at the opposite end of the line connect a temporary shorting jumper across the two line legs (leg 1 to leg 2). Measure the resistance from leg 1 to leg 2. Record this value. If it was measured and recorded prior to this, check your measurement against the recorded measurement. Remove the temporary shorting jumper.

Voltage Checks Across Line

You may wonder about the need or purpose of this test, but strange things can show up on leased control lines! This is especially true of lines that may be one pair of several hundred pairs in a large cable and most especially so if the line pair runs through the telephone company interconnect panels. Problems of this sort occur much less frequently on direct lines — that is, lines that run directly from the remote control point to the base station site without becoming integrated with the maze of other telephone wires.

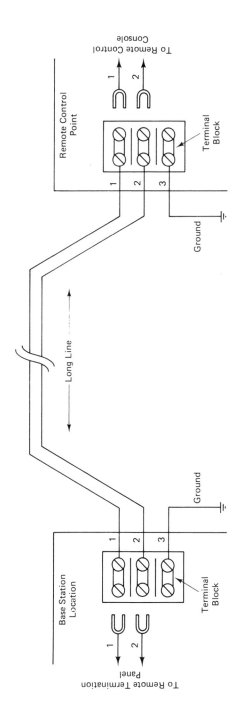

FIGURE 11-1. A typical remote control line consists of two *hot* legs. Some operations require two pair (four wires).

381

Test Procedure

1. Isolate the control line by disconnecting both legs from equipment on each end of the line.
2. Connect a voltmeter across the two line legs and check for any voltage (ac or dc). Also, check for the presence of voltage from ground to either line leg.

Frequency Response Tests

Voice-grade control lines should provide a good frequency response from 300 to 3,000 Hz. Typical frequency response for a *nonconditioned* control line may be −3 dB to +12 dB over the 300 to 3,000 Hz range. The local Telco should provide you with the minimum frequency response specification for any line they lease. Any great deviation from this specification should be reported to Telco.

Test Procedure

1. Disconnect all other equipment from the control line and hook up test equipment as shown in Figure 11-2. This procedure requires two persons, one on each end of the control line, in direct contact with each other.
2. Set the audio generator frequency to 1,000 Hz and the output level to exactly 0 dBm. The graph in Figure 11-3 correlates *rms* audio voltage across 600 Ω with the equivalent dBm value. The audio voltage at the other end of the line is noted. This will serve as the 0-dB reference level. If the audio meter has a decibel scale, use it. The dB scale will indicate the line loss directly in dB. For example, if the line loss at 1,000 Hz is 15 dB, the dB scale will indicate −15 dB (when proper correction factors are applied for the various ranges). This −15 dB reading serves as the 0-dB reference for the measurements at all other frequencies. For example, if the measurement at 2,000 Hz is −10 dB, you will record this as +5 dB since this is 5 dB *above* the 0-dB reference level of −15 dB.
3. Run the frequency response test at several frequencies from 300 Hz to 3,000 Hz and use the data to plot a response curve of frequency versus amplitude. The closer the spacing of the test frequencies, the more accurate the response curve will be. If the line is used with a tone remote system, be sure to include a response test at each tone frequency used by the tone remote in performing the various functions. Some of the more commonly used tone remote frequencies and the associated functions are listed in Table 11-1. All data obtained from such response tests should be recorded and kept for future reference.

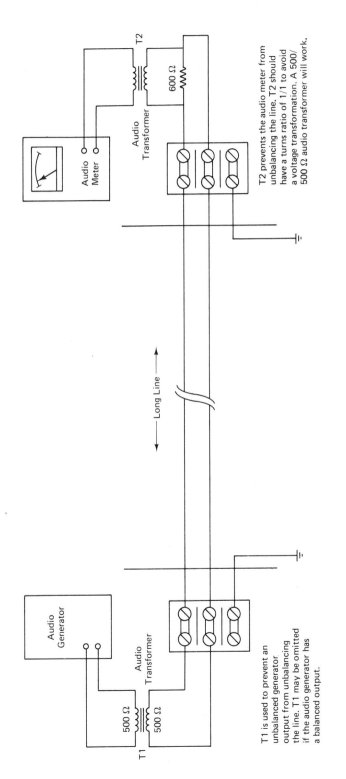

Audio Meter

T2

600 Ω

Audio Transformer

T2 prevents the audio meter from unbalancing the line. T2 should have a turns ratio of 1/1 to avoid a voltage transformation. A 500/500 Ω audio transformer will work.

Long Line

Audio Generator

Audio Transformer

500 Ω
500 Ω
T1

T1 is used to prevent an unbalancing generator output from unbalancing the line. T1 may be omitted if the audio generator has a balanced output.

FIGURE 11-2. Typical setup for making a frequency response check on a control line. This requires two persons, one at each end of the line.

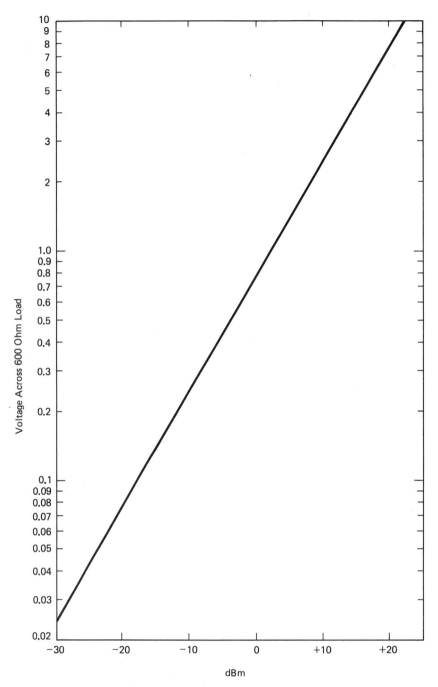

FIGURE 11-3. This graph correlates audio *rms* voltage across 600 Ω with the equivalent dBm value.

FREQUENCY	TYPICAL FUNCTION PERFORMED
1050 Hz	#2 auxiliary or wild card "OFF"
1150 Hz	#2 auxiliary or wild card "ON"
1250 Hz	#1 auxiliary or wild card "OFF"
1350 Hz	#1 auxiliary or wild card "ON"
1450 Hz	Carrier squelch, repeater "ON", or minimum squelch on Motorola equipment. Repeater "OFF", maximum squelch on G.E. equipment.
1500 Hz	Repeater "OFF" or maximum squelch on Motorola equipment. Repeater "ON" or minimum squelch on G.E. equipment.
1650 Hz	R2 unmute or receiver #2.
1750 Hz	R2 mute or receiver #1.
1850 Hz	Frequency #2.
1950 Hz	Frequency #1.
2050 Hz	Private line/channel guard monitor or disable.
2175 Hz	Guard tone.

TABLE 11-1. This table lists the various tone remote frequencies and the functions they typically perform. Courtesy Industrial Electronics Service Company.

The Three-Point Response Test   A quick three-point frequency response test can be performed by checking the low end (400 Hz), middle (1,000 Hz), and high end (2,800 Hz) of the frequency range. Again, the 1,000-Hz frequency serves as the reference for the other two. Subtract the losses (in dB) at 400 Hz and 2,800 Hz from the loss at 1,000 Hz. This will give you the *slope* at 400 Hz and 2,800 Hz. Telco often uses odd frequencies such as 404 Hz, 1,004 Hz, and 2,804 Hz for this test. A typical response curve might look like the one shown in Figure 11-4. The response curve in Figure 11-4 shows a 6-dB slope at 400 Hz and a 10-dB slope at 2,800 Hz.

Bandwidth   The bandwidth of the line is defined as the band of frequencies within which the attenuation of any frequency is no more than 10 dB as referenced to the attenuation at 1,000 Hz. Voice-grade lines of the type used for remote control lines usually have a 10 dB bandwidth of 300 Hz to 3,000 Hz. Unless the line is used for tone remotes, the frequencies between 500 Hz and 2,500 Hz are of primary importance since this frequency range contributes most to normal speech communication.

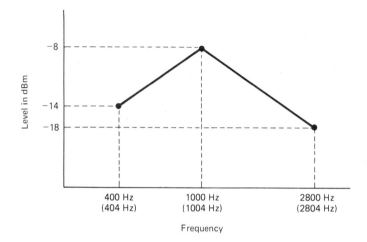

FIGURE 11-4. A 3-point frequency response test is made to find the *slope* at 400 Hz and 2,800 Hz, 1000 Hz being the reference. This response shows a slope of −6 dB at 400 Hz and a slope of −10 dB at 2800 Hz. The frequencies: 404 Hz, 1004 Hz, and 2804 Hz may be used by Telco for this test.

Signal-to-Noise Ratio

Control line noise can be defined as any signal other than the desired signal. Since we are only concerned with frequencies between 300 Hz and 3,000 Hz, the only detrimental noise is that which is located within this passband. Proper noise measurements would then require a special filter which passes the band of frequencies from 300 Hz to 3,000 Hz with little attenuation but which sharply attenuates frequencies outside this range. Telco refers to this filter as a "C-message weighting filter."

The ideal response of such a filter would look like that shown in Figure 11-5A with no attenuation between 300 Hz and 3,000 Hz and maximum attenuation to outside frequencies. The response curve of a practical "C-message weighting filter" would resemble that in Figure 11-5B.

The signal-to-noise ratio measurement can be made with the same setup shown in Figure 11-2 except that the "C-message weighting filter" is connected between the measuring device (audio meter) and the control line. The signal-to-noise ratio for any frequency from 300 Hz to 3,000 Hz can be measured in the following manner.

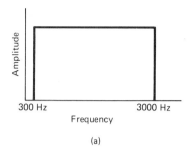
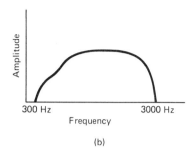

(a)   (b)

FIGURE 11-5. The response of the *ideal* "C-message weighting filter" would look like that at (A) while the response curve at (B) is more representative of a practical "C-message weighting filter."

Test Procedure

1. Hook up the equipment as decribed above.

2. With the audio generator off, note the noise level as indicated on the audio meter.

3. Tune the audio generator to the desired test frequency (between 300 Hz and 3,000 Hz) and set the audio output level to 0 dBm.

4. Note the new reading on the audio level meter. The difference (in dB) between the reading with the audio generator on and the reading with the audio generator off is the signal-to-noise ratio. Actually, to be exact, this is the *signal + noise/noise* ratio ($S + N/N$) since the noise and signal are being measured as a *composite* signal. To get a *true S/N* measurement would require that a special filter be inserted between the audio meter and the line. This filter should pass only the audio frequency used for the test. For practical purposes, however, the $S + N/N$ measurement is a good indicator of noise performance. Signal-to-noise ratios typically should be 50 dB or better, although some systems may tolerate a lower ratio than this.

If the remote control line passes all the tests described above, it must be in very good shape. Figure 11-6 shows an instrument that is *dedicated* to testing remote control lines and associated equipment. This is the model CLT-1000 Communications Line Tester manufactured by Industrial Electronics Service Company. Standard features are:

1. A high-quality line amplifier for matching or bridging telephone lines.

2. A function-tone synthesizer for generating test tones.

3. A separate guard-tone generator (2175 Hz) that is reed-controlled for high accuracy.

4. A tone burst generator to simulate the actual signal transmitted from a remote control console.

5. An intercom feature for communicating with a remote console operator, a base station with intercom feature, or a second service technician.

6. A wide-range dBm meter for measuring low-level hum or noise.

Optional features include:

1. A dc current generating module for testing dc line-current-controlled base stations.

2. A dual tone module with a sixteen-button keypad that generates standard DTMF tones.

FIGURE 11-6. The CLT-1000 Communications Line Tester can be used to run complete performance tests on the control line as well as the associated equipment. Courtesy: Industrial Electronics Service Company.

BASIC SWEEP TECHNIQUES AND TIPS

To align bandpass circuits, filters, etc., properly in communications equipment, it is often desirable to run a sweep frequency response test on the circuit or device in question. Such a test provides an instant scope display of the frequency versus amplitude response of the circuit under test.

When a very wide frequency range must be swept, a dedicated RF sweep generator is usually required. However, when a relatively narrow frequency range must be covered such as required for testing narrow band IF filters in communications receivers, it is possible to use an FM signal generator to obtain a frequency sweep. The FM generator must have an external modulation capability so that a low-frequency external signal can be used to modulate (sweep) the FM generator. This same low-frequency modulating signal must be used to either synchronize or directly drive the scope's horizontal sweep system. Depending upon the type of trace desired the modulating signal may be a sine-wave or a sawtooth waveform. Very narrow bandpass circuits must be swept at a slow rate in order to avoid amplitude distortion in the response curve. A commonly used sweep frequency of 60 Hz is used in many sweep generators mainly because this frequency is readily available from the power source. However, a much slower sweep rate must be used for narrow band filters. Typically, a 20- to 30-Hz sweep rate is used for these narrow bandpass circuits.

The Single Trace Pattern

A single trace pattern is usually used to display the response curve on the scope. The single trace pattern can be obtained by using either a sawtooth waveform or a sine-wave. Both procedures are described here.

The Sawtooth Signal   Figure 11-7 shows a typical setup in which a sawtooth signal is used to provide both the horizontal sweep for the scope and to modulate the FM generator. In this arrangement, the scope's horizontal sweep oscillator (time base generator) is switched out and the sawtooth is applied to the horizontal amplifier. Figure 11-8 shows the relationship of the sawtooth signal, the instantaneous frequency of the FM generator, and the scope display. The height of the response curve represents the amplitude of the detected output signal which is applied to the scope vertical input. At time 1, the sawtooth signal is at minimum amplitude, the FM generator instantaneous frequency is at F1, and the scope trace is just beginning at one side of the CRT. At time 2, the sawtooth amplitude is at one-half the maximum amplitude, the FM generator instantaneous frequency is at F2, and the scope trace is half-way across the CRT. At time 3, the sawtooth waveform has reached maximum amplitude, the FM generator instantaneous frequency is at F3, and the scope beam has reached to the far side of the CRT. From time 3 to time 4, the sawtooth waveform goes from maximum to minimum amplitude at a very rapid rate, causing the scope beam to return (retrace) rapidly back to the other side of the CRT to the starting point and the FM generator frequency returns to F1. This retrace happens so fast that it will not be seen on the scope, or if it is seen at all it will be very dim.

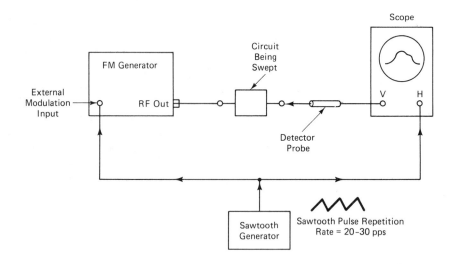

FIGURE 11-7. A sawtooth signal of low repetition rate is used to sweep modulate the FM generator and to drive the scope's horizontal input to obtain a synchronized sweep of the circuit under test.

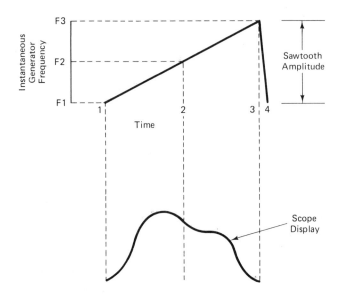

FIGURE 11-8. The relationship of the sawtooth signal, FM generator frequency, and scope display for the setup shown in Figure 11-7.

The FM generator frequency and the sweep width (modulation) should be adjusted so that the sweep range of the FM generator covers approximately twice the bandwidth of the circuit being swept with the circuit bandpass located at the center of the sweep range.

Some scopes have a sawtooth output terminal that provides a sawtooth waveform which is a sample of the internal horizontal timebase of the scope. This sawtooth waveform can be used to synchronize the FM generator sweep with the scope sweep rather than using a separate sawtooth generator.

The Sine-Wave Signal A sine-wave signal can be used to provide the sweep for the generator and the scope just as well as the sawtooth signal. The setup is the same as shown in Figure 11-7 except that a sine-wave generator (such as an audio generator) is used instead of the sawtooth generator. The manner in which the response curve is traced on the CRT is a little different with a sine-wave driving the CRT horizontal deflection plates.

Figure 11-9 shows the relationship between the sine-wave, the FM generator frequency, and the scope beam position at any given instant. When the sine-wave amplitude goes from maximum positive amplitude (point 1) to maximum negative amplitude (point 2), the generator frequency goes from F3 to F1 and the scope beam is swept across the CRT from right to left, thus tracing response curve *A*. When the sine-wave amplitude goes from maximum negative amplitude (point 2) to maximum positive amplitude (point 3) the generator frequency goes from F1 to F3 and the scope beam is swept across the CRT from left to right, thus tracing response curve *B*. Since the two traces are identical and one superimposed over the other, they appear as a single trace on the CRT.

The Double Trace Pattern

A double trace pattern results when the *direction of frequency change* of the FM generator is reversed on alternate sweeps across the CRT. That is, in one sweep across the CRT the FM generator frequency is changing from high to low and on the next sweep across the CRT the FM generator frequency is changing from low to high. The scope horizontal sweep is done with a sawtooth signal so the beam is swept across the CRT in the same direction each time. The manner in which this is accomplished is as follows.

The necessary hookup is shown in Figure 11-10. The diagram of a simple sweep modulator appears in Figure 11-11. This particular circuit appeared in the General Electric Mobile Radio Datafile Bulletin #1000-6. A 15 to 30 Hz audio signal is fed to the input of the sweep modulator. This audio signal is used to modulate (sweep) the FM signal generator. The deviation level is controlled by R4. R4 may be omitted if the FM generator has a built-in modulation control. The audio signal also is fed (through transformer T1)

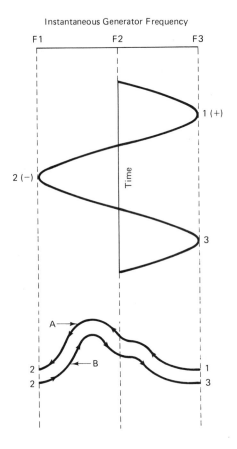

FIGURE 11-9. The relationship of the sine-wave signal, FM generator frequency, and scope waveform when a sine-wave generator is substituted for the sawtooth generator in Figure 11-7.

to a full-wave bridge rectifier. The output of the rectifier is a series of pulses. These pulses are fed to the scope's *sync input*. The phasing network on the primary side of T1 is used to control the phase of the sync pulses in relation to the sine-wave signal used to modulate the FM generator. Thus, the sync pulses can be adjusted so that the horizontal sweep of the scope starts at the proper instant.

Figure 11-12 illustrates the relationship between the sine-wave signal, the sawtooth sweep of the scope, the sync pulses, and the resulting waveform on the scope. Between time 1 and time 2, the frequency of the FM generator is being swept from high to low (F3 to F1). The sync pulse causes the scope sweep to begin at time 1 and end at time 2. Thus, during this sweep the frequency is changing from high to low and produces the wave-

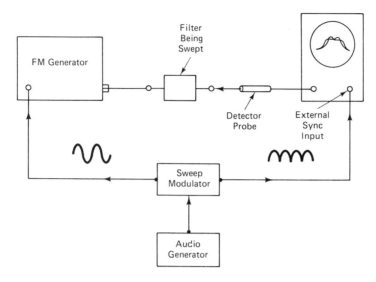

FIGURE 11-10. In this hookup a special *sweep modulator* is used in conjunction with a low-frequency audio generator to obtain a double trace sweep pattern on the scope.

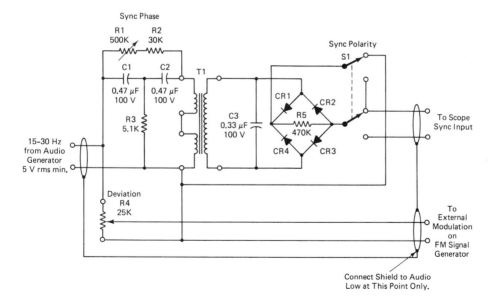

FIGURE 11-11. This simple sweep modulator can be used to form the double trace pattern in sweep tests. Courtesy of General Electric Mobile Radio Division.

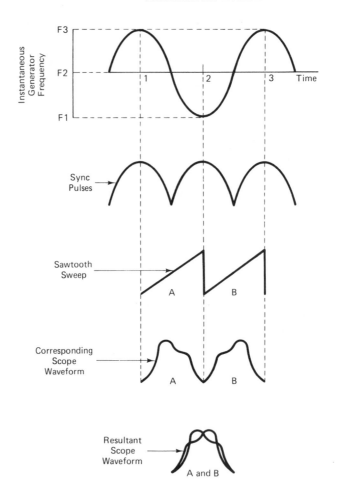

FIGURE 11-12. This illustratin shows the relationship between the audio modulating signal, the FM generator frequency, the sync pulses, and the scope waveform.

form shown in Figure 11-12A. At time 2, a second sync pulse causes a new sweep to begin, from time 2 to time 3. During this same time interval, the frequency of the FM generator is changing from low to high (F1 to F3). Since this is in the opposite direction of the prior sweep, the scope waveform (at B) will be reversed from the previous sweep. The resultant waveform shown at the bottom is called the *double trace waveform.* Each trace is produced on alternate sweeps.

It may first appear that the second trace is useless, since one trace is simply the reverse of the other. However, the double trace method has been found to offer certain advantages over the single-trace method. If the shoulders of the two traces line up as shown in Figure 11-13A, the center frequency of the FM generator is centered in the passband of the circuit being swept. If the center frequency of the FM generator does not fall in the center of the passband, the shoulders will be separated as shown in Figure 11-13B. If the sweep *rate* is too fast (the modulating frequency too high), the noses of both traces will slope in the same direction. A modulating frequency of 20 Hz is sufficiently slow for most of the narrow band IF filters found in modern communications equipment.

If the FM generator has a built-in deviation meter, the bandpass of the filter being swept can be determined by reducing the modulation level until just the bandpass curve is presented on the scope, eliminating the oversweep in both directions. And, assuming symmetrical modulation, the FM generator center frequency indicates the center of the bandpass if the generator frequency was centered properly by checking the shoulders of the double trace curve.

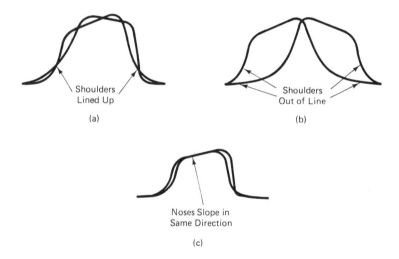

FIGURE 11-13. If the shoulders of the double trace pattern are lined up as at *A* the bandpass response is centered in the sweep range of the FM generator. If the shoulders are out of line, as at *B*, the bandpass response is not properly centered in the sweep range of the FM generator. If the sweep *rate* is too fast (modulating frequency too high) the nose of both traces will slope in the same direction, as shown at *C*.

MISCELLANEOUS OSCILLOSCOPE APPLICATIONS

The oscilloscope is one of the most valuable multipurpose test instruments available to the technician. It is impossible to cover every conceivable application of the scope in test and measurement work. However, in addition to the many uses mentioned in the preceding chapters, the following applications are of interest.

Phase Measurement

The test setup shown in Figure 11-14 is used for phase measurement tests of audio amplifiers or audio networks, filters, etc. Typical patterns and their interpretations are shown in Figure 11-15A to F. The degree of phase shift can be determined as illustrated in Figure 11-16.

Example: Suppose the measurement of $B = 6$ divisions on the scope's scale and the measurement of $A = 7.5$ divisions on the scope's scale (sin $\phi =$ $6/7.5 = 0.8$). The angle (ϕ) that corresponds to a sine of 0.8 is 53 degrees. This is found from trigonometry tables.

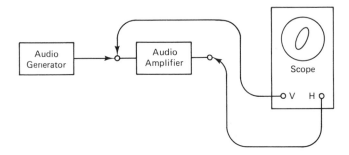

FIGURE 11-14. Typical hookup for making a phase measurement with the scope.

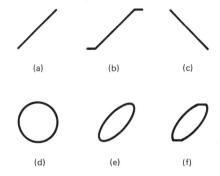

FIGURE 11-15. Typical scope displays of phase measurement. $A =$ no amplitude distortion and no phase shift. $B =$ amplitude distortion but no phase shift. $C = 180°$ out of phase. $D = 90°$ out of phase. $E =$ phase shift but no amplitude distortion. $F =$ amplitude distortion and phase shift.

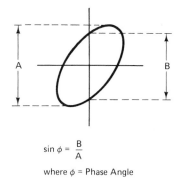

$$\sin \phi = \frac{B}{A}$$

where ϕ = Phase Angle

FIGURE 11-16. The degree of phase shift can be determined as shown here.

Square Wave Test

A practical and proven method of checking the response of wideband audio amplifiers has been to inject a square wave low-frequency signal into the amplifier input and monitor the output of the amplifier with a scope. An indication of the amplifier response can be obtained by close examination of the output waveform and comparing this to the input square-wave signal. By nature, a square wave contains a large number of odd-order harmonics. For example, a 500-Hz square-wave signal (of good square shape) may be used for testing amplifiers from 500 Hz up to the 15th, 19, or even 21st odd harmonic of 500 Hz. Square wave testing is not usually done with narrow-band speech amplifiers such as those found in communications equipment. The procedure is presented here for general information. Besides, communications technicians often find themselves working on other equipment for various reasons.

The setup for the test is shown in Figure 11-17. The square-wave signal at the input should first be checked with the scope to assure that the input signal is of good quality. Defects in the amplifier response will show up

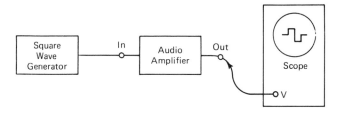

FIGURE 11-17. A typical setup for using square waves to check the frequency response characteristics of an amplifier.

as a change in the shape of the square wave (distortion) at the output of the amplifier. Typical defective waveforms and their causes are shown in Figure 11-18.

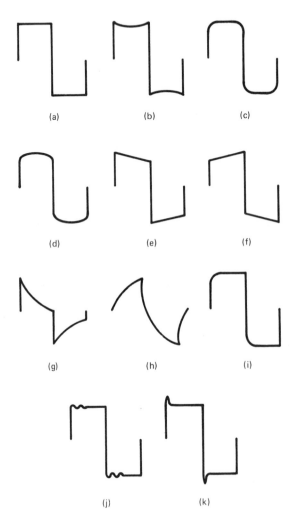

FIGURE 11-18. Typical square wave patterns are shown here (after passing through the audio amplifier). $A =$ ideal waveform (rarely seen at the output). $B =$ low frequency loss, no phase shift. $C =$ high frequency loss, no phase shift. $D =$ low frequency boost at or near the fundamental frequency. $E =$ low frequency *leading* phase shift. $F =$ low frequency *lagging* phase shift. $G =$ low frequency loss and phase shift. $H =$ high frequency loss and low frequency phase shift. $I =$ high frequency loss, and high frequency phase shift. $J =$ damped oscillation. $K =$ high frequency overshoot.

Frequency Comparison with Lissajous Patterns

By using the setup shown in Figure 11-19, an oscilloscope can be used to compare and determine frequency relationships of two signal sources. The frequencies must lie within the bandwidth of the horizontal and/or vertical amplifier of the scope, unless the signals are of sufficient amplitude to drive the deflection plates directly. There is a limitation to the ratio that can be determined in this manner since higher ratios present a very complicated pattern on the scope making it impossible to study the pattern in sufficient detail. A frequency ratio of 10 to 1 is a realistic upper limit, although it may be possible to do somewhat better in some cases.

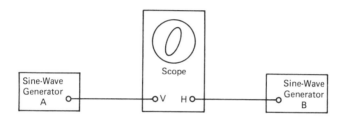

FIGURE 11-19. This setup is used to obtain the Lissajous pattern for frequency comparison.

Several different patterns are shown in Figure 11-20 along with the frequency ratios they represent. The method of determining the frequency ratio is as follows. Following a horizontal line across the pattern, count the number of times the trace pattern is crossed. Use this number to represent the vertical frequency in the ratio. Next, follow a vertical line down across the trace pattern counting the number of times the trace is crossed. Use this number to represent the horizontal frequency in the ratio. Note: It makes no difference at which point the horizontal or vertical line falls across the pattern; the result will be the same. If the horizontal or vertical lines cross an intersection or crossover point in the pattern, count this twice since two lines are being crossed at once.

Figure 11-21 should help to clarify this procedure. This procedure works equally well for both closed-loop and open-loop patterns. In Figure 11-21A, the horizontal line crosses the trace pattern ten times. The vertical line crosses the trace pattern six times. The ratio is then 10/6 or 5/3. In the equivalent open-loop pattern in Figure 11-21B, the horizontal line crosses the trace pattern five times and the vertical line crosses the trace pattern three times, giving the same 5/3 ratio.

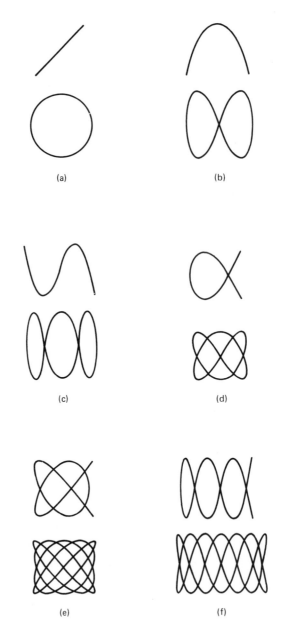

FIGURE 11-20. Typical Lissajous patterns showing both the *closed-loop* pattern and the equivalent *open-loop* pattern. The following frequency ratios (vertical/horizontal) are illustrated here: *A* = 1/1; *B* = 2/1; *C* = 3/1; *D* = 3/2; *E* = 5/4; *F* = 7/2.

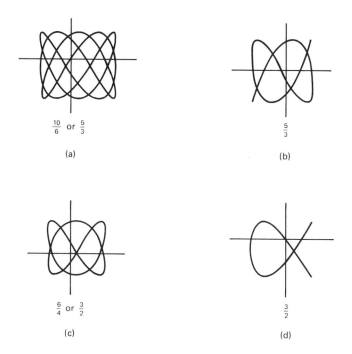

FIGURE 11-21. The procedure for determining the frequency ratio from a Lissajous pattern is illustrated here. See text for details.

BIBLIOGRAPHY

Industrial Electronics Service Company, *CLT-1000 Communications Line Tester Operating Manual.* Schaumburg, Ill., 1981.

General Electric Mobile Communications Division, "Mobile Radio Datafile Bulletin # 1000-6. Subject: Receivers, General." Lynchburg, Va., 1978.

Lenk, John D., *Handbook of Oscilloscopes.* Englewood Cliffs, N.J.: Prentice-Hall, Inc., 1968.

Philco Corporation, "Radio Communication System Measurements" ©1952 Philco Corporation.

Wolf, Stanley, *Guide to Electronic Measurements and Laboratory Practice.* Englewood Cliffs, N.J.: Prentice-Hall, Inc., 1973.

APPENDIX A

A MINI-LESSON IN LOGARITHMS

Since so many formulas used in communications work contain logarithm expressions, it is important to know how to work with them in order to apply these formulas to your particular applications. You don't have to be a mathematician to be able to work with logarithms. This appendix condenses the rules and procedures for working with logarithms in a simplified manner.

THE LOGARITHM DEFINED

Basically, the logarithm is an exponent. The logarithms you will encounter are almost always *common* logarithms. Common logarithms are based on 10. The common logarithm (hereafter called log) of a number is defined as the exponent of 10 which will yield that number. The following expression defines that statement in mathematical form.

$$A = 10^n$$

$$\log A = n$$

STANDARD NOTATION

When working with mathematical formulas, it is common practice to use a form called "standard notation" (also called "scientific notation"). Basically, standard notation is defined by the following expression:

$$A = B \times 10^n$$

where A is the number to be written in standard notation, B is a number between 1 and 10, and n is an integer.

Example: When written in standard notation, the number 1,000 becomes 1×10^3. By the definition of logs, the log of 1,000 is 3 (the exponent of 10 which yields 1,000). The exponent of 10 can be positive or negative, depending upon which direction the decimal point of B must be moved to get back to the number being represented. The value of the exponent tells how many places the decimal point must be moved. The chart in Table A-1 gives examples of this. A logarithm always consists of two parts: the characteristic and the mantissa. The characteristic is always an integer or zero. It can be positive or negative. The purpose of the characteristic is to tell you where to place the decimal point. The mantissa is always a fractional part or zero. It must always be positive. The mantissa tells you what the digits of the number are. If the expression $A = B \times 10^n$ is written in standard notation form, the log of A is equal to the log of B combined with the log of 10^n. Furthermore, the mantissa of the log of A is equal to the log of B and the characteristic of the log of A is equal to n.

You have seen from the previous example how easy it is to find the log of a number that is an integral power of 10. But what about all those other numbers that fall between those integral powers of 10? You can determine the log of any number by the following procedure:

1. Rewrite the number in standard notation ($A = B \times 10^n$)
2. Find the mantissa of B from the graph in Appendix B or from the log table in Appendix C.
3. The characteristic is equal to n.
4. The complete log of A = mantissa of B (step 2) combined with n.

Example: Find the log of 192.

Solution: 1. Rewrite the number in standard notation.
   $192 = 1.92 \times 10^2$
2. The mantissa = log 1.92 = 0.28 (from the graph, Appendix B)
3. The characteristic = 2 (from exponent of 10)

Number	0.01	0.1	1.0	10	100	1000
Exponent	-2	-1	0	1	2	3

TABLE A-1

EXPRESSING LOGS WITH A NEGATIVE CHARACTERISTIC

Logarithms of numbers which are smaller than 1 have a negative characteristic. When expressing logarithms that have negative characteristics, care must be taken to express the log properly, or the entire meaning of the log will be changed. An example will serve to clarify this.

Example: Find the log of 0.00473.

Solution: 1. Rewrite the number in standard notation.
   4.73×10^{-3}

2. Find the mantissa of 4.73 from the log table in Appendix C or from the graph in Appendix B. It is 0.675.

3. The characteristic is the exponent of 10 or -3.

4. Since the characteristic is negative, we must be careful to keep it separated from the mantissa—so we write the complete log as: $0.675 - 3$.

Never reduce a log expression such as $0.675 - 3$ to the algebraic sum of the mantissa and the characteristic because the sum of the two will be a value that is entirely negative. This goes against the law of logs that says that the mantissa can never be negative, only the characteristic can be negative. Thus, we must keep the two separated when the characteristic is negative. The form $0.675 - 3$ is perfectly acceptable for expressing the log of 0.00473, but another commonly used method is to express this as some value minus 10. Simply add 10 to and subtract 10 from the expression $0.675 - 3$ to get $0.675 - 3 + 10 - 10$. This gives $0.675 + 7 - 10 = 7.675 - 10$. The characteristic is equal to $-10 + 7 = -3$. Any number can be added and subtracted in this manner as long as the *difference* between the negative and positive whole numbers equals the original characteristic. Notice that in each of these expressions the mantissa is still intact and easily recognizable.

FINDING THE ANTILOG

Finding the antilog is the reverse of finding the log. The expression $\log A = n$ may also be written as antilog $n = A$. One is simply the converse of the other. For a given number, there is one and only one log and for any log there is only one corresponding number or antilog. In the expression $B = $ antilog A, A is a logarithm. If the value of A is given, we can determine the value of B by finding the antilog of A. In the following expression, a real value is substituted for A. Find the value of B.

Example: $B = $ antilog(0.470).

Solution: 1. Since the characteristic is 0, we know that B is a number between 1 and 10.

2. On the graph in Appendix B, find 0.47 on the log N scale (horizontal) and the corresponding number on the N scale (vertical). The antilog is 2.95 (approximately).

FINDING THE ANTILOG OF A NEGATIVE NUMBER

Once in a while when working with formulas you might end up with an expression such as $A = $ antilog -2.7000. This expression indicates that the entire log (2.7000) is negative, but it can't be. So we must change this to another form that gives us a separate negative integer (characteristic) and a positive fractional part (mantissa). The procedure is to add the next higher integer (in this case, 3) and then subtract that same integer as follows: $-2.7 = 3 - 2.7 - 3 = 0.3000 - 3$. This gives us a mantissa of 0.3000 and a characteristic of -3. The antilog can then be found. From the graph in Appendix B, the number corresponding to 0.3000 is 1.98. The characteristic tells us to multiply this by 10^{-3} to get $1.98 \times 10^{-3} = 0.00198$.

SUBTRACTING LOGS

Example: $A = \log B - \log C$, where $\log B = 0.725 - 3$ and $\log C = 2.180$.

1. Change the form of log B from $0.725 - 3$ to $7.725 - 10$.

2. Then subtract 2.180 from this value as follows:

$$
\begin{array}{r}
7.725 - 10 \\
(-)\ 2.180 \\
\hline
5.545 - 10
\end{array}
$$

3. Reduce this number to $5.545 - 10 = -4.455$.

ADDING LOGS

Example: $A = \log B + \log C$, where $\log B = 0.412$ and $\log C = 0.637 - 2$.

1. Place these in a column and add as you would any other number as follows:

$$
\begin{array}{r}
0.412 \\
(+)\ 0.637 - 2 \\
\hline
1.049 - 2
\end{array}
$$

2. Reduce this number to $1.049 - 2 = -0.951$.

DIVIDING LOGS

Example: $A = \log B/3$, where $\log B = 7.5950 - 10$.

1. Change the expression to $27.5950 - 30$ so that the negative figure (-30) can be divided evenly by 3.

2. Proceed to divide each term by 3 as follows:

$$
\frac{27.5950 - 30}{3} = 9.1983 - 10
$$

3. Reduce this number to $9.1983 - 10 = -0.8017$

MULTIPLYING LOGS

Example: $A = 3(\log B)$, where $\log B = 5.0975 - 10$.

1. Proceed as follows: $A = 3(5.0975 - 10) = 15.2925 - 30$.
2. Reduce as follows: $15.2925 - 30 = 14.7075$.

BIBLIOGRAPHY—APPENDIX A

Allied Radio Corporation (Eugene Carrington), *Allied Electronics Data Handbook.* Chicago, Ill. © Allied Radio Corporation, 1966.

Andres, Paul G.; Miser, Hugh J.; and Reingold, Haim. *Basic Mathematics for Engineers.* New York: John Wiley & Sons, Inc., 1948 (7th printing).

Juszli, Frank L., and Rodgers, Charles A., *Elementary Technical Mathematics.* Englewood Cliffs, N.J.: Prentice-Hall, Inc., 1962.

APPENDIX B

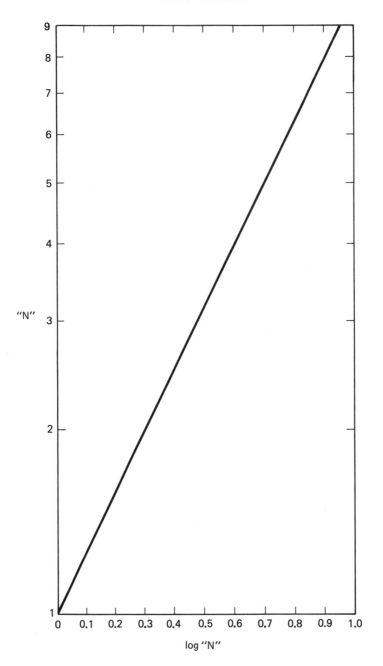

APPENDIX C

N	0	1	2	3	4	5	6	7	8	9
10	0000	0043	0086	0128	0170	0212	0253	0294	0334	0374
11	0414	0453	0492	0531	0569	0607	0645	0682	0719	0755
12	0792	0828	0864	0899	0934	0969	1004	1038	1072	1106
13	1139	1173	1206	1239	1271	1303	1335	1367	1399	1430
14	1461	1492	1523	1553	1584	1614	1644	1673	1703	1732
15	1761	1790	1818	1847	1875	1903	1931	1959	1987	2014
16	2041	2068	2095	2122	2148	2175	2201	2227	2253	2279
17	2304	2330	2355	2380	2405	2430	2455	2480	2504	2529
18	2553	2577	2601	2625	2648	2672	2695	2718	2742	2765
19	2788	2810	2833	2856	2878	2900	2923	2945	2967	2989
20	3010	3032	3054	3075	3096	3118	3139	3160	3181	3201
21	3222	3243	3263	3284	3304	3324	3345	3365	3385	3404
22	3424	3444	3464	3483	3502	3522	3541	3560	3579	3598
23	3617	3636	3655	3674	3692	3711	3729	3747	3766	3784
24	3802	3820	3838	3856	3874	3892	3909	3927	3945	3962
25	3979	3997	4014	4031	4048	4065	4082	4099	4116	4133
26	4150	4166	4183	4200	4216	4232	4249	4265	4281	4298
27	4314	4330	4346	4362	4378	4393	4409	4425	4440	4456
28	4472	4487	4502	4518	4533	4548	4564	4579	4594	4609
29	4624	4639	4654	4669	4683	4698	4713	4728	4742	4757
30	4771	4786	4800	4814	4829	4843	4857	4871	4886	4900
31	4914	4928	4942	4955	4969	4983	4997	5011	5024	5038
32	5051	5065	5079	5092	5105	5119	5132	5145	5159	5172
33	5185	5198	5211	5224	5237	5250	5263	5276	5289	5302
34	5315	5328	5340	5353	5366	5378	5391	5403	5416	5428
35	5441	5453	5465	5478	5490	5502	5514	5527	5539	5551
36	5563	5575	5587	5599	5611	5623	5635	5647	5658	5670
37	5682	5694	5705	5717	5729	5740	5752	5763	5775	5786
38	5798	5809	5821	5832	5843	5855	5866	5877	5888	5899

N	0	1	2	3	4	5	6	7	8	9
39	5911	5922	5933	5944	5955	5966	5977	5988	5999	6010
40	6021	6031	6042	6053	6064	6075	6085	6096	6107	6117
41	6128	6138	6149	6160	6170	6180	6191	6201	6212	6222
42	6232	6243	6253	6263	6274	6284	6294	6304	6314	6325
43	6335	6345	6355	6365	6375	6385	6395	6405	6415	6425
44	6435	6444	6454	6464	6474	6484	6493	6503	6513	6522
45	6532	6542	6551	6561	6571	6580	6590	6599	6609	6618
46	6628	6637	6646	6656	6665	6675	6684	6693	6702	6712
47	6721	6730	6739	6749	6758	6767	6776	6785	6794	6803
48	6812	6821	6830	6839	6848	6857	6866	6875	6884	6893
49	6902	6911	6920	6928	6937	6946	6955	6964	6972	6981
50	6990	6998	7007	7016	7024	7033	7042	7050	7059	7067
51	7076	7084	7093	7101	7110	7118	7126	7135	7143	7152
52	7160	7168	7177	7185	7193	7202	7210	7218	7226	7235
53	7243	7251	7259	7267	7275	7284	7292	7300	7308	7316
54	7324	7332	7340	7348	7356	7364	7372	7380	7388	7396
55	7404	7412	7419	7427	7435	7443	7451	7459	7466	7474
56	7482	7490	7497	7505	7513	7520	7528	7536	7543	7551
57	7559	7566	7574	7582	7589	7597	7604	7612	7619	7627
58	7634	7642	7649	7657	7664	7672	7679	7686	7694	7701
59	7709	7716	7723	7731	7738	7745	7752	7760	7767	7774
60	7782	7789	7796	7803	7810	7818	7825	7832	7839	7846
61	7853	7860	7868	7875	7882	7889	7896	7903	7910	7917
62	7924	7931	7938	7945	7952	7959	7966	7973	7980	7987
63	7993	8000	8007	8014	8021	8028	8035	8041	8048	8055
64	8062	8069	8075	8082	8089	8096	8102	8109	8116	8122
65	8129	8136	8142	8149	8156	8162	8169	8176	8182	8189
66	8195	8202	8209	8215	8222	8228	8235	8241	8248	8254
67	8261	8267	8274	8280	8287	8293	8299	8306	8312	8319
68	8325	8331	8338	8344	8351	8357	8363	8370	8376	8382
69	8388	8395	8401	8407	8414	8420	8426	8432	8439	8445

Common Logarithm Table, continued

N	0	1	2	3	4	5	6	7	8	9
70	8451	8457	8463	8470	8476	8482	8488	8494	8500	8506
71	8513	8519	8525	8531	8537	8543	8549	8555	8561	8567
72	8573	8579	8585	8591	8597	8603	8609	8615	8621	8627
73	8633	8639	8645	8651	8657	8663	8669	8675	8681	8686
74	8692	8698	8704	8710	8716	8721	8727	8733	8739	8745
75	8751	8756	8762	8768	8774	8779	8785	8791	8797	8802
76	8808	8814	8820	8825	8831	8837	8842	8848	8854	8859
77	8865	8871	8876	8882	8887	8893	8899	8904	8910	8915
78	8921	8927	8932	8938	8943	8949	8954	8960	8965	8971
79	8976	8982	8987	8993	8998	9004	9009	9015	9020	9025
80	9031	9036	9042	9047	9053	9058	9063	9069	9074	9079
81	9085	9090	9096	9101	9106	9112	9117	9122	9128	9133
82	9138	9143	9149	9154	9159	9165	9170	9175	9180	9186
83	9191	9196	9201	9206	9212	9217	9222	9227	9232	9238
84	9243	9248	9253	9258	9263	9269	9274	9279	9284	9289
85	9294	9299	9304	9309	9315	9320	9325	9330	9335	9340
86	9345	9350	9355	9360	9365	9370	9375	9380	9385	9390
87	9395	9400	9405	9410	9415	9420	9425	9430	9435	9440
88	9445	9450	9455	9460	9465	9469	9474	9479	9484	9489
89	9494	9499	9504	9509	9513	9518	9523	9528	9533	9538
90	9542	9547	9552	9557	9562	9566	9571	9576	9581	9586
91	9590	9595	9600	9605	9609	9614	9619	9624	9628	9633
92	9638	9643	9647	9652	9657	9661	9666	9671	9675	9680
93	9685	9689	9694	9699	9703	9708	9713	9717	9722	9727
94	9731	9736	9741	9745	9750	9754	9759	9763	9768	9773
95	9777	9782	9786	9791	9795	9800	9805	9809	9814	9818
96	9823	9827	9832	9836	9841	9845	9850	9854	9859	9863
97	9868	9872	9877	9881	9886	9890	9894	9899	9903	9908
98	9912	9917	9921	9926	9930	9934	9939	9943	9948	9952
99	9956	9961	9965	9969	9974	9978	9983	9987	9991	9996

APPENDIX D

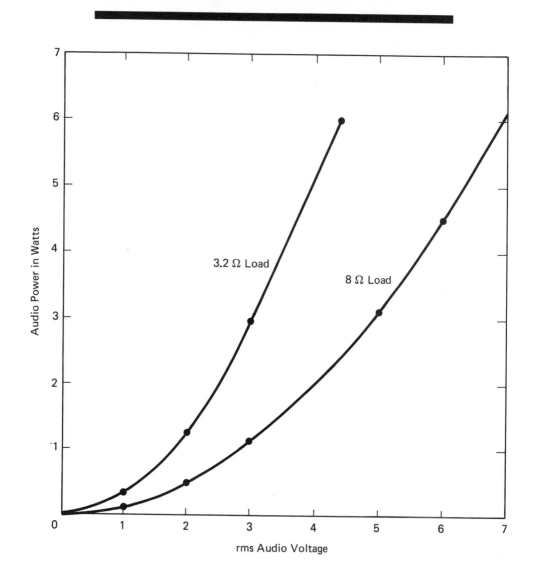

Audio Voltage Versus
Audio Power for
3.2 and 8 Ohm Load

APPENDIX E

RF Connectors
Courtesy Coaxial Dynamics, Inc.

413

APPENDIX F

To wind things up, I'd like to share with you an amusing story that was written by a friend of mine, who so generously allowed me to use it here. Although it is amusing, there is a very good message in it that is as important as anything else presented in this book.

THE "SPECTRONS"
by William B. "Ben" Langston

The "Spectrons," that's what I call the little people who inhabit my shop. I can't see them. But I *know* they're there. They *must* be there! How do I know? Simple. They borrow my tools, disturb my components, and destroy my concentration. They never steal, but they borrow with total disregard for my convenience. Some days I dare not lay down a tool for fear it will be gone when next I reach for it. They will, of course, return it undamaged in time; but meanwhile I must proceed with awkward substitutes such as using long-nose pliers for a nut driver or my pocket knife for wire cutters and even my fingernails for wire strippers!

Naturally, they are most active around a *cluttered* workbench. I have never known of them to borrow a tool from it's *proper* place. I suppose that's why there are so many of the little rascals around my shop! Last week they borrowed my long number 8 nut driver. It was three days before I found it in a box of miscellaneous resistors! That is the most infuriating part of their activities. When they do return a tool or a component, they usually hide it— under a cabinet or rag or even under the chassis that I'm working on at the time!

There are other activities that I suspect them of but cannot prove. For example, sometimes I can be thinking through a real hairy problem when suddenly, from nowhere, the solution will pop into my head; or I may be staring (with gloom and despair) at the cluttered face of a printed circuit board when suddenly the hairline crack appears where it had seemed solid before!

Over the years, we have become tolerant of each other. When they borrow my tools, I hide my exasperation and walk away for a few minutes. Usually, when I return, it is back. But starting today things are going to be different around here. To discourage their plundering, I'm going to clean up this messy shop and I'll start by sweeping the floor. Now, where is my broom?

414

INDEX